Specialty Construction Techniques for Dam and Levee Remediation

Specialty Construction Techniques for Dam and Levee Remediation

Edited by Donald A. Bruce

CRC Press
Taylor & Francis Group
Boca Raton   London   New York

CRC Press is an imprint of the
Taylor & Francis Group, an **informa** business

A SPON PRESS BOOK

CRC Press
Taylor & Francis Group
6000 Broken Sound Parkway NW, Suite 300
Boca Raton, FL 33487-2742

© 2013 by Donald A. Bruce
CRC Press is an imprint of Taylor & Francis Group, an Informa business

No claim to original U.S. Government works

Printed in the United States of America on acid-free paper
Version Date: 20120823

International Standard Book Number: 978-0-415-78194-7 (Hardback)

Visit the Taylor & Francis Web site at
http://www.taylorandfrancis.com

and the CRC Press Web site at
http://www.crcpress.com

This book is dedicated to the ladies and gentlemen with whom I have worked, for and against, during my career in dam and levee remediation, and the memories of my Mum and Dad, Wally, Ken, and Renato. Thank you for everything you did.

Donald

Contents

Preface

The aging infrastructure of North America is a topic of fundamental importance that attracts the consciousness of the general public only when a catastrophic failure occurs. Owners and operators of these assets do their best to maintain the elements of the infrastructure under their control. However, the efficiency and scope of their efforts are usually limited by financial considerations, and frequently by bureaucracy. Quite simply, our infrastructure is so huge and complex, and its maintenance needs so vast, that there are not enough resources to spare.

Some level of repair activity is always underway, yet it needs one of Nature's "wake-up calls" to trigger a major surge. One classic example is the seismic retrofit initiative for bridges and other life-line structures that has followed the 1989 Loma Prieta earthquake in California.

For those of us who work in the dam and levee communities, our most recent clamorous alarm was the impact of Hurricanes Katrina and Rita on the Gulf Shore in August 2005. Not only did this stimulate the federal government to restructure its approach to the risk analysis of its dam portfolio, but it catalyzed similar programs for levee assessments, in both the public and private sectors.

Dam remediation has been around as long as dam construction, and there has been strong, if somewhat sporadic, activity in North America since the 1970s in particular. Concrete dams have had prestressed rock anchors installed, while successive phases of interventions such as relief wells and clay blankets can be detected for seepage control in embankment dams. More recently, seepage-control principles have devolved toward the widespread use of cutoff walls, often in concert with sophisticated grouting programs. Seismic foundation remediations with deep mixing methods have become common, while the same techniques are being widely employed for cutoff wall construction in levees. Mainly as a result of federal initiatives, the years from 2006 have therefore seen an unprecedented intensity of dam and levee remediation in North America using specialty geotechnical engineering techniques such as anchors, grouting, cutoff (diaphragm) walls, and deep mixing.

Such works are being constructed by a cadre of well-resourced, experienced contractors, many of whom owe their origins and/or current ownership to countries in Europe, and to Japan. A wide range of methodologies and techniques has been developed in response to the particular challenges of the dam and levee remediation environment, and to the need to provide a competitive edge. These works are being designed and monitored by the owners and/or their consultants, using state-of-the-art methods of investigation and analyses. As a result, there is a new "Age of Enlightenment," the intensity of which is being reflected in record attendances at the various annual conferences dealing with dams and levees.

This book attempts to capture the spirit and of the New Age, through its description of the theory and practice of contemporary remedial techniques. Widespread use is made of case histories of more recent vintage, so providing a snapshot (known as a Polaroid in the Old Age) of our industry in the first decade of the twenty-first century. It is hoped this will be valuable as a source of reference and inspiration to colleagues both in North America, and throughout the world, who may not have had, so far, the privilege and pleasure of direct involvement in the dam and levee remediation market.

Acknowledgments

This book is the result of collaboration by many practicing, professional engineers who devoted precious personal time to drafting various sections and chapters. It also reflects the extraordinary skills of my personal assistant, Mrs. Terri Metz, and the Georgian understanding of my wife, Hope. I also owe a great deal to Charlie Naples. I am also indebted to the leadership and organization of all the folks at Spon Press, an imprint of Taylor & Francis, and to the extreme efficiency of my project manager at Cenveo, Chris Schoedel. Thank you all so much.

Contributors

Donald A. Bruce Geosystems, L.P., Venetia, Pennsylvania

George K. Burke Hayward Baker, Inc., Odenton, Maryland

Marcelo Chuaqui Monir Precision Monitoring, Inc., Mississauga, Ontario, Canada

Trent L. Dreese Gannett Fleming, Inc., Camp Hill, Pennsylvania

Brian Jasperse Geo-Con, Inc., Monroeville, Pennsylvania

Shigeru Katsukura Specially Operating Machinery Division KG Machinery Co., Ltd., Tokyo, Japan

Yujin Nishimura Overseas Business Division Raito Kogyo, Co., Ltd., Tokyo, Japan

Maurizio Siepi TREVI SpA, Cesena, Italy

Daniel P. Stare Gannett Fleming, Inc., Camp Hill, Pennsylvania

Ulli Wiedenmann BAUER Spezialtiefbau GmbH, Schrobenhausen, Germany

John S. Wolfhope Freese and Nichols, Inc., Austin, Texas

David S. Yang DSY Geotech, Inc., Fremont, California

Chapter 1

Background and scope

Donald A. Bruce

1.1 DAMS AND LEVEES IN THE UNITED STATES: A SITUATION ASSESSMENT

The Congress of the United States of America authorized the U.S. Army Corps of Engineers (USACE) to conduct an inventory of dams in the United States through the National Dam Inspection Act of 1972. The resultant National Inventory of Dams (NID) was first published in 1975 and updates have been made in the succeeding years, in accordance with the Water Resources Development Act of 1986. The most recent Dam Safety Act (2006) reauthorized the maintenance and update of the NID.

As described in the USACE's NID website, the NID covers dams meeting at least one of the following criteria (Ragon 2011):

1. High hazard classification—loss of one human life is likely if the dam fails.
2. Significant hazard classification—possible loss of human life and likely significant property or environmental destruction.
3. Low hazard classification—no probable loss of human life and low economic and/or environmental losses, but the dam:
   Equals or exceeds 25 feet in height and exceed 15 acre-feet in storage;
   Equals or exceeds 50 acre-feet storage and exceeds 6 feet in height.

The goal of the NID is to include all dams in the United States that meet these criteria, yet in reality the program is limited to information that can be gathered and properly interpreted with the given funding. The inventory initially consisted of approximately 45,000 dams, identified mainly from extensive record searches, although some were extracted from aerial imagery. As methodical updates have continued, data collection has been focused on the most reliable data sources, which are the various federal and state government dam construction and regulation offices. In most cases, dams within the NID criteria are regulated (construction permit, inspection, and/or enforcement) by federal or state agencies, who have basic

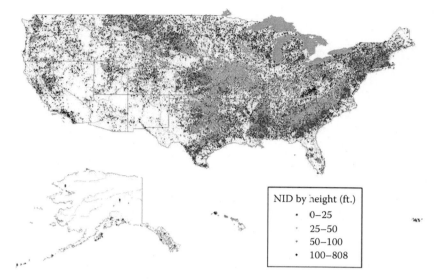

Figure 1.1 Geographic distribution of dams in the United States. (From National Inventory of Dams, *CorpsMap*, http://nid.usace.army.mil, 2010.)

information on the dams within their jurisdiction. Therein lies the biggest challenge, and most of the effort to maintain the NID, namely the periodic collection of dam characteristics from fifty states, Puerto Rico, and sixteen federal agencies. Based on the input from these sources and, in coordination with the Federal Emergency Management Agency (FEMA), more than 100,000 dams have now been reported, although only about 84,000 of these meet the NID criteria. The following statistics are freely available on the nation's dams, as a group.

Figure 1.1 gives an indication of the geographic distribution of these dams, which include almost 14,000 which are classified as "High Hazard." Texas has the most dams (7,069), followed by Kansas (5,650), Missouri (4,850), Oklahoma (4,672), and Georgia (4,158). Almost 6,200 dams are at least 50 feet high, with over 1,600 being in excess of 100 feet, while about 50 percent of the dams are less than 25 feet high. Oroville Dam, California, is the highest earthfill dam (770 feet), Hoover Dam, Nevada, is the highest concrete dam (730 feet), and New Bullards Bar Dam, California, is the highest arch dam (645 feet) in the United States. The dams have the following ownership distribution:

Private	68%
Local government	20%
State government	5%
Federal government	4%

Public utilities   2%
Unknown   1%

The federal government total of 3,075 includes those owned and/or operated by the USACE, the Bureau of Reclamation, Forest Service, Bureau of Indian Affairs, Bureau of Land Management, Fish and Wildlife Service, National Park Service, International Boundary and Water Commission, and the Tennessee Valley Authority. These agencies only began building dams in the early 1900s. Before then, private funding prevailed and design and construction methodologies were uneven and unregulated.

Over 87 percent of the total dams are primarily classified as earth embankments of some form, while no other category (including arch, buttress, concrete, gravity, masonry, multiarch, rockfills, and timber cribs) exceeds 3 percent of the total. There are estimated to be just under 1,500 RCC structures, although this total most likely includes a large number of spillway overlays.

Figure 1.2 summarizes the dam population by primary purpose: more than one-third are for recreation, while less than 3 percent are primarily for power generation. Many of the dams are multipurpose. Figure 1.3 categorizes them by completion date: about 50 percent of the nation's dams were completed between 1950 and 1979, while the median age in the year 2011 is sixty years. Fewer than 200 high dams have been completed since 1990 in the United States, although 1,372 new dams were completed between 2000 and 2005. As illustrated in Section 1.2, a very large percentage of our dams is located in areas underlain by solution susceptible rocks and/or is potentially threatened by seismicity.

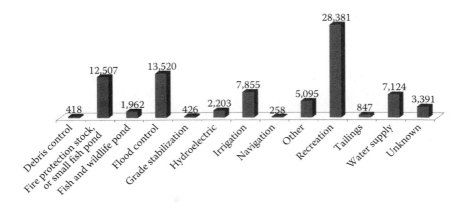

Figure 1.2 Dams by primary purpose. (From National Inventory of Dams, *CorpsMap*, http://nid.usace.army.mil, 2010.)

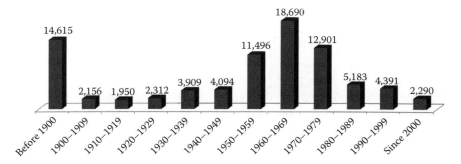

Figure 1.3 Dams by completion date. (From National Inventory of Dams, *CorpsMap*, http://nid.usace.army.mil, 2010.)

Three of the dams cited in case histories in this book impound reservoirs in the top eleven based on maximum reservoir capacity:

Dam	Reservoir	State	Maximum Reservoir Capacity (acre-feet)	Ranking
Herbert Hoover Dike	Lake Okeechobee	FL	8,519,000	7th
Sam Rayburn Dam	Sam Rayburn Lake	TX	6,520,000	8th
Wolf Creek Dam	Lake Cumberland	KY	6,089,000	11th

The largest reservoir, Lake Mead, Nevada, is impounded by Hoover Dam and accounts for 30,237,000 acre-feet of storage.

Whereas one can calculate that all the dams in the United States, if placed end to end, would form a structure about 17,000 miles long, *preliminary* estimates put the total length of levees at over 100,000 miles. ("The total number of levees across the nation is still unknown" [USACE 2009]). Only about 14 percent of this total may be regarded as "federal." The balance includes municipal, local, and agricultural structures.

It is important to note the distinction between a dam and a levee. FEMA (1998a) defined a levee as "a manmade structure, usually an earthen embankment, designed and constructed . . . to *contain, control,* or *divert* the flow of water, so as to provide protection from temporary flooding" [emphasis added].

A somewhat different approach to the distinction was proposed by Davis and Kennedy (2001), albeit in the "pre-Katrina" years. They began by noting that the federal "Hazard Potential Classification Systems for Dams" (FEMA, 1998b) defines a "High Hazard Potential" dam as one "where failure or mis-operation will probably cause loss of one or more human lives if it should fail." A "Significant Hazard Potential" dam is defined as one where failure or mis-operation results in no probable loss of human life, but can cause "economic loss, environmental damage, disruption of lifeline facilities, or can impact other concerns." High and Significant Hazard Potential dams are more highly regulated, subject to increased inspection frequency, and must meet higher design standards. Davis and Kennedy (2001) argue that the same classification system should be applied to levees, and further argue that a levee is the same as a flood-control dam ("dry dam"). In commentary to their paper, the authors also cite the USACE definitions of the time:

> Dam—an artificial barrier, together with its appurtenant works, constructed for the purpose of impounding or diverting water.
> Levee—an embankment whose primary purpose is to furnish flood protection from high water and which is subject to water loading for period of a few days or weeks per year.

According to Halpin (2010), there are levees in all fifty states and, in addition to the uncertainty regarding their number and extent, there is equal ignorance about their current condition and the threat they pose to the population and property they protect. It is clear, however, that tens of millions of people live and work in close proximity to levee systems: originally constructed to protect property, levees have often inadvertently increased flood risks by attracting greater development to the flood plain.

From the earliest days of the United States until the 1930s, levee construction was sporadic and unsophisticated without the benefit of systematic engineering or scientific expertise. Levees were considered "simple" structures and so not designed or built to contemporary dam standards. After great devastation and loss of life from the floods in the Mississippi and Ohio river valleys and in Florida, the USACE was directed, at full federal expense, to take a more active role in levee design and construction, resulting in the 14,000 miles or so of "robust" levee systems now in place. Many of these levees, which make up the backbone of the nation's levee system, are now therefore over fifty years old and, even when regularly maintained, may not have been brought up to the most recent engineering standards.

The nation's attention has been refocused on this critical part of our infrastructure by the aftermath of the floods in the Midwest (1993, 2008, and 2011), in California (1986 and 1997), and, of course, in Louisiana (2005). In 2007 Congress therefore passed the National Levee Safety Act, whereby a select committee will provide technical leadership, safety evaluations, and standardization of processes nationwide for both federal and private-sector structures and owners. The American Society of Civil Engineers estimates it will require a $50-billion investment over a five-year period to repair and rehabilitate the country's levees (Halpin 2010).

The situation was described in stark and chilling terms:

> The potential consequences of levee failure can be devastating. The situation is the result of more than 100 years of inattention to, and in some cases neglect of, levee infrastructure combined with a growing

population living behind levees and an economy and social fabric that are in a particularly vulnerable state. The current levee safety reality for the United States is stark—uncertainty of location, performance and condition of levees, and a lack of oversight, technical standards, and effective communication of risk. (Halpin 2010)

1.2 GEOLOGICAL CHALLENGES

One can always cite specific examples of dams that have proved problematical in service (occasionally to the point of abandonment or failure) as a result of specific factors relating to design, foundation, geology, construction techniques, or operational shortcomings. Focusing on the geological aspects, the United States was of course historically such a huge and open country that there was typically a "walkaway" solution to a potential foundation challenge: if the site was not favorable, the final dam location could be moved. However, after the first few decades of the twentieth century, the length of the "walk" had already been severely reduced, and the geological demons had therefore to be faced *in situ*. Some of the resultant dams could not be completed or were never able to sustain the anticipated reservoir. Hales Bar, Tennessee, is a classic example, even though heroic and pioneering seepage cutoff attempts were made in the karstic limestone under this structure over many years up to the 1940s. This project is the main subject of Chapter 8. Such cases aside, the *main and systematic geological challenges* to U.S. dams remain the threats posed (a) by solution susceptible foundations (and in particular karstic limestone terrains), and (b) by seismic activity.

Veni (2002) described the process and rationale for the development of the karst areas map of the United States (Figure 1.4). While Veni also discussed possible inaccuracies and limitations of the map, it is extremely important and simple to read, and is based on lithology rather than the locations of observed caves. It illustrates exposed and buried carbonates, exposed and buried evaporites, volcanic pseudokarst, and unconsolidated pseudokarst. Any local comparisons with the dam locations shown in Figure 1.1 must therefore be made with care. For example, while very few dams actually sit on evaporite, this material, especially as confined in other lithologies, underlies 35–40 percent of the forty-eight contiguous U.S. states (Martinez, Johnson, and Neal 1998). Furthermore, the degree of karstification varies from formation to formation across the country, although there is no doubt that the intensity is greatest in the huge belt of Ordovician carbonates that sweeps from central Pennsylvania through West Virginia, and into Kentucky, Missouri, Arkansas, Tennessee, and Alabama. It is no coincidence that many of the case histories described in this book relate to seepage cutoffs for existing embankment dams in these states.

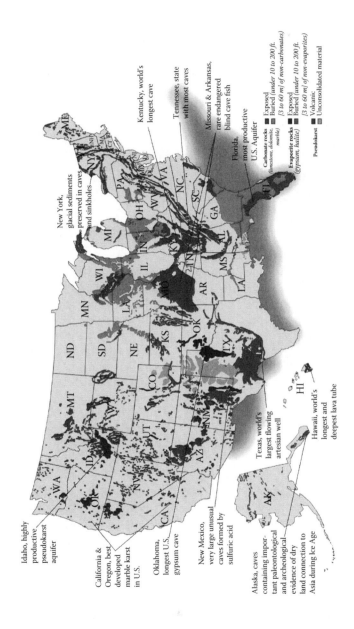

Figure 1.4 Karst map of the United States published by AGI. This map is a general representation of U.S. karst and pseudokarst areas. While based on the best available information, the scale does not allow detailed and precise representation of the areas. Local geologic maps and field examinations should be used where exact information is needed. Karst features and hydrology vary from place to place. Some areas are highly cavernous, and others are not. Although most karst is exposed at the land surface, some is buried under layers of sediment and rock, and still affect surface activities. (From Veni, G., et al., *Living with Karst: A Fragile Foundation*, American Geological Institute, 2001. With permission.)

Figure 1.5 is the U.S. earthquake hazard map for 2 percent probability of ground motion exceedance in fifty years, that is, a 2,500-year return period. A slightly different appreciation is given in Figure 1.6, although the overall message is the same: areas of extreme seismicity exist in the Mississippi/Tennessee river area (focused on the New Madrid fault system) and, of course, along the West Coast and in the eastern Rockies. The high frequency of damaging earthquakes in New England surprises many, although one must also take into consideration the larger and more intense human habitation of that area and so the high sensitivity and precision of the historical database.

The presence of high seismicity zones centered on New Madrid, Missouri, and Charleston, South Carolina, may also be surprising to some readers otherwise acquainted only with the various western seismic zones. As illustration, the New Madrid zone comprises a series of buried strike-slip and thrust faults situated under the continental crust. It is not a result of interplate actions, as is the case, for example, of the San Andreas Fault in California. The highest historical earthquake magnitude of over 8.0 on the New Madrid system was recorded in the 1811–1812 events, while the Charleston earthquake of 1886 was estimated to be of magnitude 7.5. Both the New Madrid and Charleston zones have since entered quieter phases, whereas the western zone remains active, as illustrated by the quite recent Loma Prieta (1989) and Northridge (1994) earthquakes in particular.

Very simplistically, therefore, geology and seismicity—either alone or together—pose a clear and present threat to tens of thousands of water-retention structures nationwide, but especially to those in the basins of the central Mississippi–Missouri river system and its major tributaries such as the Tennessee and Ohio rivers, and to those in the environs of the greater Rocky Mountain chain. To these concerns must be added the more transient, but equally destructive, threat posed by extreme weather events to levees all across the country, but especially in the upper Midwest, the lower Mississippi, and central California. The problem in the New Orleans area is exacerbated by the continual regional settlement of the entire delta area, estimated at 0.1 to 0.5 inches per year.

Mother Nature hates an imbalance, and water-retaining structures constitute to Her a particularly attractive challenge and a tempting target. And what better target than a defensive line composed of aging performers, of polyglot pedigree, planted on treacherous fields?

1.3 THE PATH OF REMEDIATION

Traditionally, remedial projects were initiated either in response to the early development of an obvious deficiency having the potential to threaten dam or levee safety, or in response to the later findings of structural reevaluations

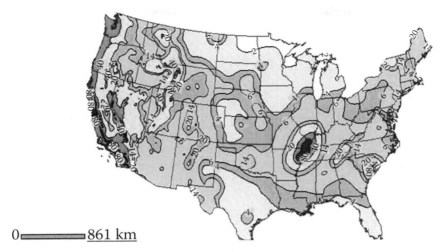

0 ▬▬▬▬▬ 861 km

Figure 1.5 Peak horizontal acceleration with 2 percent probability of exceedance in 50 years. Note: The contour values represent peak ground accelerations in percent g. (From U.S. Geological Survey, *National Seismic Hazard Maps of Conterminous U.S.*, http://earthquake.usgs.gov/hazards/apps/, 2008.)

Modified Mercalli Intensity VI–XII

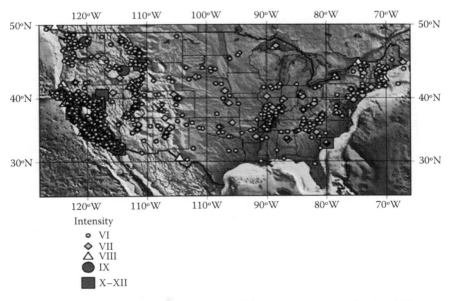

Figure 1.6 U.S. earthquakes causing damage, 1750–1996. (From U.S. Geological Survey, *National Seismic Hazard Maps of Conterminous U.S.*, http://earthquake.usgs.gov/hazards/apps/, 2008.)

based on revised design criteria, updated parameters, and new assumptions. In either case, the remediation was more reactive than proactive. Furthermore, with the finite and often limited funds available to any given owner—including the federal government—the work was spaced out as far as safely possible, especially when the said owner was funding the remediation out of operational cash flow. It was often the case that effective and accurate prioritization was defeated by the "squeaky wheel" syndrome, with the result that remediation of certain truly needy cases was deferred or overlooked.

Galvanized by the Gulf Coast tragedy of August 2005, the federal government, in the form of the USACE, developed and implemented a radically different approach to dam-remediation prioritization, supported by expertise and experience from the Bureau of Reclamation. This "risk-based" or "risk-informed" approach has since become a model for other bodies with large portfolios of dams, including the Tennessee Valley Authority and the larger utilities.

It is first insightful to review certain facts regarding the USACE (as provided in their 2009 publication) to more fully appreciate its role and responsibilities, and the evolution of its thinking:

- USACE functions within the Department of the Army and comprises approximately 37,000 civilian and military employees.
- Its management structure includes headquarters in Washington, DC, nine division offices, 46 district offices, six technical centers, including the Engineer Research and Development Center in Vicksburg, Mississippi (formerly Waterways Experiment Station).
- It owns and operates about 650 dams, 241 navigation lock chambers at 195 locations, and about 12,000 miles of commercial inland navigation channels, and has specific authorities to routinely inspect about 2,000 levees, totaling 14,000 miles, under its ongoing levee safety program.
- It maintains 926 coastal, Great Lakes, and inland harbors.
- It owns and operates 24 percent of U.S. hydropower capacity, equivalent to 3 percent of the total national electricity capacity.
- It is the nation's leading federal provider of outdoor recreation, including more than 4,200 sites, at 423 lake and river projects.
- Major lakes provide a total water-supply storage capacity of 329.2 million acre-feet.
- Given that most of the nation's infrastructure was built under previous "stimulus" packages during the Roosevelt and Eisenhower eras, most of the locks were built in the 1940s and 1950s, while many dams are several decades older and most levees are equally venerable. In effect, many structures are approaching the end of their *projected* useful lives.

The USACE's new approach, known as risk-informed decisionmaking, was launched in 2005 when the first and most basic evaluation program for individual dam safety—the Screening for Portfolio Risk Analysis (SPRA)—was conducted. This takes into account many and wide-ranging factors, including the dam's purpose, the site characteristics, construction history, historical performance, hydraulics and hydrology, the probability of a hazardous event, and the potential consequences (of all types) of dam failure. The initial phase involved about 20 percent of the USACE's portfolio, considered likely to be of highest risk. This led, in some cases, to the immediate implementation of remedial work and identified a number of dams requiring extensive and fast-track modifications. In fact, several of the case histories referred to in this book in Chapters 2 through 5 are these "Class I" structures having an "urgent and compelling" need for risk-reduction measures to be implemented. The SPRAs were completed on all USACE dams by the close of 2009 and, in addition to identifying "clear and present" dangers, has precipitated further focused evaluations on many other structures.

The next step is the preparation of an Issue Evaluation Study (IES), involving an intense technical level of study with sophisticated modeling, leading to the development of targeted, cost-effective solutions. In not all cases have these studies proved the risk to be as high as originally feared, although in other cases the true risk was calculated to be higher.

Combined with the systematic use of the Potential Failure Mode Analysis (PFMA) process and the input of independent peer review groups, it may be concluded that dams and levees in the United States are now being remediated in the correct order of priority with the most appropriate methodologies.

1.4 THE PURPOSE AND CONTENT OF THIS BOOK

The purpose of this book is to present the state of practice in dam and levee remediation in the United States, as it relates to the use of specialty geotechnical construction techniques. Other modes of risk reduction, such as permanent reservoir lowering, crest raising, or methods to combat inadequate spillway or drawdown capacity, are outside the scope. Although the focus is on the actual construction processes themselves, aspects of the design and performance of the remediations are discussed where appropriate, and especially in sections dealing with case histories. Emphasis has been placed on more recent—and in some cases current—projects given the rapid pace at which many of the subject techniques are developing and evolving.

Chapter 2 deals with the application of drilling and grouting methods, mainly for the treatment of rock masses, but also for seismic retrofits in soil foundations. Of particular importance is the section on contemporary rock fissure grouting methods, initially drafted by Daniel P. Stare and Trent L.

Dreese of Gannett Fleming, Inc.; over the last ten years or so, there has been a technological revolution in this particular topic in the United States. Practice in treating voided karstic rock conditions is also addressed at length and is especially relevant given the number of major embankments founded on such conditions. Although jet grouting has not, for many reasons, been a common method for constructing seepage-control structures, it has proved very effective when correctly employed, especially for seismic mitigation schemes. Recent case histories are described to illustrate the applicability and potential.

In Chapter 3, methods used to install what are referred to as "Category 2" cutoff structures are described. These methods each produce an *in situ* blend of the native soil, or fill, with a cementitious "binder." This blend is referred to variously as soilcrete or soil-cement. A simple but fundamental distinction is drawn between the three main categories of methods, which are the "conventional" deep mixing, the trench remixing deep (TRD), and the cutter soil mixing (CSM) methods. The primary authors of these sections are, respectively, Dr. David S. Yang and Yujin Nishimura of Raito Inc., George K. Burke and Shigeru Katsukura of the Keller Group, and Ulli Wiedenmann of Bauer Spezialtiefbau GmbH. Category 2 walls have particular importance and relevance to the construction of cutoffs through levees, and for the installation of downstream buttress structures to stabilize embankments on potentially liquefiable soils.

Category 1 structures are discussed in Chapter 4. These involve the excavation of trenches under a stabilizing fluid and the subsequent backfilling of these trenches with an engineered material, typically some type of concrete. A subclassification is based primarily on the excavation methodology. Brian Jasperse of GeoCon, Inc., leads with a discussion of longitudinally continuous walls constructed with long-reach backhoes, followed by a description of diaphragm walls constructed in discrete panels. Dr. Arturo Ressi, now of Kiewit, contributed information on certain panel wall projects. Maurizio Siepi of the Trevi Group drafted certain case histories of cutoffs constructed by the overlapping, or secant, pile method. This method is somewhat uncommon but, when it is used, it is because overriding geological or dam safety concerns absolutely rule out the use of backhoe or panel methods.

Chapter 5 describes the concept and details of "composite" cutoff walls, wherein the whole alignment of a concrete cutoff wall is systematically pregrouted to a high and verified engineering standard prior to the concrete cutoff being installed. This concept has proved especially useful and cost effective in several recent remediations of embankments on karst, and has three main technical advantages: (a) it prevents the risk of massive and sudden slurry loss into karstic and epikarstic features during wall excavation, (b) it constitutes a very detailed site investigation to allow optimization of the depth of the (expensive) concrete wall, and (c) it provides a curtain of low and engineered permeability in "clean" rock fissures below and beyond the concrete wall.

Chapter 6 focuses on the use of prestressed rock anchors for the stabilization of concrete structures. It draws heavily on the results of a National Research Project conducted in mid-decade by John S. Wolfhope, of Freese and Nichols, Inc., and the author. This chapter provides particular food for thought in that the data it contains indicate that many—and probably most—of the anchors installed in over 400 North American structures since the first anchoring project in the early 1960s simply do not meet currently recommended levels of corrosion protection. Furthermore, due to their mode of construction, it is impossible to measure their residual prestress load in the vast majority of cases.

Marcelo Chuaqui, of Monir Precision Monitoring, Inc., is the author of Chapter 7, which provides a comprehensive generic guide to dam and levee instrumentation. This chapter reaffirms the link between the various tools of risk analysis, such as the PFMA process, and the proper design and analysis of a responsive and informative instrumentation program. Such programs are essential to establish the potential need for remediation, and the actual efficiency sustainability of the remediations.

Chapter 8 is superficially a case history of a long series of remediations on one project that ultimately proved unsuccessful. Its deeper goal is to diffuse the euphoria that certain readers may begin to experience having consumed the case histories detailing unrelenting success in the intervening six chapters of this book. Caveat lector!

The reference lists are provided, for convenience, at the end of each chapter.

REFERENCES

Davis, A. P., and W. F. Kennedy. 2001. "Levees, A Dam by Any Other Name (Improving Levees with Dam Safety Practices)." Association of State Dam Safety Officials, Annual Conference, Snowbird, UT, September 9–12.

Federal Emergency Management Agency (FEMA). 1998a. *Emergency Management and Assistance*. http://www.access.gpo.gov/nara/cfr/waisidx_00/44cfrv1_00.html.

Federal Emergency Management Agency (FEMA). 1998b. "Federal Guidelines for Dam Safety." FEMA No. 333.

Halpin, E. C. 2010. "Creating a National Levee Safety Program: Recommendations from the National Committee on Levee Safety." ASDSO Dam Safety Conference, Seattle, WA, September 19–23.

Martinez, J. D., K. S. Johnson, and J. T. Neal. 1998. "Sinkholes in Evaporite Rocks." *American Scientist* 86 (1): 38–51.

National Inventory of Dams (NID). *CorpsMap*. 2010. http://nid.usace.army.mil.

Ragon, R. 2011. "National Inventory of Dams Provides Interesting Data and Statistics." *USSD Newsletter*, March, p. 14.

U.S. Army Corps of Engineers (USACE). 2009. "Serving the Nation and the Armed Forces." http://www.spa.usace.army.mil/portals/16/docs/serving.pdf.

U.S. Geological Survey (USGS). 2008. *National Seismic Hazard Maps of Conterminous U.S.* http://earthquake.usgs.gov/hazards/apps/.

Veni, G. 2002. "Revising the Karst Map of the United States." *Journal of Cave and Karst Studies* 64 (1): 45–50.

Veni, G., H. DuChene, H. C. Crawford, C. G. Groves, G. H. Huppert, E. H. Kastning, R. Olson, and B. J. Wheeler. 2001. *Living with Karst: A Fragile Foundation.* Environmental Awareness series. Alexandria, VA: American Geological Institute.

Chapter 2

Contemporary drilling and grouting methods

Daniel P. Stare, Trent L. Dreese, and Donald A. Bruce

2.1 BACKGROUND

The world of grouting is truly and indeed wide, as illustrated in a succession of recent conferences and textbooks. Paradoxically, the grouting techniques used in the United States to remediate existing dams and levees are surprisingly few, even though their annual dollar value has reached extraordinary levels in recent years: industry reviews and estimates put this figure in the $80–$100-million range in each of the years from 2006 to 2010. It is, of course, highly debatable if that level of intensity can and will be sustained in the years ahead, and indeed the debate is most likely to be lost. As introduced in Chapter 1, the current phase of dam and levee remediation represents an unparalleled intensity of activity in specialty geotechnical construction circles in the United States and it may well be unprecedented in any country so far.

So what are the applications of grouting in dam and levee remediation? Potential opportunities exist to treat zones in the embankments themselves and, in the case of certain concrete or roller compacted concrete (RCC) structures, to seal cracks that have been induced during curing or by subsequent structural distresses. In this regard, the profession has generally preferred other ways to seal embankments, as demonstrated in Chapters 3 and 4 (in particular) of this volume. Similarly, cracks or fissures induced in concrete structures are now quite routinely and effectively addressed by the use of "solution" grouts, typically of the polyurethane, acrylic, or epoxy families. Such projects in the United States have typically been of relatively small scale, notwithstanding the admirable efforts at Dworshak Dam, Idaho (Smoak and Gularte 1998); Upper Stillwater Dam, Utah; Santeetlah Dam, North Carolina (Bruce 1989); and in Arizona (Arora and Kinley 2008), as more recent examples of a grander scale.

Contemporary grouting applications have therefore focused on four main deliverables, namely:

1. Remedial grout curtains in rock under and around existing dams, and mainly in karstic limestone conditions.

15

2. Quantitatively engineered grout curtains in rock under new structure.
3. Jet grouting in soils underlying existing embankment dams and levees to form or complete seepage cutoffs or to improve the foundation's seismic response characteristics.
4. Sealing of the interface between embankment and bedrock, and treatment of the bedrock itself, to facilitate the safe construction of a subsequent "positive" cutoff wall and to treat the rock beyond and below the cutoff to a quantitatively engineered standard.

This chapter reviews the first three applications listed above. The fourth role of grouting, that is, in conjunction with the concept of "composite" cutoffs, is described in Chapter 5, in which the fundamental and oft-overlooked value of a drilling and grouting program as a definitive site investigation is illustrated.

2.2 ROCK GROUTING

2.2.1 Introduction and historical perspective

2.2.1.1 Introduction

Grouts for cutoffs are typically low-viscosity and/or low-cohesion solutions or suspensions that gel or set. In fissure grouting applications, which constitute the vast majority of grouting applications for cutoffs, the material intrudes the pore spaces or discontinuities in the foundation resulting in little or no displacement of the parent foundation materials. The goal is to homogenize the foundation materials and/or fill any discontinuities without significant disturbance to the foundation.

This section presents contemporary U.S. grouting practices for improvement of dam and levee foundations with specific concentration on the use of cementitious suspension grouts having consistencies ranging from near water-like fluids up to thick mortars. Chemical grouts, while having a well-defined niche in the grouting industry, are rarely used for large foundation improvement projects due to their high costs and concerns with durability and toxicity. Chemical grouts are beyond the purview of this chapter and the reader is referred to Karol (1990) for detailed information on chemical grouting.

2.2.1.2 Historical perspective

The use of grouting is well documented throughout the nineteenth and twentieth centuries. Prior to the 1800s grouting was mainly used to strengthen walls and other manmade structures (Houlsby 1990), but little technical documentation of these processes remains. Members of the grouting fraternity, notably Houlsby (1990), Littlejohn (2003), and Weaver and

Bruce (2007), have endeavored to document the historical use of grouting. Repeating the entirety of their historical documentation efforts here would leave little room for other information and be an injustice to their efforts. For perspective, however, it would appear that grouting in the United States dates back at least to 1893, when cement-based grout was injected into the limestone formation of a 290-foot-high dam in the New Croton Project, New York. The prime goal, according to Glossop (1961), was to reduce uplift pressures, and that therefore no attempt was made to construct an "impermeable" cutoff. The curtain was as much as 100 feet deep. Thereafter, there was a "slow advance in grouting technology throughout much of the 20th Century" (Weaver and Bruce 2007, p. 12) for reasons that reflect on technological isolation and parochialism in the face of unprecedented amounts of new dam construction and the popular belief in very prescriptive specifications.

Dam foundation grouting practice in the United States had to wait for visionaries such as Dr. Wallace Baker and for the influx of foreign ideas and concepts that began in the early 1980s through the efforts of Don Deere, Clive Houlsby, and others, before practices and attitudes changed. Construction of two-row grout curtains and grouting to standards became common practice, and the Swiss concept of multiple-row grout curtains with holes at oppositely inclined orientations was adopted for major projects. Ultrafine cements, first introduced from Japan, came into common use for treating finely fractured rock foundations. Artificial pozzolans came to be standard ingredients in cement-based grouts, as did superplasticizers and, later, stabilizing additives. The European concept of using stable grouts gradually (if grudgingly) began to be accepted, and low-mobility ("compaction") grouts also were adapted for use in remedial foundation grouting. Fear of applying injection pressures greater than conservative rules of thumb began to subside. In large part because of the efforts of innovative specialty contractors, equipment manufacturers, material suppliers, and assorted consultants with international experience, U.S. practice began to evolve rapidly and is internationally acknowledged as a source of innovation, accomplishment, and expertise—especially in the remediation of grout curtains originally constructed between 1920 and 1970. This coming of age was particularly well demonstrated during the ASCE Geo-Institute's International Conference on Grouting in 2003, and is chronicled in Weaver and Bruce (2007) and USACE (2009) in particular.

Of all the areas where technological advancement has been achieved in the grouting industry, the use of computers and electronics is likely more responsible for the recent exponential growth than any other. Our ability to measure and control equipment and display data from numerous sources has allowed us to systematically reduce data to meaningful formats that previously required such exorbitant manual effort it was rarely if ever contemplated. Four examples are notable. First, sophisticated drilling equipment is available with automatic recording of drilling parameters such as torque, rpm, weight on bit, and depth. This is often referred to as "monitoring while drilling" or drilling parameter recording and allows us to calculate the amount of energy required to produce a borehole by drilling, as well as delineate soft or permeable zones or other possible features of interest below ground surface. Second, process control systems allow accurate and quick batching of large volumes of grout at the touch of a button. We can automatically record the amount of all materials added to the grout batch as well as measure certain physical properties of the grout. Third, our abilities to "see" underground through down-hole investigation techniques have greatly advanced the understanding of fracture networks in bedrock. This technology, originally developed by the petroleum industry, has gained general acceptance as a valuable site investigation tool in advance of and during grouting. And fourth, computer systems for collection and display of data allow us to gain valuable insight into the conditions we have encountered, measure "improvement" to the ground as work progresses, and provide extensive documentation of the work performed. They allow us to measure our processes at a high frequency and produce quality control records at a pace never before possible.

2.2.2 Drilling through unconsolidated materials

2.2.2.1 Overburden drilling perspective

Unconsolidated materials consist of noncemented or nonlithified materials commonly referred to as *soils*. The term *overburden* is often used to describe these materials and is one borrowed from the mining industry: from the miner's perspective, where the ore body is rock, the material *over* the ore is a *burden* that must be removed prior to reaching the economical ore body. In the case of manmade fills, particularly for water-retaining structures, the term *embankment* is commonly used. Regardless of whether you call it soil, overburden, or embankment, if the areas requiring treatment by grouting lie beneath or within it, you must drill through it. Consider yourself fortunate if your working surface consists of bedrock as drilling through overburden often presents one of the most challenging aspects of a grouting project.

As with any drilling system, overburden drilling productivity is highly dependent upon the materials penetrated. A system designed for high productivity in soft ground conditions can be brought to a standstill upon encountering a boulder. On the other hand, systems designed for hard ground conditions can clog and jam upon encountering soft ground. Unfortunately, given the wide array of overburden conditions that can be encountered (clay, silt, sand, gravel, cobbles, boulders, or a mixture) and specific site constraints, there is no single drilling system suitable for all ground conditions, or even all ground conditions on a particular site. Couple this with the fact that engineers and contractors often do not see things from the same perspective when it pertains to drilling methods, and overburden drilling can quickly become a contentious issue on a project.

From the contractor's perspective, use of the most efficient and economical drilling system, particularly if it does not require retooling or hiring an overburden drilling subcontractor, is desired. From the engineer's perspective, protection of the foundation (and embankment if present) and access to the features requiring treatment are paramount. The best recipe for success is for the engineer to provide a specification that clearly denotes the performance requirements of the drilling method, while identifying methods and procedures that are *not* acceptable.

2.2.2.2 Overburden drilling methods

In the majority of cases, the overburden drilling method must be capable of providing a stable borehole for subsequent insertion of a temporary standpipe through which soil grouting or rock drilling and grouting will be performed. Standpipe is typically constructed of either plastic or steel and the term is specific to a permanent casing left in the ground after the drill tools are removed. In some cases, the term *casing* is incorrectly used to describe a standpipe. Casing is a part of the drill tooling (which is removed), whereas standpipe remains in the hole after the drill tooling has been removed. Cased drilling methods are necessary when hole stability issues result in hole caving after tooling extraction.

Drilling with inappropriate methods can result in damage to the embankments and soil foundations: it is therefore crucial that appropriate drilling methods be employed for both investigative and production drilling activities. Although various agencies have specific guidelines, the U.S. Army Corps of Engineers' ER 1110-1-1807 (1997), *Procedures for Drilling in Earth Embankments*, provides excellent guidance with regard to methods and materials to be used to minimize the potential for embankment damage while drilling.

Borehole stability, drilling accuracy, ground conditions, and ultimately the specific requirements of the project dictate the appropriateness of a particular drilling method. Figure 2.1 illustrates the large variety of methods available for overburden drilling, which are discussed in the following. Further details are provided in Weaver and Bruce (2007).

2.2.2.2.1 Solid-stem augers

A solid-stem auger consists of a small-diameter drill rod or pipe with helical flights welded to the perimeter. The cutting head of the auger commonly consists of hardened steel with recesses for insertion of various cutting teeth.

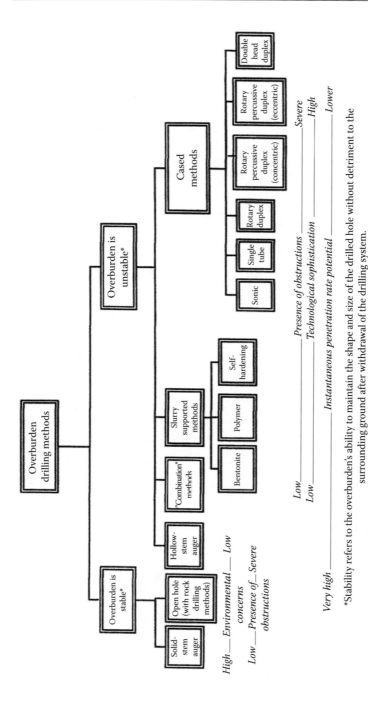

Figure 2.1 Classification of overburden drilling systems. (From Bruce, D. A., Grouting and Ground Treatment, Proceedings of the Third International Conference, Geotechnical Special Publication No. 120, edited by L. F. Johnsen, D. A. Bruce, and M. J. Byle, American Society of Civil Engineers, 2003. With permission.)

Hole advancement is achieved by rotation of the drill string and simultaneous downward force (crowd). Cuttings are evacuated to the surface by the rotational action of the helical flights. Auger sections are typically coupled by loosely fitting male/female–type connections that transmit rotational forces with through bolts or pins to resist crowd and pullback forces.

Solid-stem auger systems are appropriate for shallow holes in stable overburden. Clays or other soils that exhibit adequate standup time after tooling removal are also considered appropriate. However, shallow groundwater can greatly reduce standup time. Solid-stem auger systems can penetrate cobbles and small boulders if equipped with the appropriate cutter head, although the system is generally not appropriate for ground with frequent cobbles and boulders. Solid-stem auger systems have limited ability to "socket" the hole into bedrock as minimal—if any—penetration into bedrock is possible. Penetration in weathered bedrock is slow. Upon reaching the desired hole depth, the augers are removed and a standpipe is inserted. Due to the loose-fitting nature of the coupling between sections, the use of solid-stem augers can result in large deviations from the design borehole alignment. Angled boreholes are typically not appropriate for solid-stem augers for the same reason.

2.2.2.2.2 Hollow-stem augers

Hollow-stem augers are similar to solid-stem augers in the sense that the rotational action of the helical flights and the crowd advance the hole. The major distinction between the two is the larger-diameter hollow stem of the hollow-stem system and the use of an inner rod or plug. The cutting head of the inner rod is typically slightly ahead of the hollow-stem cutting head. Hollow-stem auger systems allow for the use of standard geotechnical sampling methods such as the Shelby tube or standard penetration test during advancement. The inner rod or plug is necessary to prevent migration of cuttings up into the hollow stem, and use of the inner rod is mandatory to produce a clean hole bottom.

Hollow-stem auger systems have similar characteristics to solid-stem systems with regard to their ability to penetrate bedrock and drill-hole accuracy. One benefit is the ability to retrieve the center plug and install standpipe while still providing a hole supported by the outer helical stem. This makes the hollow-stem auger system a superior choice where borehole stability issues hamper standpipe installation.

2.2.2.2.3 Drilled or driven casing

Drilled and driven-casing drilling methods consist of an outer casing advanced into the ground followed by a secondary method to remove cuttings from the casing interior. Casings are advanced using rotary methods

Photo 2.1 Inner auger used to clean out drilled casing. (Courtesy of Advanced Construction Techniques, Ltd.)

with casing equipped with a diamond- or carbide-tipped shoe, or driven using a large hammer or weight operating on casing equipped with a driving shoe. In either method, cuttings are fed to the interior of the casing during casing advancement. Once the casing is advanced a predetermined increment, the interior of the casing is cleaned using a variety of methods (e.g., Photo 2.1). If cobbles and boulders are penetrated, rotary drilling methods and a tricone-type bit are typically utilized. Noncohesive soils may simply be flushed from the casing. Solid-stem auger methods are appropriate in nearly all soils provided they are free of significant cobbles and boulders.

Drilled and driven casing methods typically result in reasonably accurate drill holes in both vertical and angled drilling, and casings, particularly when fitted with diamond shoes, may be seated into bedrock. On water-retaining structures there is often some concern that flushing fluids during casing cleanout will pose risks to the embankment. However, this risk can be mitigated by maintaining a soil plug of a few feet inside the casing. Standpipe placement is facilitated by removal of the inner flushing or cleanout string, insertion of the standpipe, then removal of the casing.

2.2.2.2.4 Rotary sonic

Rotary sonic or sonic drilling has gained recent acceptance as the overburden drilling method of choice on water-retaining embankments (Bruce and Davis 2005). Sonic drilling utilizes a multiple casing system with each casing equipped with a hardened casing shoe typically containing carbide button bits. The drill utilizes rotation and crowd, as well as intense vibration from the hydraulically driven oscillator mounted above the casing, to advance the hole. Typically the inner casing is advanced the full stroke of the rig, then the outer or override casing is advanced to the same depth. During advancement of the inner casing, collection of a continuous "core" of the overburden is possible. After the inner casing is removed, the sample is typically extruded by vibrating the casing while pushing out the sample with water pressure. In cohesive soils the samples are minimally disturbed in the interior of the core although samples at the beginning and end of the run may not be representative of the *in situ* condition. Some water is typically used in this drilling method, as described in Section 2.2.2.4. Sonic drilling methods are capable of advancing through hard ground conditions and even significant depths into bedrock. The vibratory action of the drill string pulverizes the rock at the casing shoe, allowing for reasonable advancement rates in soft-rock conditions. Hard-rock drilling is possible with the method, but the samples recovered are typically significantly disturbed and advancement rates are undesirably slow. Hole depths greater than 200 feet are not uncommon with the sonic drilling method, so telescoping casing is often necessary to reduce skin friction below a depth of 100 feet or so. Rotary sonic drilling methods typically produce very accurate holes in both vertical and angled drilling applications. The use of dual casings provides a stiff drill string to counteract deviation tendencies as a result of ground conditions. Sonic drilling also allows for the taking of both soil and rock samples using standard geotechnical methods (diamond rotary coring, Shelby tube, standard penetration test). Setting of standpipe is simply a matter of removing the inner casing, inserting the standpipe, then retrieving the outer override casing.

2.2.2.2.5 Casing advancement systems

Casing advancement systems comprise the widest variety of overburden-drilling methods. Since the method includes an outer casing and an inner drill string simultaneously advanced and rotated and/or percussed, it is commonly referred to as duplex drilling. With these systems a casing is advanced with or without rotation and with or without percussion depending on the specific method. The bulk of the drilling effort is handled by an inner drill string. The inner string may include a top-hole percussion system or a down-the-hole hammer (DTH). The bit on the inner string may be an underreaming-type bit to clear space for the advancing casing or nonreaming-type bit depending on the method. In some cases the DTH on the inner string locks to the casing shoe and percusses both the inner string bit and casing shoe simultaneously. A wide variety of casing advancement systems has been developed by various drilling suppliers and goes by various trade names, including the venerable Odex and Tubex and more recent

Centrex, Super Jaws, and Roto Loc. In all cases, flushing fluids are required to evacuate cuttings. Cuttings are typically conveyed to the surface in the annulus between the inner and outer string. For top-hole percussion systems water or air, or water under pressure (Section 2.2.3.2), is circulated down the inner string.

Casing advancement systems typically produce very straight and accurate drill holes. These systems are likely the best suited for difficult ground conditions where cobbles and boulders are present. With most systems, particularly those utilizing a DTH, rock penetration is possible. In fact, these same types of system are commonly used for drilling micropiles and anchors to significant depth in rock. The only drawback to these systems is the use of flushing fluids in embankments. Some systems incorporate sophisticated fluid channels within the bit and hammer, which minimize exposure of fluids to the embankment. However, partial or complete obstruction of these fluid passages, which does occur even with the best drill operator, can result in pressurization of the embankment at the bit face, so increasing the danger of hydraulic fracturing.

2.2.2.3 Standpipe and interface treatment

The particular type or configuration of standpipe and the treatment of the embankment/rock interface (if required) are vital issues in a typical dam grouting project. Proper selection or specification of the appropriate materials, diameter, port spacing, and grouting methods is essential. Ultimately, the specific goals of the project dictate the type of standpipe and any subsequent grouting thereof. If grouting of highly permeable soils is desired, then the configuration of the standpipe will be significantly different than that of a standpipe required for purely rock grouting. If damage to the soil-rock interface is suspected or the extreme top of rock otherwise requires treatment, then the standpipe must be configured to accommodate these challenges.

2.2.2.3.1 Standpipe

Standpipe typically consists of either plastic or metal pipe of a diameter suitable to facilitate future subsequent work to be performed in the hole below it. For fissure grouting applications a typical standpipe will range from approximately two to four inches in diameter. If subsequent rock drilling must be performed through the standpipe, then the bit diameter of the rock drill will control standpipe size. If soil grouting through tube-à-manchette (TAM) or multiple-port sleeve pipe (MPSP) (Bruce and Gallavresi 1988) is required, then the diameter of the packers to be utilized will control the standpipe size.

In soil grouting applications, MPSP or TAM systems allow for grout access to the soil repeatedly and at frequent intervals. The standpipe includes machined recesses at intervals of typically two or more feet along its length.

Grout access holes of suitable diameter (typically on the order of one-half-inch) are drilled through recesses through the pipe. A tight-fitting rubber or elastic sleeve is then placed in the recess. When grout is pressurized inside the standpipe, it forces open the sleeve, allowing grout access to the formation. When the fluid pressure is relieved, the sleeve retracts, preventing the injected grout from running back into the standpipe. A double packer injection system is typically used in this application. With sleeves placed at frequent intervals and the standpipe remaining open due to the one-way valve action of the sleeves, injections can be made multiple times in any particular zone as needed. The major distinction between TAM and MPSP standpipes is inclusion of barrier bags on the exterior of MPSP assemblies. Barrier bags allow for the zoning of ports whereby a fabric bag affixed to the exterior of the standpipe is inflated with grout, effectively isolating the standpipe annulus between sleeve ports (Figure 2.2).

In rock grouting applications the degree of sophistication of the standpipe is a matter of the required level of interface treatment. If no interface treatment is required and if the interface will not accept grout under static head, it may be possible to simply use solid pipe with an open bottom and tremie both the annulus and interior of the standpipe to the surface simultaneously. In this scenario it is critical, however, that the grout inside the standpipe be removed after gelling for reasons described below. An

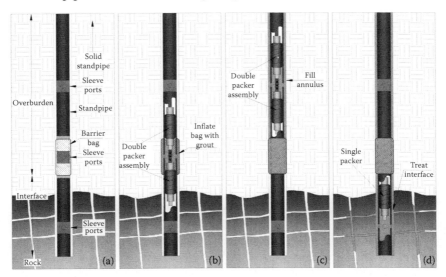

Figure 2.2 Sequence of construction for interface treatment using the MPSP system: (a) MPSP system installed in a borehole terminating in bedrock; (b) inflation of barrier bag with neat cement grout using double packer; (c) filling of casing annulus with neat cement grout using double packer; (d) treatment of the interface with appropriate grout mix using single packer. (Courtesy of Gannett Fleming, Inc.)

alternative is to provide a single sleeved section at the bottom of the stand-pipe with a closed bottom. A packer is inserted and pressurized grout opens the sleeve and fills the annulus. This minimizes the amount of grout left inside the casing, although flushing of the casing after grout gel should still be conducted.

2.2.2.3.2 Interface treatment

If interface treatment is required, it is prudent to isolate the interface from general standpipe annulus grouting and treat it as a separate operation. On previous water-retaining embankments where interface treatment was warranted but not specified, experience has shown that improper planning or contract provisions for such treatment can result in an incompletely grouted casing annulus and dam safety implications. These experiences include pumping thousands of gallons of grout during standpipe grout-ing (the interface was obviously capable of accepting large quantities of grout under static head) and finding casings that were grouted the day before later accepting significant quantities of grout (again, the interface accepted large quantities of grout simply under the static head in the annulus). Proper isolation and treatment of the interface is mandatory for all water-retaining embankments, but is often overlooked by designers of grouting programs.

The best available method for isolating and treating the soil-rock inter-face is the use of MPSP with a barrier bag placed immediately above the interface. The interface may consist of soils immediately above the top of rock that readily accept grout or otherwise necessitate treatment, weath-ered rock that was determined unsuitable for termination of overburden drilling, or freshly weathered rock in the borehole rock socket. Regardless of the reason, oftentimes overburden drilling extends through groutable conditions immediately overlying bedrock, and below the top of rock for a significant distance. Effective and systematic treatment of the materials penetrated by the standpipe is only possible using MPSP techniques.

The configuration is typically two or more ports above the top of rock or interface, the lowest of which includes a barrier bag, and at least one port below the barrier bag. The barrier bag typically consists of a tubular sec-tion of woven geotextile banded above and below the sleeve port. A double packer is utilized to isolate the port with the barrier bag, and neat cement grout is pumped into the bag. Pressure from the grout inflates the bag and firmly presses it against the borehole sidewall. The high filtration capa-bility of the woven geotextile barrier bag is exploited to pressure-filtrate the neat cement grout, resulting in dry-packing the bag. This effectively isolates the annulus between the subsequent standpipe annulus grouting operation and the interface treatment. Since dry-packing occurs, there is no need to wait for the grout to set before moving to subsequently higher ports. The double-packer system is simply moved to higher-level sleeve ports and neat cement grout is pumped into the annulus until observed at the ground surface. The remaining standpipe grouting activity is interface grouting through ports below the barrier bag, which may be conducted immediately following casing grouting or at a later time using a single-packer assembly.

The benefits of the MPSP system include the following:

- Positive and rapidly deployed isolation between the casing annulus and interface.
- Determination of whether embankment materials above the barrier bag are accepting grout beyond the theoretical standpipe annulus volume. If so, MPSP techniques may be utilized on subsequent holes to treat overburden at these locations.
- Permits systematic treatment of the interface using standard thickening procedures and the appropriate grout mixes.
- Quantities of grout accepted by the interface can be evaluated in a closure analysis to determine if project goals have been achieved.
- Provides a higher level of protection to the embankment.
- Results in fewer requirements for hole "topping off" to ensure the standpipe and overburden are fully supported over the length of the standpipe.

2.2.2.4 Specific concerns regarding embankments

Embankment dams present specific hazards with regard to overburden drilling and standpipe installation. The uncontrolled use of circulating fluids during overburden drilling can result in hydrofracturing of the embankment under certain conditions. The integrity of a standpipe can be compromised, resulting in embankment exposure to the erosive forces of flushing fluids during hole washing or rock drilling. It is prudent that these be considered during both design and construction of grout curtains through water-retaining embankments.

Again, the U.S. Army Corps of Engineers' ER 1110-1-1807, *Procedures for Drilling in Earth Embankments*, provides excellent guidance on the issue and is recommended for further reading by all grouting practitioners contemplating drilling through embankments. The ER, however, is geared toward exploratory drilling operations and the setting of temporary casing for subsequent exploration in rock. It is therefore somewhat specific to exploration methods, proper hole abandonment after sampling, and retrieval of the temporary casing. It does not specify methods for installing permanent standpipes or standpipe configurations for treatment of the interface. Those specifying methods for drilling in embankments for purposes of foundation grouting must bear this in mind and supplement the requirements of the ER with project-specific requirements. When drilling through embankment for purposes of setting a permanent standpipe, use of MPSP methods is highly recommended and provides additional protection to the embankment.

2.2.2.4.1 Hydrofracturing embankments

The term *hydrofracture* in this connotation is specific to tensile failure of the soil as a result of the application of pressurized fluid from the drilling process. Sometimes called a "frac." it can propagate from the borehole significant distances and hydraulically connect zones of the embankment, foundation, and possibly the pool and downstream areas. In the worst-case scenario it could potentially result in a direct hydraulic connection between the pool and downstream areas with little to no headloss to seepage along its length—obviously a disastrous condition for an embankment dam. Otherwise, it may locally damage the embankment, which can allow for piping, settlement, or stability issues to occur at a later date.

It is imperative that proper drilling techniques be selected when overburden drilling in embankments. Auger drilling methods and drilled or driven casing are likely the safest methods for hole advancement through overburden on embankments. However, there are obvious drawbacks to these methods, including hole accuracy, depth limitations, and productivity. Rotary sonic and casing advancement methods do not have these same limitations, although they do have their own inherent risks. The sonic drilling method is often described as being possible with no flushing fluids. While in theory the system can advance a hole using "dry-drilling" methods and has been utilized as such in the past, some water is required for two reasons. First, the water lubricates the inner drill string during advancement, increasing the penetration rate, and, second, significant heat is generated during drilling and water provides a means of cooling to prevent premature tooling failure. In some scenarios, sonic drilling has been prescribed using "minimal" amounts of water and only during certain phases of the tooling advancement. While this looks good on paper and satisfies regulatory requirements, the reality is that if any water is permitted, its use will likely be in excess of that prescribed and inevitably water will find its way into the tooling by other means not easily observed by an inspector. The authors believe that if water is utilized at inappropriate times during sonic drilling, it presents the same if not greater risks as other drilling systems that use water as a flushing fluid.

If water is utilized during sonic drilling, it should only be done so in very limited and quantifiable amounts. Ideally, water should not be used when advancing the inner casing as it results in disturbance to the recovered sample (particularly the top several feet) and likely represents the greatest potential for hydrofracturing to occur. If water use is permitted during inner casing advancement, the entire length of the inner drill string should be demonstrated to be unobstructed after removal of every sample from the sampler (material often travels up the length of the sampler and into the overlying drill rods), and water that is used to demonstrate that the casing is unobstructed should not be "accidentally" directed down the outer casing as has been observed on prior occasions. In no case should water be poured down the borehole and then a casing with a top closed to the atmosphere driven into the standing water. The top of the casing should always be open to the atmosphere to prevent the downward force of the casing from pressuring the fluid. Advancement of the outer override casing likely presents less risk with regard to hydrofracturing, provided the system is again open to the atmosphere. With the sonic method, in no scenario should pressurized water be fed down the drill string while drilling through an embankment.

For those inspecting the work and ensuring compliance with contract provisions, the best method for demonstrating that the sonic drilling process is not damaging the embankment is to observe the recovered samples. If water is utilized and more than the top one or two feet of the sample is significantly wet and disturbed (particularly so for soils exhibiting significant cohesion), then an inappropriate amount of water is likely being utilized. Water use should be cut back until minimal sample disturbance is observed. Sonic drilling is a valuable investigation tool and production standpipe installation method for grouting projects on embankments. Drilling with no water, although greatly limiting the productivity of the system, is possible. If water use is permitted, it should only be done so during certain sequences of the drilling method as described above and only in minimal amounts. Specification writers must be very specific with regard to these issues and inspectors must be vigilant in their duties when observing the process.

Casing advancement systems likely present the greatest risk for hydrofracturing, particularly those that use air-powered DTH drilling methods. Frac propagation can be maintained at pressures less than those needed to initiate the frac; therefore, a zone of compressed air with tremendous stored energy has the capability to propagate a significant frac prior to any loss of circulation being noticed by the driller. Pressurization at the bit face due to obstruction of the flushing fluid by cuttings bridging the relief ports is typically the reason for circulation loss. This is largely the reason why casing advancement systems are usually prohibited from use on embankments.

Casing advancement systems that use top-hole percussion methods and low-pressure circulation fluids present a much lesser risk with regard to hydrofracturing. Some systems that utilize drilling mud as the circulation medium explicitly meet the requirements of some regulatory agency guidelines. However, their use has continued to be prohibited as a result of previous dam safety issues with casing advancement systems in general. With the continued advancement of casing advancement systems, a system guaranteeing a low-pressure bit face and zero potential for hydrofracturing will likely be developed. Until that time it is recommended that casing advancement systems on embankments be limited to those that use top-hole percussion methods with low-pressure flushing fluids, most notably drilling mud.

2.2.2.4.2 Unsupported standpipes

In addition to the need for fully grouting a standpipe to prevent direct hydraulic connection between the embankment and the interface/foundation, a standpipe must be fully supported to withstand the loads imposed by subsequent rock drilling and grouting operations. When drilling through grout inside a standpipe, significant tensile stresses are applied to the standpipe wall. Deep standpipes where high packer pressures are utilized also result in high-tensile loads on the standpipe. If the standpipe is not fully encased by grout and free to expand in response to the loads, splitting of the standpipe is possible. If the breach in the standpipe is substantial, the embankment may be exposed to significant flushing fluids during subsequent rock drilling and hole-washing operations. Again, adequate casing grouting and interface treatment mitigate many of these concerns.

In some cases the casing may split during drilling as a result of very soft or weak soils surrounding the standpipe, regardless of whether or not it is fully grouted. In order to address this and other concerns, a good practice is to flush any grout from inside the casing after any operations that present the possibility for grout to intrude into the casing. These would include after any casing grouting, barrier bag filling or interface treatment, or after downstage drilling and grouting techniques in the zones immediately below the bottom of standpipe. While these concerns generally pertain to the use of plastic casings, drilling through grout even in steel casing is not an ideal situation as it can result in significant wear to the bit.

2.2.3 Drilling through consolidated and/or lithified materials

2.2.3.1 Rock drilling perspective

Consolidated materials consist of cemented or lithified materials commonly referred to as rock. The main distinctions between consolidated and unconsolidated materials as they pertain to drilling for grouting projects are that consolidated materials generally require the use of percussion or cutting for removal, and they generally produce a stable borehole after removal of the drill tooling.

Like overburden drilling systems, the multitude of rock drilling systems can vary in significant ways and no single system can be expected to produce satisfactory productivity in all conditions. However, unlike over-burden drilling systems, many times modification to the specific tooling being used (specifically changing the bit type) will allow for drilling through most bedrock formations. Again, the best recipe for success for owners and specification writers is to provide a specification that clearly denotes the requirements of the method and leave ultimate selection of the method up to the drilling specialist with the caveat that the specification should be precise about what is not acceptable.

2.2.3.2 Drilling methods

A multitude of drilling methods are available for the grouting practitioner. Generally speaking, two major categories exist: rotary methods and per-cussion methods. Rotary methods use cutting or shearing actions for rock removal, whereas percussion methods use impact energy to crush or chip the rock. Rotary methods can be further subdivided into high-speed and low-speed methods. Similarly, percussion methods can be further divided into top-hole and bottom-hole percussion methods. With regard to the depths to which we grout, practically speaking, there is no depth limitation with any of the above drilling methods with the exception of top-hole percussion as discussed below.

For the majority of grouting projects, self-contained, self-propelled drilling rigs are the norm. This minimizes the site infrastructure required to support the equipment. These are gasoline- or, more typically, diesel-driven machines having sophisticated hydraulic systems. Tracked or rubber-tired truck configurations are common. Skid-type rigs are still available. Where drilling must be performed indoors or within confined spaces, smaller electrically powered drills are used.

2.2.3.2.1 High-speed rotary

High-speed rotary drilling methods are commonly associated with coring or diamond drilling. This drilling method utilizes a high rotation speed, of the drill tooling and typically a coring-type bit to cut the rock, although plug-type bits are available that result in no sample recovery. Rotation rates range from several hundred to one thousand RPM for larger-diameter holes to several thousand RPM for smaller diameter holes. Minimal crowd and torque is necessary to advance the hole in comparison to other methods. Water is typically used as the flushing fluid. Most commonly a cored sample of the rock is recovered in this method. A wide variety of bits is available, but the majority of high-speed rotary methods use either impregnated or surface-set bits.

Both impregnated bits and surface-set bits typically use diamond as the cutting agent. The business end of an impregnated bit is a matrix of diamonds set in powdered metal. The powdered metal wears over the life of the bit, exposing additional diamonds to aid in cutting. In some cases intentional wear of the bits' powdered metal matrix is induced to expose more diamonds, which results in faster penetration rates. Surface-set bits differ from impregnated bits in that a surface-set bit contains only a single layer of diamonds on the working surface of the bit. Diamonds are typically set in a predetermined pattern to ensure that cutting agents are located over the full face of the bit-cutting area. Within both the impregnated and surface-set types, various other options, including matrix hardness, diamond size, and bit shape, are available in consideration of the nature of the rock to be cut, including abrasiveness, hardness, and particle size.

In addition to the configuration of the bits, the configuration of the rotary tooling itself and method in which rock samples are retrieved can be highly variable. Simple methods require that the entire drill string be removed in order to remove the rock sample from the core barrel. After sample removal the entire string of tooling must be reassembled and deployed down the hole. These methods typically result in significant sample disturbance and are generally limited to holes where only a few runs at shallow depth are necessary to achieve hole bottom. More sophisticated methods utilize a complicated core barrel assembly whereby the drill tools are left in the hole and a separate winching system, commonly referred to as a wireline, retrieves a detachable inner core barrel containing the sample. This speeds production and results in minimal disturbance to recovered samples.

For many decades high-speed rotary drilling was the method of choice for grouting projects. This is due to the use of water as the flushing fluid, the ability to drill deep holes with smaller rigs, the relatively high degree of accuracy achieved, the smaller-sized cuttings produced that generally do not clog the majority of rock fractures of interest, and the ability to recover a rock sample for subsequent inspection. High-speed rotary drilling is generally considered to be a more expensive and less productive drilling method in comparison to other presently available techniques. However, its use is still common on many grouting projects as an investigation and verification tool, and use of the method as a production drilling tool will typically provide equivalent or superior hole characteristics in comparison to other available methods.

2.2.3.2.2 Low-speed rotary

Low-speed rotary drilling is not as common as other methods in grouting applications as it is typically reserved for hole sizes larger than those necessary for grouting. Low-speed rotary drills employ high torque and high crowd to cut the rock. This method typically requires a larger and heavier drill rig to develop adequate crowd on the bit. Depending on the exact configuration of the drill bit, shearing or crushing action excavates the materials penetrated. Flushing fluids are typically used to remove cuttings from the bit face.

The bits typically used are of either tricone or drag-type configuration. Tricone bits employ conical-shaped rollers typically constructed of hardened steel or faced with carbide for particularly hard formation. Tricone bits use a crushing-type action to advance the hole. Drag-type bits are also similarly constructed of hardened steel or faced with carbides. The bit cutters penetrate the rock and the torque applied by the drill locally shears the rock at the bit face, resulting in hole advancement. A bit type more common in the petroleum industry, which utilizes similar shearing action, is the polycrystalline diamond compact bit or PCD. PCD bits consist of synthetically grown diamond discs mounted to the face of the bit and are found in a wide variety of configurations throughout the petroleum industry.

2.2.3.2.3 Top-hole percussion

Top-hole percussion methods utilize either a pneumatically or hydraulically activated hammer, sometimes referred to as a drifter, to advance the hole. The hammer is located at the top of the drill string and moves up and down the mast of the rig. Slow rotation of the drill string is provided to facilitate cleaning at the bit face and ensure that the hammer energy from subsequent strikes is delivered to intact rock rather than the cuttings from the previous strike. Typically water is used as a flushing medium.

Due to their high production rates, top-hole percussion methods are extremely economical choices for shallow depth curtains or blanket treatments. As hole depths increase, drilling production rates will decrease to the point where the method is no longer economically viable. The depth limitation to top-hole percussion rigs is attributed to the energy lost at joints between the individual drill rods. As more rods are added to the drill string, more energy is lost at the joints, reducing the impact energy delivered to the bit. This limits the method to a practical depth of approximately 150 feet. For deeper holes, alternate drilling methods should be considered.

One of the major drawbacks to the top-hole percussion method is hole deviation. The location of the hammer at the top of the drill string, couplers between rod sections, and smaller diameter of the rods in comparison to the bit all allow for sometimes significant deviations to occur regardless of good rig alignment at the surface (Weaver and Bruce 2007). One method for improving hole accuracies with top-hole percussion methods is to use a guide rod or tube located directly above the bit. A guide rod's outside diameter is just slightly smaller than that of the bit and therefore the bottom several feet of the drill string is unable to significantly deviate. Adequate flow channels are provided in the guide tube to facilitate transport of cuttings along its length. A guide tube is an inexpensive and standard drilling tool for top-hole percussion rigs and its use is recommended in grouting applications.

Top-hole percussion bits are available to handle a wide variety of ground conditions. The controlling factors for bit selection are typically penetration rate and wear rate. These factors are controlled by modifying the geometry of the bit, and particularly the arrangement and shape of the carbide buttons. The shape, number, and size of these buttons can be modified to accommodate various ground conditions. Additionally, removal of cuttings is facilitated by the number, size, and location of flushing ports on the bit.

2.2.3.2.4 Bottom-hole percussion

Bottom-hole percussion is also known as down-the-hole (DTH) drilling. As the name implies, the method utilizes a percussion hammer located at the bottom of the drill string. Slow rotation of the drill string is provided by the rotary head mounted on the drill mast above. The method by which rock is excavated is identical to that of the top-hole percussion. The majority of bottom-hole percussion hammers are pneumatically actuated by high-pressure air fed down the drill rods. Since use of air flush is prohibited on fissure grouting projects, bottom-hole percussion methods were not common for rock fissure grouting applications until recently. With the development of the water-powered hammer, bottom-hole percussion drilling on grouting projects has become a viable and attractive option.

While there are distinct differences in depth capacity between pneumatic hammers and water hammers that warrant discussion for other applications, for grouting projects and the depths to which we treat they are practically equivalent. The same applies to production rates as both methods are considerably more productive at greater depths than top-hole percussion methods. The main distinction between the two methods is the fluid used to deliver energy to the hammer and subsequently used to flush the hole. The required air-flow rates and pressures for holes in the four-inch-diameter range are typically on the order of several hundred to generally less than one thousand cubic feet per minute with air pressures typically ranging from about 150 to 350 psi. In comparison, water hammers of similar size use water at rates of approximately 20–30 gallons per minute and at delivery pressures of 2,000–2,500 psi. In both systems, the flushing fluids exit at the bit face and carry cuttings to the surface through the annulus between the drill rods and the borehole sidewall.

The fundamental difference between air- and water-hammer systems is the compressibility of the flushing fluids. Air is a compressible fluid and even after significant pressure is lost through the hammer, stored energy remains present in the spent air and is necessary to convey such a large volume of fluid up the annulus. The issues regarding this are threefold. First, it can cause hydrofracturing, either due to the presence of soft zones or due to obstruction of the annulus, causing pressures to rapidly build in the borehole below the obstruction. Second, the high velocity of the flushing fluid scours the sidewalls of the borehole and results in an irregular borehole section, particularly in softer rocks, which can present challenges with setting and sealing packers. And third, as it expands, the pressurized air can drive the cuttings into the fractures we wish to treat. For these reasons, air flush is prohibited in grouting applications, which functionally prohibits the use of air-powered DTH systems.

At first glance, water-hammer systems may appear to present even higher risks due to the higher initial pressure of the drilling fluid. However, the majority of this pressure is consumed by the hammer and the pressures exiting the hammer are similar to those in traditional top-hole percussion systems that use water flush. The other distinguishing factor is that water is an incompressible fluid and therefore does not present the same risks due to its inability to store energy (pressure). Hydrofracturing, while a risk with any flushing fluid, is no more likely to occur with a water hammer than with other traditional grout hole drilling methods such as top-hole percussion or core drilling with water flush. Significantly lower volumes of flushing fluid are required as cuttings are more easily evacuated with water in comparison to air. The velocity of the flushing fluid and corresponding velocity of the cuttings is therefore much lower, resulting in minimal scour of the borehole sidewalls. This also greatly reduced the likelihood for cuttings to be driven into the fractures we wish to treat. Water hammers are a significant and recent technological achievement in drilling for grouting applications and provide the same benefits as other water-flush methods typically permitted, while simultaneously resulting in the ability to drill deeper and more accurate boreholes at high penetration rates.

Top-hole percussion and water-activated DTH are the primary drilling methods employed in the grouting industry for production grout hole drilling in rock. Both methods produce satisfactory results. For maximum depths of approximately 150 feet top-hole percussion methods will likely prove suitable. Beyond this depth top-hole percussion methods are ineffective and DTH or high-speed rotary are commonly used.

2.2.3.3 Circulating fluids

Use of air as a circulating fluid in rock fissure grouting applications is prohibited due to the tendency for air to drive cuttings into the fractures, so effectively obstructing them against subsequent grout penetration. Subsequent flushing of the hole with water will typically not remove these cuttings. Air circulation also presents particular risks for hydrofracturing in embankments, as discussed above, possibly even when drilling in an underlying rock foundation if fractures allow for transmission of pressurized air to the interface. Air flush is a viable alternative when drilling through voids or cavities where permeation of fine fractures in the rock is not a requirement, and no collateral damage will be caused by air escaping from the drill hole. Thus water flush is the medium of choice and practice.

2.2.4 Hole washing

2.2.4.1 Necessity

Washing of grout holes is necessary for a variety of reasons. First and foremost, washing is necessary to scour the rock sidewalls of the borehole to remove any remnant cuttings, joint gouge, or infilling that otherwise could obstruct grout penetration. Hole washing is also the first activity after drilling and provides valuable insight into the stability of the hole (i.e., hole collapsed or open and stable) and the consistency of the materials penetrated (i.e., competent rock or highly weathered material in flush return). Washing of grout holes is also necessary to remove fresh grout from the standpipe installation or from the rock portion of the borehole if downstage techniques are utilized.

2.2.4.2 Washing tools

Washing is not a complicated process. It relies on the use of a wash bit (sometimes called a torpedo or wash wand) to direct radial jets of water against the borehole sidewall. Typically, wash bits are manufactured by the contractor out of standard pipe sections. A typical wash bit will be two to four feet in length with a series of small-diameter holes (approximately one-eighth inch) drilled around the circumference. It is common to have a slightly larger diameter hole at the bottom of the tool that points directly down the hole, since after drilling and during washing, cuttings not evacuated from the hole will fall to the hole bottom. The bottom discharge port on the wash bit agitates the material on the bottom of the hole as it approaches, allowing it to be flushed to the surface. The diameter of the bit should also be considered. It is undesirable for the body of the bit to be significantly smaller than the borehole diameter in order to minimize the column of water between the radial jets and the borehole. The diameter of the wash bit should be between one-half and one inch smaller than that of the borehole. Connection to the water supply hose or pipe is made at the top of the wash bit.

2.2.4.3 Pressure and flowrates

It is common practice to specify a minimum water pressure and flowrate to be delivered to the hole collar for hole washing. Typical specifications will include a minimum rate of 10–20 gallons per minute at a pressure of 100–150 psi. This range of values is adequate for the majority of applications.

The greatest concern regarding the pressure and flowrate is that it be delivered to the appropriate wash bit; it is desirable for the bit to produce pressurized streams of water in a radial fashion. This can only be produced if pressure is allowed to build inside the wash bit. The ability to develop pressure inside the wash bit is a function of the size and quantity of radial discharge holes drilled through the bit. If too many holes are drilled, no pressure will build in the bit but the flowrate will be significant. If too few holes are drilled, the pressure will be adequate although the flowrate will be minimal. Prior to utilizing a new wash bit, it should be demonstrated that the appropriate jetting action will be produced by the pump and bit combination. If a bit performs unsatisfactorily, it can be modified by adding more radial holes or closing existing holes.

2.2.4.4 Field observation

The washing operation typically starts at the top of the hole and progresses downward. The operator in the field visually monitors the turbidity of the water and any materials evacuated as the wash tool is lowered. Zones of the hole that produce particularly turbid water and result in accumulation of sediments at the surface should be washed thoroughly until clear. In some cases where flush water does not clear after a reasonable amount of effort, washing should continue to deeper zones as continued scouring of the hole will result in difficulties setting and retrieving packers. The wash tool should be raised and lowered a few feet off the bottom of the borehole to be sure any accumulated cuttings or sediment are flushed to the surface.

It is important that washing operators keep a log of the materials and conditions encountered, particularly if difficulties were encountered with collapsing hole conditions. This is a strong indication that packers may be lost in the hole during subsequent operations. Accurate washing records allow for setting of packers in consideration of the locations where difficulties were encountered.

2.2.5 Water-pressure testing

2.2.5.1 Purpose and applicability

Water-pressure testing is conducted to systematically evaluate the permeability of the rock in each stage. Given the inability to directly observe the effectiveness of a grouting program in the subsurface, and the inherent difficulties in quantifying seepage sources and flow paths, pressure testing is a vital element in assessing and guiding the performance of the work. Pressure testing is therefore mandatory for any grouting application for seepage reduction. Pressure testing is applicable for nearly any rock foundation, although certain precautions must be taken if the project includes remedial efforts in embankment dam foundations. This is due to the potential for the injected water to flow to the top of the rock and potentially damage the embankment: a properly treated interface greatly reduces the risk of hydrofracturing or piping during pressure testing.

2.2.5.2 Methods

Three primary methods are used to pressure test grout holes at depth. These include the single-packer method, the double-packer method, and the wireline method. The single- and double-packer methods are the most common methods for pressure testing grout holes, while wireline packer systems are typically reserved for difficult ground conditions. The configuration of and materials used to construct the packer vary between manufacturers, but all include an inflatable bladder allowing for isolation of a zone in the grout hole such that water may be introduced into the stage. Typically, at least two connections, either by hose or pipe, are made to the packer. A small-diameter inflation line allows for inflation and deflation of the packer. Compressed gas or liquids may be used to inflate packers. However, it is recommended that only compressed gas be utilized as it allows for more rapid inflation and deflation. The second connection is to the water source. Packers used in grouting applications are typically on the order of two to three feet in length. Longer packers, which are recommended, routinely provide a better seal with the borehole sidewall.

Single-packer systems are used to pressure-test the bottom of an upstaged grout hole or all stages in a downstaged grout hole. This packer arrangement simply isolates the portion of the borehole below the packer and water is injected below the body of the packer. Double-packer systems are used to test stages above the bottom of an upstaged grout hole. Double-packers isolate the zone between the upper and lower packer, and typically a perforated pipe or hose is located between the packers to allow water to access the stage. The bottom of the lower packer is capped in a double-packer arrangement. In some formations caving of grout holes after drilling is common, resulting in difficulties setting packers or even the loss of packers. In these situations the wireline packer system proves most suitable. Examples of these three packer arrangements are shown in Figure 2.3.

Wireline packer systems must be used in conjunction with high-speed rotary drilling methods (coring), the reason being that the drill rods act as a temporary casing, preventing caving of the hole. After making a core run, the drill tools are pulled up the length of the run plus the length of the wireline packer. With the inner core barrel removed, the wireline packer system is inserted through the tooling until the packer exits from the bit

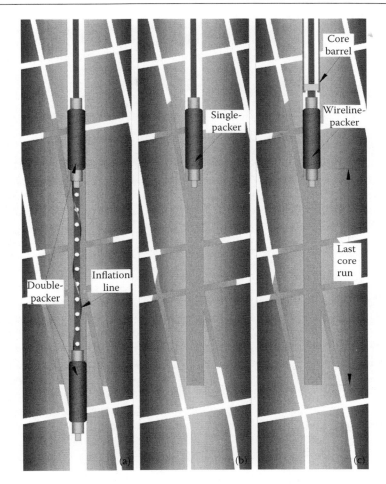

Figure 2.3 Packer systems: (a) double-packer system; (b) single-packer system; (c) wire-line-packer system. (Courtesy of Gannett Fleming, Inc.)

face. Typically the geometry of the wireline packer system only allows the packer to exit the bit. The packer is inflated and the recently drilled core run is pressure tested. The packer is then removed and drilling continues. There are various configurations of wireline packer systems available, but all are deployed through the drill tooling. Wireline packer systems offer distinct operational advantages over traditional single- and double-packer systems during site investigation or early phases of grouting in unstable ground.

In some instances, particularly for shallow curtains where the downstage methodology is applied, the use of a mechanical packer set at the ground surface can be both effective and economical. A mechanical packer simply uses an expandable plug or ring surrounding the injection pipe. As clamping force is applied to the ring it expands, sealing against the sidewall of the borehole. In this method it is assumed that any measured permeability is only associated with the ungrouted stage at depth, that is, no water take is assumed to occur in the previously grouted stages. (This is, however, usually not an accurate assumption, unless the stage in question took no water before grouting.) Both water-pressure testing and grouting can be performed with this method. Mechanical packers are limited in the amount of differential pressure they are capable of sustaining, more so than pneumatic packers due to their short gland length. Therefore, this method is only applicable to shallow-depth curtains. It is also critical that the pressures applied are not detrimental to the shallower stages since there is effectively no isolation between the high pressure necessary at depth and the stages grouted at lower pressure above.

2.2.5.3 Lugeon value

The unit of permeability measurement typically used in rock fissure grouting is the Lugeon, and it is unique to the grouting industry. A Lugeon is defined as one liter per minute of flow injected into a one-meter length of borehole at a pressure of ten bars. Two Lugeons would therefore be a flow of two liters per minute for the same isolated length of borehole and pressure. Practically speaking, it is uncommon to isolate a one-meter length of borehole just as it is uncommon to test all stages at a pressure of ten bars as was the original modus operandi of the good M. Lugeon himself. In order to address the varied injection pressures and stage lengths the equation is simplified to accommodate these variables as follows:

$$\text{Lugeon Value} = \frac{Q \times 10}{L_{STAGE} \times P_{EFF}}$$

Where Q is flowrate in liters per minute
L_{STAGE} is the stage length in meters
P_{EFF} is the effective pressure in bars

The conversion from Lugeon value to other permeability units is not exact, but is readily accepted by the industry as 1.3×10^{-5} cm/s.

2.2.5.4 Effective pressure

Effective pressure is the actual fluid pressure achieved in the isolated stage. Effective pressure is typically how injection pressures are specified, although it is uncommon to directly measure them. While packers are available with integral pressure transducers that do allow direct measurement of effective pressure in the stage, they still remain uncommon in the grouting industry for a variety of reasons. Pressures are typically measured at the ground surface and effective stage pressures are calculated based on the variables in the hydraulic system.

Calculation of effective pressures is a simple matter of algebraically summing all of the head losses and head gains. Static considerations include the inclination of the borehole, the proximity of groundwater to the stage, the specific gravity of the injected fluid, and the proximity of the pressure-sensing device to the borehole collar; dynamic considerations include head losses associated with the injection system. It is common to calculate effective pressures at the midpoint of the stage. However, special consideration must be given to the shallowest stages if they are overlain by embankment, particularly if long stage lengths are utilized, as pressures higher than are safely warranted may be generated if the stage midpoint is the basis for the calculation of allowable pressure.

The following are general guidelines, together with an example calculation for determining effective pressures with application to the majority of grouting projects. The selection of the actual targeted maximum pressure is addressed in Section 2.2.7.2. Special consideration must be given for uncommon conditions such as grouting a borehole drilled in an upward direction and for artesian groundwater conditions.

- The head applied by a one vertical foot column of water is the unit water head, or 0.433 psi/ft. This is derived by dividing the unit weight of water of 62.4 lbs/ft^3 by 144 in^2/ft^2. Water has a specific gravity of 1, and the head applied by other fluids (e.g., grout) is calculated by multiplying the specific gravity of the fluid by the unit water head.
- If the pressure-sensing device is above the hole collar, the vertical distance between the collar and the pressure sensor is a head gain. If it is below the collar, it is a head loss.
- The static head between the hole collar and the stage midpoint generated by the pressure of the fluid in the injection system must be calculated in consideration of the borehole inclination. The vertical distance between the collar and stage midpoint is to be used.
- Groundwater must be considered in similar fashion to the static head. The vertical column of groundwater above the stage midpoint must be measured. If groundwater is below the stage midpoint it is not considered in the calculation.
- Dynamic losses in the injection equipment will vary depending on the equipment type, size, capacity, and condition. Specific head-loss models should be developed for different types of injection equipment.

Example (using the system shown in Figure 2.4)

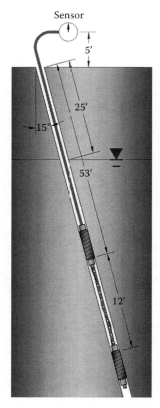

Figure 2.4 Details of the system used in the worked example. (Courtesy of Gannett Fleming, Inc.)

- Calculate the required pump pressure for the pressure test example assuming an allowable effective pressure of 1.0 psi/ft of depth. Ignore dynamic losses. The required effective pressure is the vertical distance from the hole collar to the stage midpoint times allowable effective pressure.

Effective Pressure $= 1.0 \, \dfrac{\text{psi}}{\text{ft}} \times (53 \text{ ft} + 6 \text{ ft}) \times \text{COS } 15° = 57.0 \text{ psi}$

- Calculate head losses and gains.
  Gauge is above collar; therefore, it is a head gain:

Gauge Head $= 0.433 \, \dfrac{\text{psi}}{\text{ft}} \times 5 \text{ ft} = 0.2 \text{ psi}$

Correct for vertical component of borehole angle, use stage midpoint as point of interest:

$$\text{Static Head} = 0.433 \; \frac{\text{psi}}{\text{ft}} \times (53 \; \text{ft} + 6 \; \text{ft}) \times \text{COS} \; 15° = 24.7 \; \text{psi}$$

Groundwater is above stage midpoint so it is a head loss, correct for vertical component of groundwater above stage midpoint:

$$\text{Groudwater} = -0.433 \frac{\text{psi}}{\text{ft}} \times (53 \; \text{ft} + 6 \; \text{ft} - 25 \; \text{ft}) \times \text{COS} \; 15°$$

$$= -14.2 \; \text{psi (head loss)}$$

- Calculate total head in system

  Total Head $= 2.2 \; \text{psi} + 24.7 \; \text{psi} - 14.2 \; \text{psi} = 12.7 \; \text{psi}$

- Calculate required pump pressure.

  Pump Pressure $= 57.0 \; \text{psi} - 12.7 \; \text{psi} = 44.3 \; \text{psi}$

2.2.5.5 Frequency and interpretation

Pressure testing should be performed in advance of grouting any grout stage in any series of holes. Attempting to economize the cost of a grouting project by reducing the frequency with which pressure testing is conducted is not recommended. At every step of the grouting program the permeability of the foundation should be determined for planning purposes, the overall economy of the project, and the evaluation of the conditions encountered. The cost to perform pressure testing is minimal compared to the costs associated with drilling and grouting the grout hole, and may be inconsequential relative to the cost of additional works that may have been instructed to (unqualified) concerns about the effectiveness of the treatment.

Given the outstanding capability of computer grouting systems to rapidly collect data from the field, computerized monitoring of flow and pressure during water-pressure testing is now considered standard practice. A typical precomputer monitoring age pressure-testing system included a standard household water meter and dial pressure gauge: the data collected using these types of systems provided a relatively low-resolution data set. As such, the ability to interpret such data was limited and test durations were relatively long. With computerized data collection, the duration of pressure tests can be significantly reduced in comparison to manual methods, and a very high resolution data set is collected allowing for rapid interpretation.

Regarding interpretation, of greatest concern for remediation projects is the frequency and magnitude of erosion that can be inferred from the data. Erosion is an increase in permeability of the formation (e.g., Figure 2.5) with respect to time in response to the flow applied by the test. It is commonly attributed to weathered materials present in the rock joints or fractures, which erode under flow, resulting in a larger conduit in which to transmit water. The signature of such an event is a decrease in pressure at

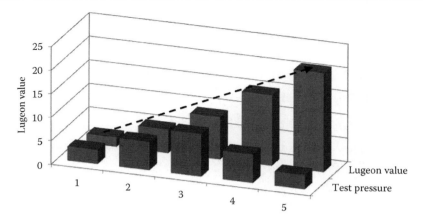

Figure 2.5 Erosion trend signature—increasing Lugeon value during stepped pressure test. (Courtesy of Gannett Fleming, Inc.)

a constant flow rate, or increase in flow rate at a constant pressure, both of which result in the Lugeon value increasing. The concern regarding erosion is future piping or transport of materials in the rock fractures after impoundment and zones of the foundation that consistently illustrate an eroding trend may require special or more intensive treatment.

Other interpretations of pressure-testing trends include dilation (elastic expansion of the fracture in response to the applied head) and infilling (decreasing permeability with time, opposite of erosion). Houlsby (1976) also discusses laminar versus turbulent flow in fractures, but with regard to treatment using grouting methods, these are of little consequence. Multistep pressure tests using varied injection pressure are typically conducted in investigatory and verification test holes; single pressure tests are usually conducted in routine grout-hole stages.

2.2.5.6 Field procedures

Pressure testing should be conducted as soon as practical after hole washing to minimize the chances that the hole will cave and thereby trap the packer. In downstage applications the bottom of the packer is set at the bottom of the previously grouted stage. In upstage applications the hole bottom is tested with a single packer. Typically the single packer is set above the hole bottom a distance equivalent to that between the packers for subsequent double-packer tests. This provides a stage length consistent with the stages above. The lone exception to this is the top stage of the hole where overlap with the stage immediately below is typically necessary.

After inflation of the packers, the supply valve on the header is gradually opened, generating pressure in the stage. This pressure should be increased at such a rate that the target injection pressure is reached within approximately one minute. Achieving the target pressure occurs rapidly in stages with little take. Test durations will vary with the conditions encountered. If steady flow and pressure are achieved rapidly, the test can be terminated after a minute or two of data are collected. If erosion is occurring and the condition is not deemed detrimental to the structure, it is recommended the test continue until a stable Lugeon value is achieved, or some maximum specified duration is achieved. This aids in removal of materials from the joints, which may later be filled with grout.

During pressure testing it is not uncommon for water to bypass the packer and be observed in the standpipe, or for water injected during pressure testing to connect to adjacent grout holes. In each circumstance, the connection should be noted in the records as similar difficulties may be experienced during grouting, and the calculated Lugeon value is likely higher than if no connection had occurred. If it is critical that an accurate permeability test be conducted, it is possible for packers to be set in the connected holes, or if water is bypassing the packer in the test hole the packer can be moved to different depths.

2.2.6 Grouts and grout injection

The quality and effectiveness of a grouted cutoff is controlled by numerous factors, many of which have been standardized in the current practice of recent years. However several factors not currently standardized in North American grouting practice can significantly impact the final quality or the cost to achieve a verified residual rock mass permeability.

2.2.6.1 Refusal criteria

The refusal criterion for a grout application is the injection rate at which grouting is stopped: it is one of the most important factors impacting the achieved residual rock-mass permeability or the number of holes required to achieve a desired permeability reduction. The specified value for refusal varies widely in current North American grouting specifications and ranges from near zero or absolute refusal, to a more traditional take of one cubic foot or less in ten minutes (equivalent to 0.75 gallons per minute [gpm]). Specifying absolute zero take is not recommended as this requirement exceeds the accuracy of flow measuring equipment. However, a very low stage refusal criterion such as 0.1 gpm over a period of five minutes or less is recommended because each fracture is only intercepted a given number of times and it is necessary to effectively and thoroughly fill all intercepted or connected fractures when they are encountered if a low permeability cutoff is to be constructed.

Imagine a site with vertical fractures and holes drilled on 20- and 10-foot centers. For a hole angle of 15 degrees from vertical, a vertical fracture will only be intercepted once every 75 feet vertically for holes at 20-foot spacing and once every 37 feet for a 10-foot hole spacing. This consideration also has implications to the final specified hole spacing where joints might not be persistent or beds are not massive. Another key consideration regarding refusal is the case where a relatively tight bedding plane fracture is intercepted that is connected to a large solutioned joint or pipe. A high refusal criterion might result in refusal being declared almost immediately and the pipe or larger opening will remain unfilled unless it is intersected directly by a future hole. With final holes spacings of 5–10 feet being typical, a very large foundation defect could go untreated.

2.2.6.2 Specifying grouting method (upstage versus downstage)

Upstage grouting methods involve drilling a hole to final depth, washing the entire hole, water-pressure testing the hole in stages using a double-packer, followed by grouting the hole in stages from the bottom up in increments of 10 to 20 feet. When using downstage grouting techniques, each 10- to 20-foot stage is drilled, washed, water tested, and grouted. The grout is allowed to take initial set before the hole can be deepened to the next stage. In general, upstage grouting is only appropriate at sites with very high quality rock that is not prone to hole collapse or after a project has proceeded to the second or third hole series and significant rock-mass improvement has been achieved.

Downstage grouting is generally recognized as the technically superior grouting method. However, it is generally only specified for difficult or problematic ground conditions due to the perceived cost savings associated with upstage grouting. Many notable authors have previously commented on the pros and cons associated with these methods, but the use of upstage grouting as the principle specified method remains common:

> This is the cheapest method on sites where all goes well but not where they don't. Its apparent lower cost is often an attraction to specification writers who are trying to minimize costs and are keeping their fingers crossed that all will go well and holes won't collapse too often. (Houlsby 1990, p. 130)

> There are also substantial technical shortcomings to this progression…. Obviously, a greater amount of drill cuttings will find their way into higher joints and defects. Significantly, this is usually the zone containing the largest number of defects and where the highest quality of work is needed because of contact with the dam body. (Warner 2004, p. 322)

> It is generally applicable where minimal problems are encountered with seating packers, where the bore holes are mechanically stable, or with

grout bypassing the packer through rock. A hole may collapse before or after the packer is introduced, leading to incomplete treatment. (Weaver and Bruce 2007, p. 333)

As has been the case with many procedures having direct impact on the economics of a project, the choice of whether to use upstage or downstage techniques is one where dollars unfortunately often trump simple common sense. Specifying (or attempting) upstage techniques in ground conditions clearly not suitable for such a method provides no benefit beyond the initial *perceived* cost savings. It is often the case that these initial perceived savings quickly disappear for a variety of reasons.

Attempting upstage techniques in ground conditions that result in frequent hole collapse results in misleading data. There is simply no way to quantify where the grout injected is traveling due to it being unknown whether the obstruction is partial, therefore allowing grout to pass the obstruction, or complete, meaning all take is between the packer and obstruction. For example, if a borehole was drilled to a depth of 250 feet, and the obstruction occurred at 100 feet, there is a significant amount of borehole (and uncertainty) below the obstruction that could have accepted the large injection volume. Depending on what assumptions are made regarding the location of the grout take, significant features could exist below the obstruction that warrant additional treatment, but go untreated due to the misleading data. These situations can also result in stages being grouted at pressure less than desired, which may necessitate the installation of additional holes to treat to the desired intensity. Sequencing among the various drilling and grouting operations is difficult to track and forecast in a meaningful way under such situations.

Ultimately, specifying advanced equipment, materials, and data-collection methods are only of benefit if sound grouting fundamentals are applied. There is no sense in specifying high-performance tools and people if the process by which the project will be executed is fundamentally flawed.

However, the use of "practical downstaging" *is* recommended. Practical downstaging consists of always starting a project utilizing downstage techniques if any questions regarding hole stability exist. This is not necessary if a site or geology is known to be stable for upstage procedures. When performing practical downstaging, downstage procedures are continued until the observed water losses and hole stability are such that upstage drilling and grouting is clearly possible.

A special consideration when performing remedial grouting through existing embankments is that the top two stages (i.e., the uppermost 10–30 feet) should always be downstaged and the entire split-spaced series of holes completed on each stage prior to advancing the hole deeper. The purpose of this is to create an improved rock mass immediately below the embankment to protect the embankment soils from potential erosion during deeper drilling and grouting activities where higher pressures will be used. Treating the top two stages with holes on a maximum of 10-foot centers greatly reduces the likelihood that no untreated vertical joint is in direct contact with the embankment.

2.2.6.3 Grouting pressures

The standard practice in North America continues to be based on rules of thumb from the U.S. Army Corps of Engineers as specified in the 1984 edition of EM 1110-2-3506, *Grouting Technology*, although it is fair to say that this issue is still extremely contentious. This document specified that a maximum pressure of 0.5 psi per vertical foot of overburden and 1.0 psi per vertical foot of rock were to be used when grouting in poor or unknown subsurface conditions. Although higher pressures were not prohibited and other guidance was provided for various types of sound rock, the 0.5 and 1.0 psi per foot rules have become standard practice regardless of rock-mass quality. In general, these adopted rules of thumb are considered to be overly conservative for most rock masses.

The grout injection pressure, the pressure-filtration coefficient of the grout (Section 2.2.5.4), and the specified refusal criterion in combination are the three biggest factors that impact the number of holes required to achieve a desired permeability reduction. Low pressures in combination with a high refusal criterion could result in the need for a final hole spacing of 2.5 feet on a project that could have otherwise been achieved with holes on 5- or even 10-foot centers. This substantially increases the final project cost and doubles the number of penetrations required through the embankment, and is a common observation when analyzing "traditional" grout curtain records, especially pre-1970.

The grout injection pressure utilized should be based on the zone being grouted, its proximity to the embankment, and the susceptibility of the embankment material to erosion or hydrofracturing, the results of water-pressure testing and ongoing experience gained at the site. Lower pressures (i.e., rule-of-thumb pressures) might be appropriate when grouting the first two stages below the embankment. Lower pressures might also be appropriate where the initial permeability is substantial to reduce the grout travel distance. However, higher pressures should often be utilized to maximize the benefit obtained from each grout hole. Pressure exceeding the conservative rules of thumb can be identified and safely utilized given the ability to detect hydrofracturing or hydrojacking when utilizing real-time computer monitoring. Due to the specific gravity of grout being higher than that of water (typically 1.4 versus 1.0), there are circumstances in embankments with significant depths to water where simply the static head of the grout exceeds the permitted injection pressures. In these cases it is common practice to utilize some nominal gauge pressure, say 5–10 psi, to ensure that positive flow is achieved in the injection system.

2.2.6.4 Grout hole spacing, orientation, and depth

An in-depth understanding of the regional and site-specific geologic setting is essential when laying out a grouting program. The depth of the curtain should be based on the geology and seepage assessments and not based on the structure height. The poor performance of hanging or partially penetrating cutoffs is thoroughly discussed in Cedergren (1989). The orientation of grout holes and the estimated final hole spacing should be based on the strike, dip, and spacing of the prevalent joint sets and selected to maximize the number of joint intersections while also considering equipment limitations and productivity impacts for holes more than 30 degrees off of vertical. Where multiple line curtains are planned, it is now standard practice that the lines are oriented in opposite directions. The final hole spacing will often be determined based on permeability reductions as the split spacing of hole proceeds, but designers must also consider the frequency of joint intersections and the likelihood of missing a feature or defect such as a pipelike solution feature in karst.

For treatment of rock foundations angled holes are *necessary* to ensure that upstream/downstream–oriented joint sets are intersected by the drill holes. The angle of the drill hole must be selected in consideration of the capabilities of commonly available drilling equipment: both overburden- and rock drilling rigs. Angles that suit the majority of available equipment are typically 15 to 30 degrees from vertical.

2.2.6.5 Types of grouts and their properties

Generally speaking, the materials range from very low viscosity water-like grouts (high-mobility grouts) to stiff, mortar-like grouts (low-mobility grouts).

Virtually all rock grouting for dams is conducted with cement-based grouts of different types. Only in very rare and unusual cases are other families of grouts employed and, even then, they are used in conjunction with cement-based grouts (Bruce and Gallavresi 1998). Cementitious grouts, as the name implies, are formulations that include cementitious materials, typically Portland cements. The vast majority of rock grouting is performed using Portland cements, typically Type I, Type II, or Type III, and subsequent discussions are based on specific experience with these materials. While other types of cements can and have been used, site-specific requirements are typically the driving factor for choosing something other than Portland cement (e.g., brackish water may require Type V cement for additional resistance to sulfate attack), and therefore their use is somewhat limited. For additional information on the various cements and their properties the reader is referred to Bye (1999).

Cementitious grouts are divided into three categories, these being high, medium, and low mobility (HMG, MMG, and LMG, respectively). The "mobility" descriptions used herein reflect the ease at which material can be conveyed and are a reflection of the viscosity, cohesion, and/or friction of the grout. These rheological properties reflect its ability to penetrate fractures, the pressures required for injection, and its ability to remain where placed after injection is terminated. Whether it be for permanent permeability control or temporary stability control for subsequent excavations as in prestabilization for cutoff wall construction, this wide variety of materials can be both a blessing and curse as achievement of project goals and objectives, both technical and economically speaking, is largely driven by selection of the appropriate materials. It is prudent to assume that multiple materials may be required for any rock-grouting project, particularly those projects with significant geologic uncertainties such as karst.

The majority of nonkarstic rocks can be adequately treated to provide permeability reduction using high-mobility grout. Medium-mobility and low-mobility grouts are typically not necessary unless large defects such as significantly enlarged joints, cavities, or other features that readily accept grout with no sign of refusal are encountered (Figure 2.6). In many cases, the presence of these defects is not known, or may be known to a limited or great extent, but the project owner may attempt to treat them using only high-mobility grout. In these cases, it is common for the major defect to be detected as a result of the drilling and grouting efforts associated with the high-mobility grout. When encountered, these defects should be thoroughly investigated by additional drilling in and around the defect such that the required level and type of appropriate treatment may be determined. It is ill advised and uneconomical to initiate or continue pumping high-mobility grout with no indication of refusal when other more suitable materials are available. Best economy is achieved by treating the feature with other methods, such as medium-mobility, or more likely low-mobility grout, then continuing with high-mobility treatment as required to produce the desired

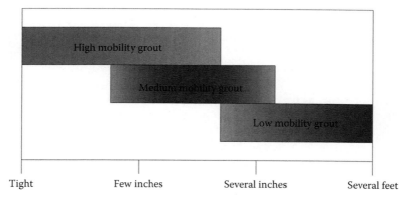

Tight   Few inches   Several inches   Several feet

Figure 2.6 Conceptual applicability of the various cement grouts based on rock fracture aperture. (Courtesy of Gannett Fleming, Inc.)

residual permeability. Karst foundations are particularly prone to such occurrences.

2.2.6.5.1 High-mobility grouts

High-mobility grout (HMG) is what has been traditionally referred to as a cement slurry. These are the "thinnest" of the cementitious grouts and therefore highly mobile. HMGs are used to penetrate the finest fractures and require relatively low injection pressures to initiate movement but can remain somewhat mobile after injection ceases.

Further refinement of the category includes neat cement grout (cement and water) and balanced-stable grout (cement, water, and admixtures). Neat cement grouts were the norm prior to the advent in the mid-1990s of balanced-stable grouts and are now considered unsuitable for rock-grouting applications due to their propensity to bleed and pressure filtrate. As illustrated in Figure 2.7, this results in undesirable rheological changes to the grout both during injection (pressure filtration—viscosity changes, bridging of fractures and small fracture penetration, fouling of equipment, and so on), and after injection (bleed—incompletely filled fractures).

Balanced-stable grouts are vastly superior in performance to neat cement grouts in that the physical properties of the grout change minimally both during and after injection. Through the use of admixtures, the detrimental properties of neat cement grout can be overcome. Balanced-stable grouts are easily formulated, do not require significant and sometimes any additional costs, and result in higher-quality grout curtains. For these reasons balanced-stable grouts should be specified for all HMG grouting projects. The main properties to be balanced are apparent cohesion (which relates

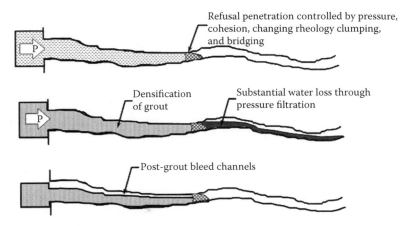

Figure 2.7 Performance of neat cement grouts during and after injection. (Courtesy of Gannett Fleming, Inc.)

to pumpability and extent of travel) and the pressure-filtration coefficient (which governs the ability of the grout to retain water during its pressurization into a fissure). This was first illustrated by De Paoli et al. (1992) in their classic graph showing the relationship between these two fundamental desiderata (Figure 2.8); regrettably, in North America, many "specialists" still do not understand—fully or at all—the significance of pressure filtration in the cost effective penetration of fissures.

What characterizes a balanced-stable grout is no or minimal bleed and a high resistance to pressure filtration. Bleed is also a significant consideration for grout selection. It is a result of material separation and sedimentation of the liquid and solid portions of the grout constituents. Photo 2.2 illustrates the difference in performance of neat cement and balanced-stable grouts having similar viscosities with respect to bleed potential: this is nothing to do with "thick" versus "thin" grouts. Typically bleed is of greater concern for less-cohesive mixes. Mixes with high water:cement ratios have considerable

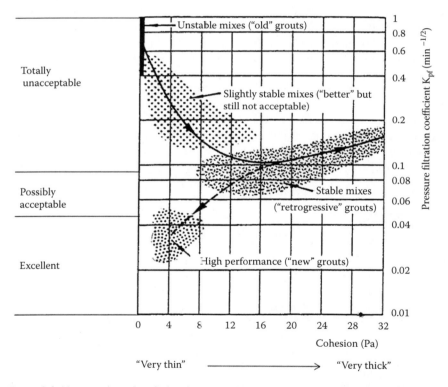

Figure 2.8 Historical path of development from unstable mixes to contemporary balanced multicomponent mixes. (Modified after De Paoli, B., B. Bosco, R. Granata, and D. A. Bruce., *Grouting, Soil Improvement and Geosynthetics*, Proceeding of the ASCE Conference, 1992. With permission.)

Neat cement grout	Balanced-stable grout

Photo 2.2 Comparison of bleed potential between neat cement and balanced-stable grouts. (Courtesy of Gannett Fleming, Inc.)

bleed potential and up to two-thirds of the injected mix may be lost as water due to bleed. In comparison, a balanced-stable mix of similar viscosity will have no observed bleed, and so the volume of grout injected will remain, resulting in a completely full fracture. A balanced-stable grout is typically defined as having less than 5 percent bleed as measured by ASTM C940, although zero bleed is an achievable target.

As noted above, resistance to pressure filtration is a vital consideration. Pressure filtration occurs during injection and results in the mix water being pushed out of the mix, functionally separating the liquid and solid components as a result of the applied pressure. This densifies the grout and changes the viscosity, resulting in bridging of fractures and reduced fracture penetration, both of which are obviously undesirable. Pressure filtration is measured using a pressure-filtration press and in accordance with API RP 13B-1 (Chuaqui and Bruce 2003). For a balanced-stable grout, the pressure-filtration coefficient (k_{PF}) should be less than 0.05 min$^{-1/2}$, and results significantly less than this are readily achievable. To contrast with the poorly performing traditional grouts of Figure 2.7, the superior grouts illustrated in Figure 2.9 have less pressure filtration and minimal

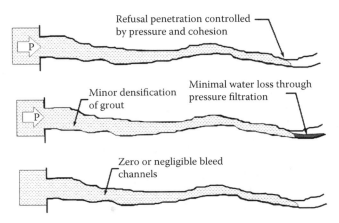

Figure 2.9 Comparative performance of balanced-stable grouts during and after injection. (Courtesy of Gannett Fleming, Inc.)

bleed, leading to completely grouted fissures over longer distances from the borehole.

Balanced-stable grouts are typically formulated with water to cement content ratios (all cementitious materials inclusive of flyash or silica fume) in the range of 0.6 to 1.5 (Weaver and Bruce 2007).

2.2.6.5.1.1 MEANS AND METHODS

High-mobility grouts are mixed using high shear mixers, sometimes referred to as colloidal mixers. These mixers consist of a cylindrical tank with a rotor and stator at the bottom (Photo 2.3). The action of the rotor withdraws grout from the bottom of the mixer and discharges it through conduits near the top of the mixing tank, resulting in the formation of a vortex. The close tolerance between the rotor and stator of the mixer in conjunction with the high rotational speed of the rotor (typically over 1,500 rpm) results in rigorous mixing action. Grouts with water-to-cement ratios as low as 0.4 by weight can be effectively mixed with such equipment.

After mixing, the grout is pumped using the action of the rotor and typically a three-way valve arrangement, to an agitator tank. Agitator tanks should be capable of holding several batches of grout at any one time. The agitator is commonly a vertical cylindrical tank with a vertical paddle operating at a low rotational rate (about 60 rpm), which provides continuous agitation to the mix. The formation of vortexes should be prevented by placement of baffles vertically around the tank perimeter. Additionally, one agitator paddle should be located as close to the tank bottom as practical.

Grout typically exits the agitator tank from the bottom and is conveyed to the pump. For HMG operations employing low to moderate pressures, say less than about 400 psi, progressive cavity pumps are the norm. Sometimes called helical-rotor pumps, or by the trade name "Moyno," they consist of a helical rotor within a stator (Photo 2.4). The number of flights of the helical rotor defines the number of stages or cavities for the particular pump. On these style pumps, each stage imparts an additional pressure of approximately

Photo 2.3 Rotor of high-shear mixer with housing cover removed. (Courtesy of Advanced Construction Techniques, Ltd.)

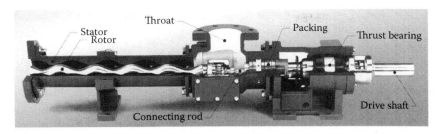

Photo 2.4 Cutaway section of progressive cavity pump. (Courtesy of Moyno, Inc.)

85 psi. Therefore, a four-stage progressive cavity pump has the capability to produce approximately 340 psi of pressure. The pressure developed is also dependent upon the viscosity of the grout. Higher-viscosity grouts will result in higher pump pressures, while lower-viscosity grouts result in lower developed pressures. Progressive cavity pumps are typically specified due to the constant nature of the developed pressure. With such pumps there are no pressure surges or cycles as is the case with piston-type pumps. This constant pressure output allows for maintenance of a near-constant pressure within the grout stage. This is a desirable trait for fissure grouting to refusal.

Grout exits the pressure side of the pump and is conveyed to a grout header located near the grout hole by a circulation loop. Two lines therefore connect the plant to the header. The supply line provides pressurized grout supply, and through manipulation of valves on the header the pressure and rate of injection into the grout stage can be controlled. Grout not conveyed to the grout stage by the header then flows back to the plant through a return line, which typically exits at the agitator. A circulation loop allows for constant movement of the grout within the grout lines and is preferred over direct injection through a single line. Depending on the location of the plant relative to the grout hole and header, circulation lines may be hundreds of feet in length and significant pressure from the pump may be required to circulate thicker grouts at the required flow rates over such distances. Circulation lines are typically constructed of flexible hose to facilitate movement and relocation in the field and must be rated to accommodate the required pump pressures. Use of exposed black hose as circulation line is discouraged as it results in significant solar heating of the grout. In cold climates protection from freezing may necessitate insulating the circulation lines.

A grout header is functionally a three-way valve assembly with pressure- and flow-measurement capability. Pressurized grout is supplied to the system and typically two valves are used to manipulate the flow of the grout, either through the instruments and to the grout stage, or back to the plant through the return line. Diaphragm-style valves are recommended in lieu of ball- or gate-style valves as they allow for more control of the injection process and do not wear nearly as quickly. Headers are typically operated through manual means and careful attention is required in order to ensure appropriate pressures are applied.

Grout injection is conducted with single-packer assemblies similar to those used during water-pressure testing. Grouting typically initiates with the lowest viscosity mix and, if conditions are appropriate, a thickening sequence is initiated. It is common practice to inject a predetermined volume of the first grout mix, say 200 to 400 gallons, prior to considering thickening. This ensures that fractures in the immediate vicinity of the grout stage that will only accept such grouts are full prior to injection of more viscous mixes. If no sign of grout refusal is noted using the starting mix, a thickening procedure is initiated, again with predetermined volumes of each subsequently thicker mix injected. If at any time during injection a refusal trend occurs, the current injected mix should be maintained until completion of the stage. Refusal trends include the following: decreasing flow rate at a steady maximum injection pressure or increasing pressure at a steady injection flow rate. Each is indicated by calculation of the apparent Lugeon value, that is, grout is used as the test fluid (instead of water) and a correction is made for the difference between the Marsh value of water and that of the grout. In most contemporary practice, the apparent Lugeon value is continuously calculated and displayed.

2.2.6.5.1.2 QUALITY-CONTROL METHODS

Grout quality-control methods rely predominantly on field methods developed using American Petroleum Institute procedures for drilling muds, and testing methods developed from the concrete industry. The first step in a HMG quality-control program is a design, testing, and categorization program in advance of production activities. This should include full-scale trials of the mixing equipment utilizing the materials to be provided for the production grouting activities. A multitude of mixes with varying ingredient dosages are attempted and tested. Based on the testing results, a suite of suitable mixes is selected. A typical HMG mix suite includes up to five mixes; the first mixes have Marsh funnel times ranging from about 35 to 40 seconds, while more viscous mixes may not continuously flow through the Marsh funnel. The viscosities range is selected based on the conditions to be treated.

The intensity of testing required varies depending upon the type of test. During the mix-design program, intensive testing is required to gain an understanding of the variation in grout mix physical parameters with modification to the ingredient dosages. Once a suite of mixes is selected, tests are conducted on a much less frequent basis (Table 2.1).

During production grouting activities, testing for viscosity and specific gravity are required daily for each mix type batched. Specific gravity testing using the Baroid mud balance provides highly repeatable results and is

Table 2.1 Recommended methods and minimum testing frequency for high-mobility grout

Parameter	Equipment	Specification	Frequency
Apparent viscosity	Marsh funnel	API 13B-1	Once per mix per day
Viscosity, cohesion, gel times	Rotating cylinder viscometer	Per manufacturer	Varies
Specific gravity	Mud balance	API 13B-1	Once per mix per day
Bleed	Graduated cylinder	ASTM C940	Once per mix per week
Pressure filtration coefficient	API filter press	API 13B-1	Once per mix per week
Set time	Vicat needle	ASTM C191-92	Mix-design testing program
Strength	Grout cubes	ASTM C942	Mix-design testing program

the most important test parameter for verifying appropriate dosage of the major grout mix constituents. Variability by about 0.01 to 0.02 units can be expected based on accuracy of the batching equipment and variability in the test procedure. Consistent deviations above 0.02 units should be investigated as this is a strong indicator that the batching equipment requires attention.

Apparent viscosity measurement using the Marsh funnel is common. Unfortunately, the Marsh funnel provides notoriously variable results especially for mixes of efflux time over 60 seconds. The test method requires the filling of an inverted cone 12 inches high and 6 inches wide at its top. A small orifice (diameter 3/16th of an inch) is located at the cone apex and plugged during filling. Once filled to the bottom of the screen, the plug is removed and the time in seconds required for a specified volume of grout to exit via the orifice is measured. This time is known commonly as the Marsh funnel time or Marsh funnel viscosity, and the specified volume is typically one quart in U.S. practice (water about 26 seconds) and one liter in European practice (water about 28 seconds). The inherent problem with the Marsh funnel lies with the diameter of the orifice. For grouts with significant cohesion and high viscosity, this orifice diameter is prohibitively small for repeatable and reliable quality control. Reliance upon such a variable test method solely for purposes of determining suitability of grout for injection is imprudent. While the test can be performed rapidly, and generally does provide for reasonable repeatability with less-viscous grout mixes, for thick grouts other test methods are warranted: grouts with Marsh funnel viscosities exceeding about 60 seconds should be tested either using a nonstandard Marsh funnel with variable orifice diameter (a commercially available although nonstandardized alternative), or using rotating cylinder viscometers.

2.2.6.5.1.3 ADMIXTURES

Admixtures allow for the development of balanced-stable grouts (appropriate rheological, bleed, and pressure-filtration properties) and less commonly modify other characteristics of the mix (such as set time, air entrainment, and antiwashout). Common admixtures and their typical dosage are included in Table 2.2.

Not all admixtures are required to produce a balanced-stable grout. At a minimum, bentonite, viscosity modifiers, and dispersants are typically required, and the necessary dosages are determined during the mix-design testing. It is particularly important to perform a mix-design testing program for every project. Changes in material suppliers, and in particular water chemistry, can drastically change the required admixture dosages. If material suppliers do happen to change during the course of a grouting project, it is necessary at a minimum to perform testing on all grout mixes to verify the effects. In some cases, another round of complete mix-design testing may be required.

Retarders and, more frequently, accelerants may be used to control set times and penetration distance. Available accelerants include calcium chloride, sodium hydroxide, sodium silicate, plaster of Paris, and sodium carbonate, among others (Weaver and Bruce 2007). Accelerants, when used in high concentrations, are capable of setting within a matter of seconds, which is advantageous for controlling flowing water conditions. This rapid set time presents a challenge to contractors as this greatly increases the likelihood of a "flash set" occurring in the injection system. In order to overcome this challenge, mixing of the HMG with the accelerant can occur either immediately at the end of the packer (twin-stream grout and accelerant through separate conduits in packer), or by injection of grout and accelerant in adjacent boreholes penetrating the same subsurface feature. The need for accelerants is typically limited to applications for control of high water-velocity conditions.

2.2.6.5.2 Medium-mobility grouts

Medium-mobility grout (MMG) is composed of thick high-mobility grouts with the addition of sand. The addition of sand will typically increase the apparent viscosity of the grout by providing some internal friction to the mix due to sand particle to particle contact. If used, sand should be fine, well-graded, consist of rounded rather than angular particles, and contain appreciable nonplastic fines. Common manufactured aggregates such as ASTM C33 (concrete sand) and ASTM C144 (masonry sand) do not provide the appropriate gradation and often natural sands not conforming to any standards are utilized (Warner 2004). The sand portion of the mix can range from 0.5 to 2 times or more the weight of cement. The key is that the sand remains in suspension during and after injection: mix-design testing is required in order to define the appropriate ratio.

MMGs are generally pumpable using the same techniques as the more viscous HMGs. Pressures required are typically higher due to the increased

Table 2.2 Common grout admixtures

Additive	Beneficial effects	Adverse effects	Typical dosage
Flyash, Type C or Type F	Improves grain size distribution of cured grout Cheap filler with pozzolanic properties Can be used as a replacement for some of the cement and reacts with the free lime resulting from the cement hydration process Increases durability and resistance to pressure filtration	Increases viscosity and cohesion	Typically 10–30% by weight of Portland cement. For Type C, concentrations higher than 20% may cause expansion and reduced durability
Bentonite	Reduces bleed and increases resistance to pressure filtration Slight lubrication and penetrability benefits	Increases viscosity and cohesion Weakens grout	Typically 2–5% by weight of all cementitious material (cement, flyash, and silica fume)
Silica fume	Fine-grained powder which improves pressure filtration resistance and reduces bleed Improves water repellency and enhances penetrability Improves grain size distribution of cured grout	Increases viscosity and cohesion	Typically 5–10% by weight of Portland cement
Viscosity modifiers (welan gum)	Makes the grout suspension more water repellant Provides resistance to pressure filtration, and reduces bleed	Increases viscosity and cohesion	Typically 0.1–0.2% by weight of cementitious material
Dispersants or water reducers (superplasticizer, fluidifier)	Overprints solid particles with a negative charge causing them to repel one another Reduces agglomeration of particles thereby reducing grain size by inhibiting the development of macro-flocs Reduces viscosity and cohesion	Depending on chemistry chosen, may accelerate or retard hydration process. This is not necessarily negative.	Typically 1.5–3% by weight of cementitious material

Source: Wilson, D. B., and T. L. Dreese. "Grouting Technologies for Dam Foundations," *Proceedings of the 1998 Annual Conference Association of State Dam Safety Officials*, Paper No. 68, 1998.

viscosity and friction of the grout. Great care must be taken to ensure that the sand stays in suspension after mixing; this requires a proper mix design and appropriate field methods. Provided the conditions are appropriate for such use, the injection of a sanded grout results in decreased penetration into fractures and a grout-filled fracture that is more resistant to immediate washout in comparison to HMG.

Sanded grouts have been used successfully on numerous projects, although there remains somewhat of a stigma regarding their use principally due to concerns about accelerated wear in the grouting equipment. Many grouting texts, notably those by Houlsby (1990) and Warner (2004), indicate no ill effects when using sanded grouts and the current authors have used MMG successfully. They consider sand to be simply a filler in the mix. When properly proportioned, the use of sand can result in a satisfactory mix and a lower-cost grout. However, the use of sanded grouts introduces a wide range of other considerations, inclusive of additional wear on equipment, decreased efficiencies due to breakdowns and dealing with obstructions in equipment, and the need for additional equipment or labor to handle an additional product. MMG should be considered for appropriate conditions only if economy of materials is an important consideration, and only so if an appreciable quantity of MMG is required such that dedicated equipment can be provided for such an operation. The use of MMG in a standard HMG thickening procedure without injection interruption is technically feasible, but is not economical or practical. MMG injection should be a separate operation from the normal HMG thickening procedure to allow for staging and setup of the appropriate equipment and materials, and proper cleaning and charging of lines. There are also limitations regarding the capability to convey the grout from the plant and to the grout hole that must be considered; higher viscosities require higher pumping pressures and sometimes decreased length of circulation loops. Very thick HMGs can be formulated having viscosities similar to MMGs, and substitution of a thick HMG for an MMG should not adversely impact a project from a technical perspective. Considering the equipment available, if coarse sand can be mixed and pumped satisfactorily, then significantly reduced penetration into even coarse fractures can be achieved. However, for all practical matters, if the use of MMG is under consideration, the subject hole will likely have consumed significant quantities of grout or a void requires filling; in either case, a major defect in the foundation has been encountered and requires specific attention. Most often low-mobility grout will prove both more effective in limiting runaway takes and more economical of a treatment method. Whether the feature is filled with MMG or thick HMG, additional treatment with HMG using the standard thickening procedures is likely required to "tighten" the area and achieve the desired residual permeability.

2.2.6.5.3 Low-mobility grouts

2.2.6.5.3.1 PERSPECTIVE

Compaction grouting, described by Baker, Cording, and MacPherson in 1983 as "this uniquely American process," has been widely used in the United States since the early 1950s and continues to attract an increasing range of applications (Warner 1982, 2003; Warner et al. 1992). The 2005 definition of compaction grout produced by the Grouting Committee of the Geo-Institute (ASCE) is as follows:

> Grout injected with less than 1 inch (25 mm) slump. Normally a soil-cement with sufficient silt sizes to provide mobility together with sufficient sand and gravel sizes to develop sufficient internal friction to cause the grout to act as a growing mass as injection continues under pressure. The grout generally does not enter soil pore (except, perhaps, where open-work boulder gravels are present) but remains a homogeneous mass that gives controlled displacement to compact loose non-plastic soils, gives controlled displacement for lifting structures, or both. (ASCE 2005, p. 1535)

It was in 1997, however, that Byle articulated a truism that many practitioners, especially in the dam grouting field had long realized—"The term 'compaction grouting' is frequently a misnomer. There are many applications of 'compaction grout' which have nothing to do with compaction. What is commonly known as compaction grouting is really just a subcategory of the broader family of limited displacement (LMD) grout" (p. 32). He specifically referenced the application of the compaction grouting *procedure* for sealing voids in karstic limestone terrains. He further opined that the intent of all LMD grouting is to inject grout that stays where it is injected and displaces a portion of the material into which it is injected. The purpose is to not permeate soils or penetrate fine fissures.

Through common usage, the term had been changed to low (or limited) -mobility grout by the time of the 2005 Geo-Institute definitions: "Low slump grout, such as a compaction-type grout, that does not travel freely and that becomes immobile when injection pressure ceases" (p. 1538). The term "LMG" therefore contrasts with the term "HMG" (high-mobility grouts) as coined by Chuaqui and Bruce (2003), and previously known as "slurries."

Compaction grouting, *sensu stricto*, has been used to improve density and so reduce liquefaction potential on hydraulic structures, and the work undertaken in the foundation sands at Pinopolis West Dam, SC, is an excellent example (Baker 1985), while the 1997 remediation of sinkholes in the core of WAC Bennett Dam in British Columbia is another (Garner, Warner, Jefferies, and Morrison 2000). Warner (2003) also described the use of compaction grouting to find and remediate weakened soils that had led to open voids in an earth embankment in the California aqueduct scheme. He notes that "compaction grouting was selected as the safest (investigatory and remedial) method, because it would provide for the greatest control of the grout deposition area, and properly performed, would not result in hydraulic fracturing of the embankment." Cadden, Bruce, and Traylor (2000) also outlined compaction grouting for seismic retrofits at the Croton Dam, New York, and at Chessman Dam, Minnesota. However, such examples of true compaction grouting for dams and levees are relatively rare, as other techniques and concepts have been judged progressively more reliable.

On the other hand, contemporary U.S. dam remediation practice makes frequent use of LMG materials and methods, but principally to treat karstified carbonate terrains and to investigate and treat embankment/rock contact zones, as a precursor to the installation of, say, a "positive" concrete diaphragm (Chapter 4, this volume). In this regard, the significance of the verb "investigate" should not be overlooked: the controlled travel characteristic of a properly formulated LMG ensures that "runaway" takes cannot occur, so avoiding wasteful and potentially damaging injections. In short, hydrofracture cannot occur, and the body of LMG injected acts like a pressure cell, as if enclosed in a rubber membrane.

2.2.6.5.3.2 MEANS, METHODS, AND MATERIALS

In very general terms, grouts are either prepared on site using auger or pugmill mixers, or they can be prepared in a readimix facility and trucked to the point of injection. It is typical to find that the former method provides more consistent and homogeneous grouts, especially those with slumps of less than two or three inches and for smaller volumetric demands. Grouts are transferred to the point of injection via "slick lines," that is, flexible or rigid pipes or hoses with no internal reductions in diameter, and no sudden changes of direction. Joints are important to keep watertight and special fittings are necessary on surface connections for safety reasons. Such pipework typically ranges from two to four inches in diameter, dictated by contractor preference and site conditions and requirements.

There are various commercial suppliers of concrete pumps, most of which are equipped with swing-tube values. These employ variable-pitch hydraulic pumps to closely control grout injection rate. Although they can operate slowly, the efficiency of these pumps deteriorates rapidly as the rate descends to within the lower 2 percent of their normal operating range. Standard pumps have pump cylinders from three to six inches in diameter: the optimal diameter is dictated by the intended injection rate, with typical LMG rates ranging from a few "strokes" per minute (one cubic foot per minute) to over 200 cubic yards per hour in "flat-out, void fill" situations.

Injection work for dams is invariably conducted in ascending stages, through the drill casing previously drilled or placed. Again, given the considerable depths involved in most dam applications, the drilling rig is used to withdraw it, in stages whose length is determined on a project-specific basis, but is usually in the range of two to ten feet.

Stages are grouted to refusal determined by one or a combination of factors, but especially a limiting pressure, a limiting volume or some adverse reaction by some type of embankment or foundation instrumentation (principally piezometers or inclinometers). Again, the exact refusal criteria are project specific, with foreseen values being very carefully verified (and modified as appropriate) in restricted test sections of the project.

The choice of grout materials and grout compositions is a subject of continuous debate among LMG specialists, each arguing from his or her own application-based experience. However, there seems general concurrence that a high-quality LMG should have the following characteristics:

- The aggregate should respect the boundaries of Figure 2.10, and preferably be rounded. If coarse, it should hug the lower limit line.
- A substantial, rheological advantage is gained if fine gravel is incorporated (3/8 inch) and if at least 20 percent is retained on the #4 sieve.
- Silts typically comprise 10–25 percent of the sand portion, but can be up to 35 percent if the silt is coarse.
- Any particles smaller than the #200 sieve should be nonplastic, and less than 1 percent clay is desirable.
- Warner (2003, p. 12) states that "poor performance will occur should clay, clay like materials, or most concrete pumping aids be included."

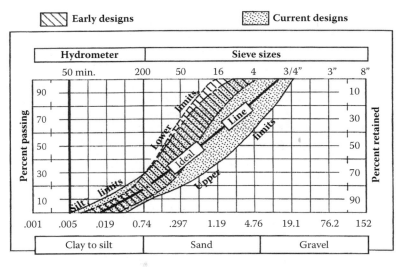

Figure 2.10 Aggregate gradation limits for LMG. (From Bandimere, S. W., *Grouting: Compaction, Remediation and Testing*, edited by C. Vipulanandan, ASCE Geotechnical Special Publication No. 66, 1997. With permission.)

This is interpreted to mean that the LMG will have unacceptable rheological properties, and will be prone to hydrofracture if used in a true compaction grouting role.

- Antiwashout properties can achieved by using high replacement proportions of flyash or other pozzolans and/or special admixtures. The use of polypropylene fibers has also been found beneficial in this regard.

Further details and excellent reviews may be found in the works of Warner (1982, 2003, and 2004), Warner, Schmidt, Reed, Shepardson, Lamb, and Wong (1992), Byle (1997), and Bandimere (1997), among the most notable.

2.2.6.5.3.3 QA/QC AND VERIFICATION

Bearing in mind that the LMG technique depends on fairly robust and traditional equipment, the means used for routine QA/QC testing have been similarly straightforward. Injected volumes have been measured by counting the strokes on the pump (each of a known volume of discharge), with reconciliation on a daily or weekly basis against the number of readimix truck deliveries or the volumes of materials passed through the on-site batching plant. Pressures have been shown on a digital or analog pressure gauge and recorded manually. Slump should be measured both at the pump and at the injection point with the ASTM slump test, while cylinders of grout are taken to give seven-, fourteen-, and twenty-eight-day strengths.

However, in recent years, automated monitoring systems have been increasingly used to record LMG injection parameters. Notwithstanding the outstanding accuracy and functionality of these systems with HMGs, the peculiarities of LMG are such that truly accurate automated injection records, especially of rate of injection, can be awkward to create. Furthermore, even accurate records can provide a source of endless debate regarding transient "peaks" or "perturbations" in the data, only revealed by the automated data collection. It must be remembered that such LMG peaks are so transient that the potential for them to cause damage to an overlying embankment dam is cosmically remote. In such cases, practical common sense must prevail in the analyses.

2.2.6.5.4 Switching from HMG to MMG to LMG

Switching from one grout material type to another rapidly is commonly specified but is not easily achievable. Switching from HMG to MMG is possible if a second grout plant is available and staffed for the purpose of adding sand to already prepared HMG. However, this does require that the owner be willing to incur the costs for the equipment to be on standby when not being used and to pay for the labor to be available to operate the equipment when it is required. However, it is not possible to easily change from grouting with HMG or MMG to LMG. Injecting LMGs requires completely different equipment and delivery systems. Recommended practice is to break out a LMG program as separate items where possible (such as a known cave feature) or to quarantine holes meeting the requirements for LMG and wait to perform the LMG injection until a number of holes requiring such treatment are available, making it cost effective to set up and perform the LMG injection.

2.2.7 Recent case histories

2.2.7.1 HMG

2.2.7.1.1 Patoka Lake Dam, Indiana (2000)

The Patoka Lake Seepage Remediation project was undertaken after a flood event resulted in sinkholes near the right end of the existing cutoff wall in the emergency spillway. The emergency spillway is located in a low area separated from the dam by a ridge in the left abutment of the dam (Figure 2.11). During the flood event, significant seepage discharges (up to 3,400 gpm) occurred from the Robert Hall cave, located well downstream from the dam. At normal pool level, total discharge from the cave entrance was typically in the range of 100 gpm, much of which could be attributed to collection of rainfall infiltration rather than seepage. The entire ridge

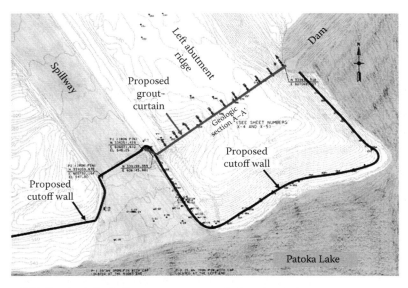

Figure 2.11 Patoka Lake Spillway seepage remediation alternatives.

area extending from the left abutment of the dam to the spillway cutoff wall was determined to be a source of potentially hazardous uncontrolled seepage, and it had not previously been treated.

The seepage barrier, as originally conceived, was planned as a cut-off wall that would follow along the shoreline of the lake. Later, it was concluded that a grouted cutoff in the karst limestone would adequately reduce risks for unusual and extreme events. The grouted cutoff would follow a shorter, direct route across the ridge and directly connect the abutment of the dam with the spillway cutoff wall. The design was a three-line curtain with the intent to achieve three Lugeons or less residual permeability.

The Patoka Lake Seepage Remediation project was important for a number of reasons: it utilized grouting in karst as the intended long-term solution; it was the USACE's first use of balanced-stable grouts, computer monitoring of injection, and defined performance criteria; and it was the introduction of best value selection to procure the grouting contractor (Hornbeck, Flaherty, and Wilson 2003).

Grouting was completed in 2000 after drilling more than 50,000 linear feet (L.F.) of bedrock. It was verified that the average residual permeability of the grouted zone was approximately one Lugeon (Figure 2.12). Verification testing included determination that the grouted zone could withstand pressures in excess of the expected applied heads without hydro-fracturing through soil-filled seams within the grouted mass.

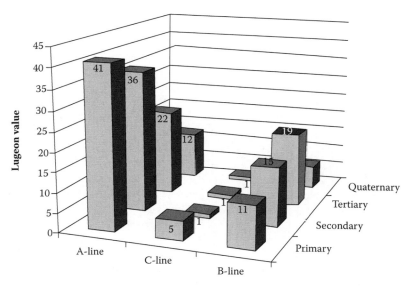

Figure 2.12 Patoka Lake Dam water test results by grout line and hole series. (Courtesy of Gannett Fleming, Inc.)

2.2.7.1.2 Mississinewa Dam, Indiana (Foundation remediation, 2002–3)

Mississinewa Dam is a 140-foot-high 8,100-foot-long earthfill dam located near Peru, Indiana, and is operated by the U.S. Army Corps of Engineers. During original construction in the mid-1960s, the right abutment was founded on glacial outwash materials overlying limestone. The glacial material was never removed, leaving the underlying bedrock untreated (Henn, Hornbeck, and Davis 2000). A foundation grouting program was implemented from 2002 to 2003 after rapid and complete slurry losses occurred during the preliminary testing of a proposed slurry wall, designed to address seepage issues along the right abutment and a 2,600-foot section of the dam embankment and foundation. The seepage issues and corresponding slurry losses were the result of solutioned limestone of the Liston Creek Formation. The foundation remediation program involved the use of balanced-stable grouts injected at pressure to seal the solutioned rock in order to reduce the overall permeability of the foundation rock to permit the safe construction of a slurry cutoff wall. Completion of the grouting program resulted in approximately 75,000 L.F. of overburden drilling and close to 27,000 L.F. of production rock drilling. The target residual permeability of 10 Lugeons was chosen such that the slurry loss, which could be anticipated in a 12-foot-long panel, would be in the order of 50–100 gpm, a value considered manageable from a construction viewpoint.

After completion of the foundation grouting program, the cutoff wall construction continued and was completed in 2004, with a return to normal pool by the spring of 2005. During the entire construction of the cutoff wall there were no observed significant slurry losses, verifying a successful grouting program.

This project marked the first time the USACE implemented a combined grout curtain and cutoff wall solution for the rehabilitation of a dam, and later became the model for many other future composite cutoff wall projects for critical high-risk dams. This project also featured the use of enhanced state-of-the-art grouting technologies, including balanced-stable grouts and real-time computer monitoring *and control* systems, which would also become the model for future grouting projects performed by the USACE. Further description of the history, foundation grouting program, and the grouting technologies implemented are provided in Section 5.5.1.

2.2.7.1.3 Clearwater Dam, Missouri (Major Rehab Phases I and Ib, 2006–9)

Clearwater Dam is a 154-foot-high, 4,225-foot-long rolled earthfill dam located in southwest Missouri, and is operated by the U.S. Army Corps of Engineers primarily for flood control along the White and Lower

Mississippi river basins. Investigations began when a sinkhole developed on the upstream slope of the dam shortly after the record pool event in 2002. A sinkhole investigation and remediation program, conducted from 2003 to 2005 discovered the existence of a large solution feature (25 feet wide and 170 feet deep) in the foundation rock just below the dam that was identified as the likely cause of the sinkhole. The feature also coincided with a pair of solutioned rock features uncovered during the excavation of the foundation during the original construction of the dam. The twin features followed the two main near vertical joint sets that existed at the site and extended almost perpendicular to the dam alignment, running completely upstream to downstream (Figure 2.13). Preliminary grouting of the solution feature was performed during the sinkhole investigation program as an interim risk-reduction measure. However, a major rehabilitation program was developed to address the issue of the solution feature as well as the overall seepage issues occurring at the dam.

Clearwater Dam is underlain by dolomite from the Potosi Formation. The formation is of Cambrian age and is extremely broken with various stages of dissolution occurring at the project site, including the solution feature located at the sinkhole, and a completely weathered residual soil layer and epikarst layer that exists along the left abutment of the dam. Clearwater Dam was one of the original six dams operated by the USACE to be designated DSAC 1 during an initial portfolio screening and dam assessment culminating in 2007. The proposed remediation of the seepage at Clearwater Dam consisted of the construction of a composite cutoff wall and grout curtain. The proposed cutoff wall was planned to extend approximately 40 feet into rock, with a maximum of 60 feet into rock, to create a cutoff through the solution feature that was discovered during the

Figure 2.13 3D representation of Clearwater Dam Foundation grade and solution feature. (Courtesy of Gannett Fleming, Inc.)

sinkhole investigation. A grout curtain was planned to precede the cutoff wall and provide a seepage barrier to reduce significant slurry losses during the construction of the cutoff wall. The grout curtain was also planned to extend between 60 and 80 feet below the proposed cutoff wall as an additional seepage barrier for the dam foundation. The grout curtain was also extended a further 120 feet below the proposed cutoff wall, along the location of the solution feature for added seepage protection. The phase 1 program consisted of an exploratory drilling and foundation grouting program that was used to both characterize the *in situ* conditions of the foundation rock below the dam along the entire length as well as treat the rock to reduce slurry losses during cutoff wall trench excavation and to reduce seepage below the proposed cutoff wall (Photo 2.5).

Once the dimensions of the composite cutoff wall were determined, using the information collected from the phase 1 work, the remaining grout curtain and solution feature treatment was completed during the phase 1b contract. The completed grout curtain consisted of two lines, both 4,100 feet long and up to 325 feet in depth. Holes on each line were located on 10-foot centers and were drilled at 15 degrees in opposing angles to increase the number and intensity at which the two existing near vertical joint sets within the formation were grouted. Overburden holes were drilled in the embankment to the top of the rock with rotary duplex drills, and combination PVC and MPSP casing was installed to both isolate the embankment and core from the foundation and to treat the existing epikarst and residual soils zones along the left and right abutments. Grouting was performed using balanced-stable grouts, state-of-the-art drilling equipment for fast and accurate drilling of holes in rock, state-of-the-art grouting equipment for rapid stage setup within the hole, and with a real-time computer

Photo 2.5 Drilling and grouting operations at Clearwater Dam. (Courtesy of Advanced Construction Techniques, Ltd.)

monitoring system capable of recording multiple grouting operations and producing as-built drawings of grouting results and up-to-date analyses for operations planning and external review. During phase 1 construction of the grout curtain, a combination of several types of balanced-stable grouts was used, including a suite of six HMGs, an MMG or sanded grout, and an LMG. The conditions of the karstic Potosi Formation changed frequently and rapidly from hole to hole and stage to stage, and the appropriate type of grout to use for each stage was determined by a combination of water-testing analysis, review of drilling logs, analysis of core samples, review of borehole televiewer information, and analysis of prior grout takes in the area according to updated as-built grouting results (Hockenberry, Harris, Van Cleave, and Knight 2009).

An additional LMG treatment program of the original solution feature was also included in phase 1 operations. The LMG treatment consisted of a single line of holes, located along the known limits of the solution feature. The single-line LMG treatment was also located between two previous LMG grout lines constructed during the sinkhole repair project from 2003 to 2005. The new LMG line extended to a maximum depth of 220 feet below top of rock (355 feet total depth).

The solution feature was then further grouted with a combination of HMG and MMG grouts using the two-line grout curtain holes extended to a total depth of 325 feet. The phase 1 and 1b grout curtain was completed in 2009 and included more than 100,000 L.F. of drilling through overburden, and approximately 120,000 L.F. of production rock drilling. Phase 2 cutoff wall construction began in 2009 and is currently under construction (Harris, Van Cleave, Howell, and Piccagli 2011). To date, no significant slurry losses have occurred during the construction of the cutoff wall.

2.2.7.1.4 Wolf Creek Dam, Kentucky (Foundation grouting 2007–8)

Wolf Creek Dam (Photo 2.6) is a 5,736-foot-long composite concrete gravity dam and rolled earthfill embankment, and is located on the Cumberland River in south central Kentucky. Completed in 1952, the dam is 258 feet in height and is capable of impounding 6,089,000 acre-feet of water at its maximum pool elevation of 760 feet. The reservoir created by the dam, Lake Cumberland, is the largest reservoir east of the Mississippi and the ninth-largest reservoir in the United States (Zoccola, Haskins, and Jackson 2006). Wolf Creek Dam provides approximately 40 percent of the total flood storage for the Cumberland River Basin and saves an estimated $34 million in annual flood damage. In addition, the dam provides $77 million in annual hydroelectric power generation, and $159 million in annual revenue to the local economy in form of recreation provided by the reservoir.

Since the first filling of the reservoir, the dam has had a history of seepage and stability issues. A system of caves and solution features was uncovered

Photo 2.6 Wolf Creek Dam, Kentucky, embankment foundation grouting project. (Courtesy of Advanced Construction Techniques, Ltd.)

during the original excavation of the embankment core trench (Photo 2.7) (Kellberg and Simmons 1977). The system lies below the transition of the concrete gravity structure and the earth embankment section and is the location of where the majority of the seepage issues have occurred. The system of caves and solution features is the result of karstic limestone of the Liepers and Catheys formations, both of Ordovician age. Several attempts have been made to mitigate the seepage occurring at the dam, including

Photo 2.7 Photo of the system of caves uncovered during excavation of the dam foundation, Wolf Creek Dam, Kentucky. (Courtesy of U.S. Army Corps of Engineers.)

an emergency grouting program from 1968 to 1970 and a concrete cutoff wall, constructed from 1975 to 1979 (Bruce, Ressi di Cervia, and Amos-Venti 2006), which included a foundation grouting program using the pilot holes drilled for the drilled shaft elements of the cutoff wall.

Current work to address the seepage issues includes the construction of a composite cutoff wall and grout curtain set even deeper than that of the 1970s cutoff wall, and which will also extend the entire length of the embankment section of the dam. In 2006, the foundation pregrouting contract was awarded. The purpose of the foundation grouting program was to construct a two-line grout curtain along the proposed length of the cutoff wall to be later constructed. An additional one-line grout curtain section was also constructed along a 263-foot section of the gallery of the concrete gravity portion of the dam near the system of caves at the transition to the embankment section. The two-line curtain was drilled up to 350 feet in depth through both embankment material and the underlying rock foundation. The section of the grout curtain from the soil/rock interface to elevation (EL) 475 feet was constructed to reduce significant slurry losses during cutoff wall construction. The section of the grout curtain from EL 475 feet to 425 feet was constructed to act as a stand-alone grout curtain to reduce overall seepage 50 feet below the final depth of the proposed cutoff wall. Since Wolf Creek was also designated a DSAC-1 structure, the work schedule for the foundation grouting contract was accelerated and an additional requirement of the program was to provide one of the grout curtain lines as an interim risk-reduction measure against seepage. Both grout curtain lines were to be completed in the shortest amount of time possible in order to expedite the start of the second phase of work, construction of the cutoff wall. The interim risk-reduction measures and the accelerated schedule were accomplished with the use of state-of-the-art drilling and grouting equipment as well as a real-time computer monitoring system capable of monitoring multiple grouting operations simultaneously and with the ability to provide detailed analyses and as-built drawings on demand, for rapid dissemination of grouting information to the project team for quick response and decisionmaking.

The two-line grout curtain was 3,840 feet long and approximately 325 feet deep, with some boreholes extending up to 350 feet in depth. The two grout lines, set 24 feet apart and each 12 feet on either side of the proposed cutoff wall alignment, were drilled at 10-degree angles (from vertical) and were oriented in opposing directions to increase the probability and frequency that existing near vertical joint sets and solution features in the rock would be intersected for grout injection. The upstream line contained holes spaced on 5-foot centers (Tertiary series) with additional split-spaced holes added as determined from the grouting results. The downstream line contained holes on 10-foot centers (Secondary series) with additional split-spaced holes added as determined from the grouting results. Drilling and

Photo 2.8 Drilling with down-the-hole rotary percussion water hammer, Wolf Creek Dam, Kentucky. (Courtesy of Advanced Construction Techniques, Ltd.)

grouting operations for the two-line curtain were conducted from a work platform constructed along the upstream slope of the earth embankment at EL 748 feet (Photos 2.8 and 2.9). A total of 1,317 holes were drilled on the upstream and downstream grout lines. At total of 192,644 L.F. of overburden drilling, 238,625 L.F. of rock drilling, 818,317 gallons of injected grout, and 5,623 hours of grouting were completed in a twenty-month schedule of operations including twenty-four-hour drilling and grouting. This contract was performed from January of 2007 to August of 2008.

State-of-art-drilling equipment included DTH rotary percussion hammers with water flush. The drilling equipment allowed for the grout holes to be drilled to the depth and angle required quickly, accurately, and within

Photo 2.9 Multiple drilling and grouting operations at Wolf Creek Dam, Kentucky. (Courtesy of Advanced Construction Techniques, Ltd.)

the alignment tolerances specified in the contract. Specially made grouting equipment allowed for the quick insertion of packers and proper injection of the grout to the required depths and pressures along the hole. Real-time computer monitoring of grouting operations and on-demand production of result analyses and as-built drawings allowed for the project team to make quick decisions for the location and depths of additional holes required to meet the project goals. Only additional holes in discrete, specified locations where closure was required were added, thereby reducing the total production quantities and creating an efficient grout curtain. The relative speed at which the drilling, grouting, and analyses could be performed was a major factor in the ability of the contract to be completed within the timeframe allotted, allowing for the start of the cutoff wall phase of construction in 2009.

2.2.7.1.5 Center Hill Dam, Tennessee (Foundation grouting 2008–10)

Center Hill Dam is a 250-foot-high and 2,100-foot-long combined earth-fill and concrete gravity dam located in Lancaster, Tennessee. Center Hill Dam was also designated as a DSAC 1 dam during the initial screening of the U.S. Army Corps of Engineers' dam portfolio. The foundation of the dam and left rim consists of the karstic limestone of the Catheys and Cannon formations. From 2008 to 2010, phase 1 foundation grouting was performed as both interim risk-reduction measures and as part of a proposed composite barrier cutoff wall and grout curtain, designed to reduce overall seepage and instability at the site (Adcock and Brimm 2008). Phase 1 work consisted of constructing multiple two-line grout curtains along the 800-foot-long main dam embankment, a 700-foot-long section of the left abutment groin, and along a 2,700-foot-long section of the left rim of the dam (1,800 feet was single line only), which also contained a 130-foot-deep open cavernous solution feature. Phase 2 of the project involves the construction of the concrete barrier wall along the embankment section of the dam and is ongoing at the time of publication.

Each two-line grout curtain section was drilled at 10 degrees in opposing directions to increase the effectiveness of grouting vertical fractures and solution features. The main dam embankment section contained holes for the upstream line on 5-foot centers with additional split-spaced holes added as needed, and holes for the downstream line on 10-foot centers with additional split-spaced holes added as needed. Holes on the groin and left rim were on 10-foot centers with additional split-spaced holes added as needed. Over 95,000 L.F. of overburden drilling and approximately 115,000 L.F. production rock drilling was completed. Drilling and grouting operations were performed with a combination of existing and state-of-the-art technologies including DTH hammer drills with water flush, balanced-stable grouts, real-time computer monitoring and control systems, high-resolution borehole imaging, geophysical electrical resistivity surveys, environmental control systems to process and treat drill water and grout wastes, and a real-time automated instrumentation system for site safety and dam safety monitoring and analysis.

The real-time computer monitoring and control system used during production was capable of monitoring multiple grouting operations at a time, and producing on-demand grouting as-built drawings as both hard copies and as electronic files uploaded to a website for technical review. The additional information collected by electrical resistivity surveys and high-resolution borehole imaging was used to further assess the subsurface conditions of the rock. The wealth of data collected during grouting operations and the speed at which the data were provided allowed the project team to make rapid but informed technical decisions and program modifications, including the addition and deletion of holes based on grouting results and known subsurface conditions. An automated instrumentation system was incorporated into the grouting program that consisted of several vibrating wire piezometers installed into new and existing piezometer standpipes along the embankment. The automated system was able to monitor the piezometric surface within the embankment during grouting operations. The incorporation of an automated instrumentation system allowed for the real-time display and analysis of the piezometric response of the subsurface foundation to the drilling and grouting operations.

2.2.7.2 LMG

The following two case histories illustrate the use of LMG in major dam remediations. In the case of Tims Ford Dam, Tennessee, a remedial grout curtain was formed in karst using LMG as the prime component, while at Wolf Creek Dam, Kentucky, an LMG program was used to investigate and treat the bedrock/embankment interface prior to the construction of a diaphragm wall through that potentially critical horizon and into the underlying bedrock, after the rock-grouting project described in Section 2.2.7.1, above.

2.2.7.2.1 Tims Ford Dam, Tennessee

Tims Ford Dam is a 175-foot-high embankment structure constructed on the Elk River approximately nine miles west of Winchester, Tennessee (Hamby and Bruce 2000). This water-regulating Tennessee Valley Authority (TVA) structure is about 1,520 feet long with the crest at EL 910 feet. The right (west) abutment of the dam intersects orthogonally a natural ridge running nearly north-south, and consisting of clay and weathered chert overburden overlying a karstic foundation of various limestone formations. The crest of this right rim abutment varies in elevation from 942 feet to about 958 feet with the top of rock generally around EL 900 feet. The maximum pool elevation is at EL 888 feet.

In May and June 1971, two leaks designated leaks 8 and 6 appeared on the downstream side of the right rim during initial filling to EL 865 feet. Leak 8 was approximately 140 feet upstream of the dam baseline. Exploratory drilling and dye testing were performed along the right rim for a distance of 200 feet upstream of the dam baseline. This work led to grouting a line of holes using cement-based slurry grouts (limited to 200 bags per hole per day) containing 4 percent calcium chloride accelerator to withstand the water-flow velocity. At that time, dye connection times from the curtain to Leak 6 were recorded in the range of four to eight hours. No attempt was made to seal it. The major outflow from Leak 6 emitted from two vertical features at EL 852 feet, some 950 feet upstream of the dam baseline, and formed an unnamed stream traveling approximately 3,000 feet to the Elk River. An outflow monitoring program was begun and data from that program showed that the outflow varied directly with reservoir level. During the period 1971 through 1994, Leak 6 peak outflow volume slowly increased to about 3,500 gpm. In 1994, however, following record drawdown of the reservoir, the Leak 6 outflow volume increased dramatically in 1995 to over 8,000 gpm (Figure 2.14) and a large slump failure occurred in the hillside around the leak (Figure 2.15). TVA determined that remedial grouting should be performed to reduce the Leak 6 outflows to less than 1,000 gpm at maximum pool.

An exploratory drilling program was performed during February–April 1997 to better define the existing foundation conditions and provide information necessary to design the remedial grout curtain. This program

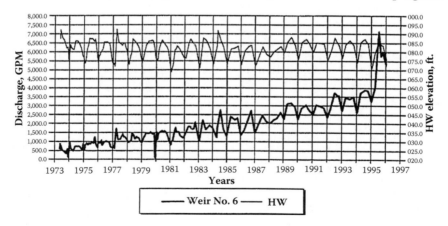

Figure 2.14 Right rim discharge at Leak 6 through 1995, Tims Ford Dam, Tennessee. (From Hamby, J. A., and D. A. Bruce, "Monitoring and Remediation of Reservoir Rim Leakage at TVA's Tims Ford Dam." U.S. Commission on Large Dams (USCOLD) 20th Annual Meeting, 2000. With permission.)

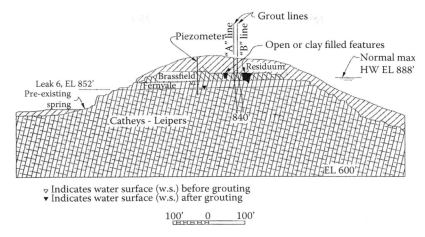

Figure 2.15 Section A-A of right rim at Leak Area 6. (From Hamby, J. A., and D. A. Bruce, "Monitoring and Remediation of Reservoir Rim Leakage at TVA's Tims Ford Dam." U.S. Commission on Large Dams (USCOLD) 20th Annual Meeting, 2000. With permission.)

consisted of drilling a total of 20 vertical and inclined holes, testing permeability in stages, and testing with a dye to develop flow connection times and paths to Leak 6. The exploratory program provided the following conclusions:

1. Progressive erosion of collapsed and/or desiccated karstic feature infill material was the likely cause of the increased seepage. These features were controlled by solutioning along bedding planes and vertical or near vertical joint sets. Open features in excess of 20 feet deep were detected. Several dye-test connection times of only minutes were encountered to the seep.
2. The bottom elevation of the remedial grout curtain as indicated by the geology and the permeability testing was estimated as EL 840 feet.
3. The southerly extent of the remedial grout curtain was geologically well defined.
4. The middle and north ends of the exploratory area were less uniform with high water takes, cavities and open features, very fast dye-connection times, and the possibility of an undetected open channel to Leak 6. (The possibility of an open channel was reinforced by the occurrence of low permeability areas near the north end on either side of a high permeability area, thus leaving the location of the north end of the curtain somewhat questionable.)
5. There was strong evidence that there would be substantial water flow through the features of the foundation rock during remedial grouting.

A multirow remedial grout curtain was designed, approximately 800 feet long, running along the rim. The holes were inclined at 30 degrees to the vertical to encourage intersection of (sub)vertical features and were oriented in opposite directions in the two outside rows. Primary holes in each row were placed at 40-foot centers, with conventional split-spacing methods to be employed (to 10-foot centers). The central, tightening, row was vertical. The grouting was to be executed between ELs 888 and 840 feet—locally deeper if dictated by the results of the stage permeability tests conducted prior to the grouting of each stage.

Because of the suspected high-velocity flow conditions, the downstream curtain row holes that encountered voids and active flow conditions were designated to be grouted with fast-setting (one to three minutes set time) hydrophilic polyurethane resin to provide an initial semi-permanent flow barrier. Holes that did not encounter voids or active flow were to be grouted with HMG. Upon completion of the downstream row it was anticipated that the active flow conditions would be mitigated, thus allowing the entire upstream row followed by the third, central, closure row to be grouted with HMG to form a permanent and durable grout curtain. The grouting was designed to be performed using upstage methods although it was anticipated that poor foundation conditions could locally require utilization of downstage methods. The grout holes were to be cased through the overburden from the surface to the top of the curtain.

The specifications contained provisions that required monitoring and limitations to outflow pH and turbidity to protect the downstream environment. TVA agreed to draw down the reservoir to EL 855 feet (10 feet below minimum normal pool) to minimize hydraulic gradient and flow velocities through the rim. The curtain was to be constructed by first grouting the far ends, so conceptually channeling the flow through a middle zone that would then be sealed. However, as the work progressed, specific geological and hydrogeological conditions caused modifications to the plan, including:

- When drawdown of the reservoir reached EL 859 feet the outflow from Leak 6 completely and naturally stopped. As a consequence, much of the grouting work could be done in "no flow" conditions, therefore eliminating the need for the polyurethane grouts and extending the applicability of cement-based formulations.
- Larger than anticipated open or clay-filled features were encountered especially in the upper 20 feet or so of the curtain. For technical, commercial, environmental, and scheduling reasons, such features were therefore treated with LMG, which had a minimum slump of two inches and contained water-reducing and antiwashout admixtures.
- A suite of HMGs was developed to permit the appropriate match of mix design and "thickening sequence" to the particular stage

conditions as revealed by drilling and permeability testing (both multi- and single-pressure tests).

- In response to conditions revealed during the treatment, observations of the seepage and further dye testing, extra groups of holes were added at the north end of the curtain, including 11 orthogonal to the original curtain, to allow specific treatment of key features.
- In early 1998 the reservoir level had to start being raised, and on February 16, 1998, the lake had risen to EL 869 feet. While injecting LMG in a certain quinary hole that had shown a strong dye connection to the leak, rim leakage decreased in the course of several hours from over 1,000 gpm to less than 60 gpm. Piezometric levels on the rim side of the cutoff dropped about 2 feet (to about EL 855 feet) and ceased thereafter to be influenced by reservoir elevation changes.
- About 2,100 cubic yards of LMG, 400 gallons of polyurethane, and 790 cubic yards of HMG were injected into a total of 250 holes (comprising 11,150 linear feet of rock drilling). Grout reduction ratios were 36 percent (primary/secondary), 49 percent (secondary/tertiary) and 37 percent (quaternary, in the middle, tightening row.)

Throughout the work, closest attention was paid in real time to data from the drilling, water-testing, and grouting activities in addition to information from leak monitoring, piezometers, and interim dye testing. The curtain was thus brought to an engineered refusal. A July 1998 reading, with the lake at EL 888 feet, indicated a seepage of around 400 gpm (net of surface runoff contributions)—about 5 percent of the flow at the equivalent lake elevation prior to grouting. Routine data from piezometers and dye testing support the existence of an efficient and durable curtain, although it is reasonable to expect some increase in flow with time as clay-filled pockets remnant in the curtain are gradually eroded.

2.2.7.2.2 Wolf Creek Dam, Kentucky

This 5,736-foot-long, homogeneous, silty clay embankment has a concrete section on the left abutment and rises a maximum of 258 feet above its Ordovician karstic limestone foundation (Spencer 2006; Fetzer 1988). It impounds the largest manmade reservoir east of the Mississippi River, namely Lake Cumberland, and the nearest population center is Jamestown, Kentucky. Owned and operated by the U.S. Army Corps of Engineers, it has become probably the most famous (or infamous) structure in dam-remediation history since its completion in 1951. A massive emergency grouting operation from 1968 to 1970 arguably saved the dam from a piping-induced failure through its foundation adjacent to the concrete section; sinkholes had appeared on the downstream side of the dam, and muddy water was observed in the dam's outflow channel. The pioneering 1975–79 concrete cutoff wall installed under a design-build contract by the ICOS Corporation, reached a maximum depth of 280 feet, including a maximum of 100 feet into karstic limestone, and totaled about 531,000 square feet of cutoff (Bruce et al. 2006). In recent years, this dam was allocated top-priority remediation status (DSAC-1) by the government following careful evaluation of a wide range of dam safety monitoring data and observations. Spencer (2006) provides a most comprehensive description of the geological and construction database.

The competition to build a new deeper and longer wall, with almost 1 million square feet of cutoff 4,000 feet long, extending from the concrete section, was won in 2008 by the TreviICOS-Soletanche JV. As a key risk-management strategy, the JV elected to explore and pretreat the embankment/interface in those sections of the dam judged likely to have been affected by seepage; if left untreated, such zones would have had the potential to cause massive and sudden slurry loss during the excavation of the diaphragm wall panels (using techniques described in Chapter 4, this volume). Such losses had been recorded on other concrete wall projects, the most notable being at Mississinewa Dam, Indiana, and the potential risk to dam safety was untenable.

Holes were located in two rows, each about 7 feet outside the planned location of the diaphragm wall. They were drilled and grouted in primary-secondary sequence, the interhole spacing in each row of production holes being 10 feet. The holes were drilled with rotary duplex methods, with a 4½-inch outside diameter (o.d.) casing and polymer flush. Drilling techniques, including restrictions on penetration rates, were in conformance with USACE's ER-1110-1-1807. In any one area, downstream holes were completed before upstream holes were installed. Each hole's progress, during drilling and grouting, was recorded by automated systems, with manual logs maintained for further support. A 1 percent deviation tolerance was set on each drill hole.

Initially, 19 exploratory holes at about 200-foot spacing were installed penetrating from 5 to 15 feet into rock (average 6.3 feet). The subsequent 248 production holes (including 58 tertiaries in critical areas) penetrated from 1 to 9 feet into rock (average 3.2 feet). Pressure grouting was conducted from these depths to about 10–15 feet above rockhead, in 1- to 2-foot ascending stages, through the drill casing. A further 8 LMG holes were drilled to treat specific features.

Three mix designs were experimented with prior to works commencing (Table 2.3), with mix 1 eventually being used throughout. The target twenty-eight-day UCS was 500 to 1,000 psi, but this was well exceeded in practice. The slump was set at 4 inches, ±1 inch.

Preproduction testing also identified that at a pump rate of 2.5 cubic feet per minute (i.e., about four pump strokes per minute), the line loss was 120 psi for the LMG. The maximum gauge pressure at refusal was therefore set at 320 psi, but was later reduced to 300 psi. The 2.5 cubic feet per minute pump rate was chosen as a conservative rate during pressure grouting as this would act

Table 2.3 LMG grout mixes tested prior to the start of work, Wolf Creek Dam, Kentucky

Components	Mix 1	Mix 2	Mix 3
Cement	300.0 lbs	250.0 lbs	200.0 lbs
Flyash	800.0	650.0	500.0
Natural sand	1112.5	1112.5	1112.5
Manufactured sand	1112.5	1112.5	1112.5
Bentonite	5.6	5.0	5.0
Water	375.3	334.0	334.0

against the possibility of damage from pressure spikes and allow for quick dissipation of soil pore pressures. In this regard, a "spike" could develop in the course of one pump stroke which, in the worst case, would mean the injection of 0.6 cubic feet of grout in the 10-second period of pump operator reaction. This rate was tripled during simple backfilling of the hole in the embankment above the pressure grouted zone, during casing extraction.

Refusal criteria included the limiting gauge pressure, a volume of 10 cubic feet per foot (but increased up to 5 cubic yards in obvious void filling situations, which were very infrequent), grout escape to the surface (not recorded), and exceedance of threshold limits on automated inclinometers and piezometers within close proximity to the point of injection. The performance of these instruments in particular was especially carefully monitored, in real time, while frequent readings were made manually of crackpins and settlement monuments. In the vast majority of stages, refusal was reached on the pressure criterion, although in the most critical areas of the foundation the reaction of the vibrating wire piezometers and inclinometers proved the controlling factors. No structural movement was recorded at any point.

Average grout takes ranged from 1.0 to 3.7 cubic yards of LMG per hole, and this was, together with the drilling data, interpreted as indicating relatively tight interface conditions. The operation also confirmed the integrity of the embankment itself. The operation involved a total of 20,187 feet of drilling and the placement of 237 cubic yards of LMG, 54 percent of which was used to backfill each hole above the pressure grouted zone.

The success of this operation is reflected in the fact that no sudden slurry losses have occurred during subsequent diaphragm walling operations at the interface.

2.3 JET GROUTING

2.3.1 Perspective

Due principally to the very aggressive promotional efforts of the respective specialty contractors, the various types of jet grouting have become very popular since the mid-1980s in the United States. Comprehensive reviews have been provided by Xanthakos, Abramson, and Bruce (1994), Kauschinger, Perry, and Hankour (1992), Shibazaki (2003), and Burke (2004), among numerous others. Paradoxically, this very high-energy approach to disaggregating soils and cementing them *in situ* has come to be regarded by certain clients as somewhat of a commodity.

Case histories of jet grouted remediations for dams can be cited around the world, and indeed Canada has an impressive record (Pacchiosi Drill SpA 2011) stretching back to the plastic cutoff installed at John Hart Dam, British Columbia, in 1988. Examples in the United States have been fewer, largely due to commercial competition from other technologies, but also due to particularly keen concerns for the potential dangers that the use of jet-injection pressures of over 5,000 psi can cause in already delicate embankment structures.

Nevertheless, certain sets of circumstances have occasionally conspired to render jet grouting as the most cost-effective technical solution. Three case histories are presented in the following sections, although the reader should also be aware that jet grouting was trialed at Mormon Island Dam, California, in 2008 for a potential seismic retrofit. No information has been published on the test, which was of relatively limited extent.

2.3.2 Case histories

2.3.2.1 American River Levee, Sacramento, California

From April 2002 to July 2003, jet grouting was used to form 100-foot-deep cutoffs on five separate areas on the USACE levee system, to close "windows" in the cutoff where other techniques could not be used for logistical reasons (Pacchiosi promotional information). The foundation principally comprised sand, clay, and silt, but there were occasional but substantial very dense beds of gravel and cobbles at depths of over 55 feet beneath crest level. The ground varied both vertically and laterally. In addition, there were frequent underground service ducts, and traffic considerations dictated night and weekend work.

A very intensive field-test program was conducted to verify design concepts and details. Five elements were installed with columns of 13-foot diameter, thin panels of over 20 feet in length, and semicircular "segments" of internal angle 30 degrees and 120 degrees. All panels were installed with the contractor's preferred three-fluid system. The program involved tests of homogeneity, column deviation (before grouting), permeability (a target of 5×10^{-5} cm/s), and unconfined compressive strength. Continuous real-time automated monitoring was conducted.

The consequence of the program was that the cutoff would be built with columns of 8-foot diameter at 6.5-foot centers. This generated about

Photo 2.10 Exposed test columns, Sacramento, California. (Pacchiosi Promotional Information, www.pacchiosi.com)

65,000 square feet of wall. Columns were vertical except for some inclined columns and panels under the Union Pacific Railroad line. Photos 2.10 and 2.11 show exposed test columns and panels, respectively.

2.3.2.2 Wickiup Dam, Oregon

Wickiup Dam, dikes, and appurtenant structures were constructed for the Bureau of Reclamation between 1939 and 1949, southwest of Bend, Oregon, on the Deschutes River (Bliss 2005). The dam is a rolled earthfill embankment with a 3:1 riprap upstream face and a 2:1 rockfill downstream face. The dam has a main river embankment section with a structural height of 100 feet and a crest elevation of 4,347 feet with a normal water-surface elevation of approximately 4,337 feet. The foundation of the main embankment section consists of basalt and mudflow debris. The left abutment of the dam transitions into an approximately three-mile-long dike with a maximum height of approximately 40 feet. The dike

Photo 2.11 Exposed test panels, Sacramento, California. (Pacchiosi Promotional Information, www.pacchiosi.com)

foundation consists of horizontal layers of fluvial sands and gravels, volcanic ash, and lacustrine silts and clays. The majority of the 13,860-foot length of the dam (left of Sta. 15+00) is commonly referred to as the left-wing dike.

The right abutment of the dam is founded on basaltic lava flows and overlying mudflow debris. The channel section and left abutment are founded on interbedded fluvio-lacustrine and alluvial sediments, which include layers of sand (Qfs), and gravel (Qfg), volcanic ash (Qfv), diatomaceous silt (Qfd), dense silt and sand (Qfds), and clay and silt (Qfc). A surficial layer of Mazama Ash (Qma) ranging from about three to five feet thick blankets the area. These layers are generally horizontally bedded and separated by well-defined boundaries.

An analysis indicated that several of the foundation soils were susceptible to liquefaction due to ground shaking from the design earthquakes and lesser events. The potentially liquefiable soils include the Mazama ash (Qma), volcanic ash (Qfv), both upper and lower diatomaceous silt layers (Qfd), and some isolated locations in the upper sand (Qfs). The Mazama ash was not a concern since it was a surficial layer that was removed from the footprint of the existing dam and from part of the new construction at the downstream toe. Furthermore, the results of postearthquake stability analyses indicated that remedial actions were necessary to prevent an overtopping failure of the left-wing dike embankment between approximate Sta. 12+00 and 48+00 during the design earthquakes.

A corrective action study indicated that either jet grouting or excavation and replacement of the foundation materials at the downstream toe were the least expensive, technically acceptable remediation methods. Both options were approximately equal in cost. However, jet grouting had important advantages over excavation and replacement. The design team estimated that the excavate-and-replace option would likely require a two-season reservoir restriction that would severely impact the water users. The reservoir restriction would have been required to assist the difficult dewatering operation for the excavation and provide a factor of safety against a slope failure of the excavation. Jet grouting did not require a reservoir restriction, nor did it require as much downstream tree removal, which was an important issue for bald eagle nesting in the vicinity.

The excavate-and-replace option would also require additional expensive drainage zones to account for reduced permeability of materials recompacted into the excavation. Without these zones, inadequate drainage of the large volume of foundation seepage transmitted through the upper foundation zone (Qfs) could cause increased pore pressure in the embankment.

A jet grouting test section contract for Wickiup Dam required the construction of both large (approximately 14.5 feet) diameter columns (L-01 through L-12) and conventional (approximately 5.5 feet) diameter columns (S-01 through S-07) in groups and as individual columns using various grout mixes and cement types.

Cement contents were varied in some of the columns in order to evaluate the effect of cement content on column strength. In addition, one column was constructed within a circular array of six cased and grouted boreholes spaced on twenty-five-foot diameters in order to perform crosshole tomography.

Four different grout mixes were utilized:

- A water:cement ratio of 1:1 by weight was used initially. Type III cement was originally specified in the contract due to the need to obtain strength results quickly from core drilling and sampling.
- A second mix was used in five columns consisting of a 1:1 water:cement ratio using Type I–II cement. The change from Type III to Type I–II cement was made, during the contract period, due to high strengths that were developing very early in the test section construction.
- The third mix was used in one column and comprised a 1.15:1 water:cement ratio using Type I–II cement.
- The fourth mix was used in one column and comprised a 1.25:1 water:cement ratio using Type I–II cement.

Based on the test section results, a grout mix of 1.15:1 water:cement ratio, using Type I–II cement, was specified for the contract. No indications of ground fracturing heaving or uplift were observed during the jet grouting work.

Both visual and chemical testing of the upper sand and gravel confirmed that grout was not contaminating this zone. In addition, monitoring of the Deschutes River was undertaken during the grouting to ensure that no grout reached the river, which, in this area, is designated as a Wild and Scenic River with zero tolerance for contamination.

Field confirmation investigations included electronic cone penetration tests (ECPT), core drilling (17 holes), and geophysical crosshole tomography testing. Verification laboratory testing consisted of unconfined compressive strength (UCS) testing of wet slurry samples obtained during construction of the columns and UCS testing of drill cores obtained after completion of the test section. In addition, some core samples were selected for cement content testing. Unit weight values were determined for select drill core samples. Nineteen core holes were drilled in the columns for a total of 940 feet. Samples of the core recovered were selected from each geologic unit and tested for UCS. UCS test values on 28-day slurry samples ranged from 260 to 3,620 psi, and UCS values on 28-day neat grout samples ranged from 1,050 to 3,500 psi. Based on UCS tests on 33- to 55-day core samples, column strength ranged from approximately 300 to 1,580 psi. In the design of the test section, a minimum 28-day UCS of 200 psi had been the target strength.

The twelve large-diameter columns were constructed (1) to achieve the largest practical diameter using large diameter grouting methods, and (2) to investigate the column spacing required to achieve closure of adjacent columns. They were constructed in an array with center-to-center spacings ranging from 11.5 to 14.5 feet. The "Superjet" system was used, basically an enhanced two-fluid system. Based on the results from core drilling, there were indications that unmixed zones or zones with decreased cutting diameter at the specified spacings were confined to the upper dense silt and sand (Qfds) layer, and to a lesser extent the lower diatomaceous silt (Qfd). Generally, the volume of unmixed material at the intersections and interstices was negligible compared to the overall mass of the soilcrete columns.

The seven smaller-diameter columns were constructed in order to achieve the largest diameter practical using conventional jet grouting methods and to investigate the column spacing required to achieve closure of the adjacent columns. The conventional diameter soilcrete columns were spaced 3.5, 4.5, and 5.5 feet from center to center. As with the large diameter columns, the top of the column stopped approximately 3 feet into the sand (Qfs) and extended to a bottom depth approximately 3 feet below the lower diatomaceous silt (Qfd) into the dense silt and sand (Qfds). The operating parameters for the conventional diameter columns were significantly different than those for the large diameter columns. The difference in lift rate, rotation speed, and grout flow resulted in lower overall cutting distances, reaching only about 5 feet diameter into the upper, less dense, geologic units and reducing to about 3.5 feet diameter in the upper Qfds and lower Qfd materials.

Based on the successful performance of the test section, jet grouting was selected for final designs. The final design geometry for the modified embankment section was largely dependent upon results from dynamic deformation analyses using the computer program FLAC. The use of FLAC also helped to confirm two other elements of the design: (1) portions of the dike could be remediated with a berm alone, and (2) the dike embankment beyond Sta. 48+00 did not require any treatment. The final design incorporated jet grouting foundation treatment and a downstream berm between Sta. 12+00 and 37+50 and a downstream berm only from Sta. 37+50 to 48+00.

In general, the estimated, postmodification seismic-induced deformations ranged from 2 to 7 feet. In order to account for uncertainty in the FLAC calculated value of deformation, the modified sections were designed to accommodate twice the calculated vertical deformation, plus 3 feet, for a minimum width of 20 feet on the deformed section (Figure 2.16).

Jet grouting was completed in September 2002, almost six months ahead of schedule. All remaining construction earthwork was completed in October 2003. The first complete filling of the modified embankment

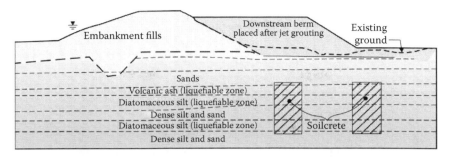

Figure 2.16 Section view of left abutment dike. (Hayward Baker promotional information.)

occurred in the spring of 2004. Extensive monitoring of piezometers, toe drains, and drainage inspection wells has been conducted to confirm satisfactory performance.

During the production work, illustrated in Photo 2.12, a total of 854 Superjet columns were installed along a 2,250-foot stretch of the dam. These typically were spaced at 13-foot centers and had a measured diameter of about 14 feet. They ranged from 20 to 87 feet deep, and required 41 million gallons of grout to create 201,000 cubic yards of soilcrete. All operations were controlled in real time by computerized systems. Coring provided the data of Figure 2.17 and confirmation that the scattered, unmixed particles were within the 1- to 3-inch range.

Photo 2.12 Production jet grouting, Wickiup Dam, Oregon. (Hayward Baker promotional information.)

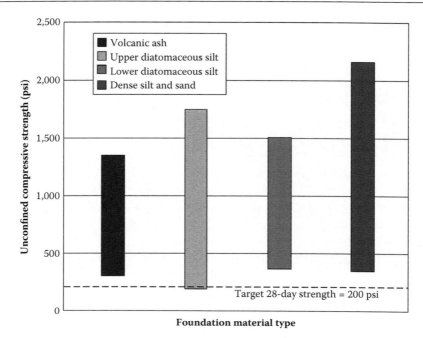

Figure 2.17 UCS strength data on cored jet grouted samples, Wickiup Dam, Oregon. (Hayward Baker promotional information.)

Bliss concluded that the benefits of the jet grout option at Wickiup Dam included:

- focused treatment of specific soil horizons;
- elimination of need for reservoir restrictions;
- minimal environmental impact (bald eagles and the Oregon spotted frog); and
- no adverse impact on "normal" seepage under the dam.

In addition, the large volume of jet grout spoils was found acceptable as fill material for the new stability berm, when blended with soils from another source.

2.3.2.3 Tuttle Creek Dam, Kansas

Tuttle Creek Dam was constructed in the 1950s by the USACE and is located five miles north of Manhattan, Kansas (Mauro and Santillan 2008). It provides for flood control, recreation, fish and wildlife conservation, water-quality control, and navigation supplementation. The main embankment is a rolled-earth and rockfill structure (Figure 2.18) about 6,000 feet long and about 180 feet high over the original valley of the Big Blue River.

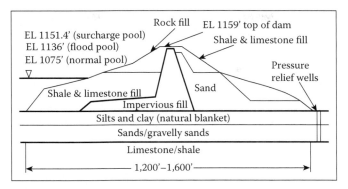

Figure 2.18 Typical cross-section, Tuttle Creek Dam, Kansas. (From Mauro, M., and F. Santillan, "Large Scale Jet Grouting and Deep Mixing Test Program at Tuttle Creek Dam." *Proceedings of the 33rd Annual and 11th International Conference on Deep Foundations,* 2008. With permission.)

A site earthquake study of 1999 postulated an earthquake of Richter magnitude 6.6 occurring twelve miles east of the dam as the maximum credible earthquake. Given the high risk associated with loss of the dam, the government authorized a multiyear remediation program featuring stabilization of foundation soils.

A large-scale jet grouting and deep mixing test was first conducted, downstream of the dam, between April 2006 and February 2007, without drawdown of the reservoir. Jet grout columns of 8 to 10 feet were targeted with a minimum UCS of 170 psi. Jet grouting was anticipated as the optimum technique for installing an upstream cutoff in the embankment, given the perceived difficulties other techniques would experience in penetrating the shale and limestone fill. The test program featured both double- and triple-fluid methods, while jet-assisted DMM columns (WJE) were also installed targeting columns of 3- to 6-foot diameter. (These are not discussed further in this review.)

Table 2.4 summarizes the main design parameters of the jet grout columns that were installed in patterns as shown in Figure 2.19.

The specific energy (E_s) per unit length of column was first described by Tornaghi (1989), and is calculated as follows:

$$E_s = \frac{P \times Q}{V_t} \, (MJ/m)$$

where:
P = pressure of injected fluids (water and grout) (MPa)
Q = flow rate of injected fluids (water and grout) (m³/hour)
V_t = jet withdrawal speed (m/hour)

Table 2.4 Design parameters for the jet grout test columns, Tuttle Creek Dam, Kansas

Jet grouting technology	Target diameter	Specific energy	Theoretical cement content
3-Fluid	8 feet	130–170 MJ/m	670–1,260 lb/cy
3-Fluid	10	240–300	760–1,420
2-Fluid	8	130–180	650–1,100

Source: Mauro, M., and F. Santillan, "Large Scale Jet Grouting and Deep Mixing Test Program at Tuttle Creek Dam." *Proceedings of the 33rd Annual and 11th International Conference on Deep Foundations*, 2008. With permission.

Column diameter is related to specific energy for any given soil type. For this trial, the water:cement ratio was varied between 0.75 and 1.25. Totals of nine double-fluid and eighteen triple-fluid columns were installed in groups of three to depths of around 45 feet below ground surface. Nine different combinations of parameters were therefore tested. As shown in Figure 2.19, a 32-inch-thick cement-bentonite cutoff wall was installed around the site, and keyed into the bedrock. This would permit later excavation of the site and exposure of the columns, to a depth of 26 feet with minimal dewatering, bearing in mind that the groundwater was about 8 feet below the ground surface.

The fluvial deposits at the test site (Figure 2.20) included a 10- to 14-foot layer of lean clay and clayey silt, underlain by medium and gravelly sand, with clayey lenses. CPT data from five holes are shown in Figure 2.21.

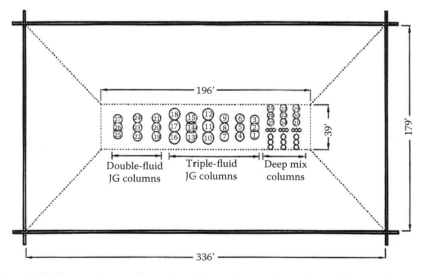

Figure 2.19 Test area layout, Tuttle Creek Dam, Kansas. (From Mauro, M., and F. Santillan, "Large Scale Jet Grouting and Deep Mixing Test Program at Tuttle Creek Dam." *Proceedings of the 33rd Annual and 11th International Conference on Deep Foundations*, 2008. With permission.)

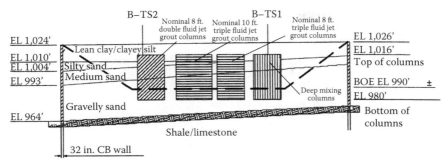

Figure 2.20 Cross-section at the test area, Tuttle Creek Dam, Kansas. (From Mauro, M., and F. Santillan, "Large Scale Jet Grouting and Deep Mixing Test Program at Tuttle Creek Dam." *Proceedings of the 33rd Annual and 11th International Conference on Deep Foundations*, 2008. With permission.)

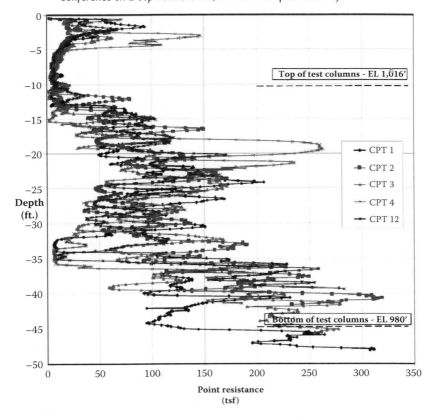

Figure 2.21 CPT point resistance at the test area, Tuttle Creek Dam, Kansas. (From Mauro, M., and F. Santillan, "Large Scale Jet Grouting and Deep Mixing Test Program at Tuttle Creek Dam." *Proceedings of the 33rd Annual and 11th International Conference on Deep Foundations*, 2008. With permission.)

Ground instrumentation included:

- surface deformation ("heave") points;
- vibrating wire piezometers, with temperature probes, fixed at EL 1,010 and 990 feet;
- deep settlement points, fixed at EL 995 feet;
- observation wells inside and outside the wall; and
- inclinometers.

Negligible ground deformations, either of the surface or at depth, were recorded. Although "a significant increase" in pore pressures was recorded in real time during the installation of the columns, this dissipated at a rate depending on the soil permeability. A "noticeable" temperature rise around the hydrating columns was also recorded.

Cores were taken from each group of three columns, one at the center, and one at an overlap. Strength test results are shown in Figure 2.22. Excavation then proceeded in steps, the columns being pressure-washed throughout. Photos 2.13 and 2.14 show the exposed columns. Diameter varied with depth varying with the soil conditions (Figures 2.23 through 2.25). For the three-fluid columns the target was generally met or exceeded, whereas for the two-fluid columns the target was met in the cohesive soils, but in granular layers the diameters were 6 to 9 feet. Three horizontal tree trunks (Photo 2.15) were found in one group of two-fluid columns, and were noted to have prevented the complete formation of columns in that area due to the "shadow effect."

Thereafter, two sets of jet grout columns were saw cut over their upper 10 feet (Photo 2.16). This exercise demonstrated good column interlock, but also sizable inclusions of soil (2–24 inches) due to localized collapses of surface soils into the freshly grouted underlying soils.

This extraordinary test program had a somewhat bittersweet, ironic conclusion: further analyses, modeling the site-specific soils, showed that the site was not as vulnerable to liquefaction as originally estimated. The upstream (jet grouting) work was therefore deleted, and a new remediation design involved the construction of a series of shear walls in the toe area. These were constructed as cement-bentonite trenches using very similar means, methods, and materials to those previously used to build the test area cutoff wall around the test columns.

2.4 CUTOFF THROUGH LANDSLIDE MATERIAL: THE CASE HISTORY OF HOWARD HANSEN DAM, WASHINGTON

As discussed in Section 2.1, and illustrated in Sections 2.3 and 2.4, it is unusual for a remedial cutoff in materials other than rock to be installed

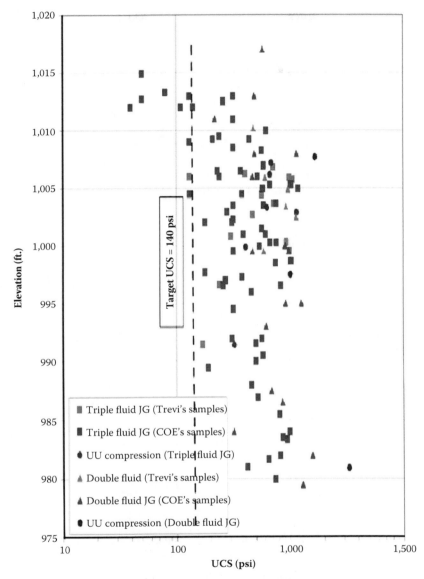

Figure 2.22 UCS and UU test results on the jet grout core samples, Tuttle Creek Dam, Kansas. (From Mauro, M., and F. Santillan, "Large Scale Jet Grouting and Deep Mixing Test Program at Tuttle Creek Dam." *Proceedings of the 33rd Annual and 11th International Conference on Deep Foundations*, 2008. With permission.)

Photo 2.13 Exposed triple-fluid columns, Tuttle Creek Dam, Kansas. (From Mauro, M., and F. Santillan, "Large Scale Jet Grouting and Deep Mixing Test Program at Tuttle Creek Dam." *Proceedings of the 33rd Annual and 11th International Conference on Deep Foundations,* 2008. With permission.)

using drilling and grouting methods (Danielson, Ebnet, Smith, Bookshier, and Sullivan 2010). However, a most significant remediation of a dam abutment comprising mainly landslide material has recently been conducted and merits close evaluation.

Howard A. Hanson Dam, built in 1962, is an earth embankment dam on the Green River in western Washington. The dam embankment is 235 feet high (crest elevation 1,228 feet) and 675 feet long. It is founded on volcanic bedrock on its left abutment and foundation. The right abutment foundation is partially bedrock and partially unconsolidated fluvial, galacio-fluvial, and landslide material. Materials in the dam consist of sandy gravel with less than 10 percent fines upstream of a vertical coarse gravel and cobble central chimney drain. Downstream of this drain is rolled rockfill and the upstream and downstream faces of the dam are covered with volcanic rip-rap. It is a multipurpose dam in which the future target

Photo 2.14 Aerial view of the test area from the west, Tuttle Creek Dam, Kansas. (From Mauro, M., and F. Santillan, "Large Scale Jet Grouting and Deep Mixing Test Program at Tuttle Creek Dam." *Proceedings of the 33rd Annual and 11th International Conference on Deep Foundations,* 2008. With permission.)

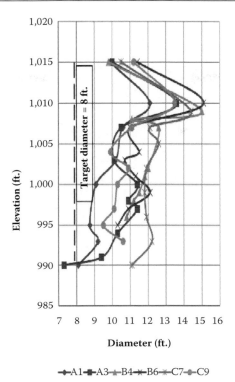

←A1–■A3–▲B4–✳B6–✳C7–●C9

Figure 2.23 Diameter of the triple-fluid columns with Es = 130–170 MJ/m, Tuttle Creek Dam, Kansas. (From Mauro, M., and F. Santillan, "Large Scale Jet Grouting and Deep Mixing Test Program at Tuttle Creek Dam." *Proceedings of the 33rd Annual and 11th International Conference on Deep Foundations*, 2008. With permission.)

summer conservation pool is 1,177 feet. This compares with an elevation of 1,075 feet (essentially empty) during the annual late October–end February flood season.

The geomorphology of the site is very complex, but of prime significance is the interglacial rock slide consisting of blocks of broken andesite with dimensions of over 20 feet. As shown in Figure 2.26, this overlies a (mainly) laterally continuous lacustrine silt deposit at about EL 1,050 feet, which hydraulically separates the landslide material from a lower glacio-fluvial aquifer.

The upper aquifer has been interpreted as having a significant seepage and internal erosion problem, especially in the "short path" zone within about 200 feet of the dam's right embankment. This is due to the nature of the materials (silt, sand, and gravel layers in contact with very high permeability fractured bedrock and landslide debris), the potential for high exit

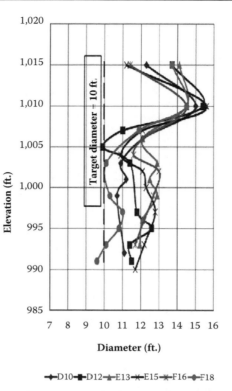

Figure 2.24 Diameter of the triple-fluid columns with Es = 230–300 MJ/m, Tuttle Creek Dam, Kansas. (From Mauro, M., and F. Santillan, "Large Scale Jet Grouting and Deep Mixing Test Program at Tuttle Creek Dam." *Proceedings of the 33rd Annual and 11th International Conference on Deep Foundations*, 2008. With permission.)

gradients at unfiltered or improperly filtered downstream slopes, and the short distance between upstream and downstream faces in this area.

The possibility of seepage was, however, identified during the design of the dam and, over the years, various defenses were introduced when the rate proved "excessive." These measures included a single row, 300-foot-long grout curtain adjacent to the embankment in the right abutment in 2002. This curtain was not brought to refusal at that time and did not tie into the embankment.

A record pool (EL 1,188.8 feet) occurred in early 2009, and during drawdown symptoms of potential piping/internal erosion were observed. Analysis of instrumentation and dye-testing results led the USACE to design and implement interim risk-reduction measures including repairs to the right abutment, including a grout curtain. The entire design, procurement and execution period of the grouting occupied six months, and was completed by the start of the flood season.

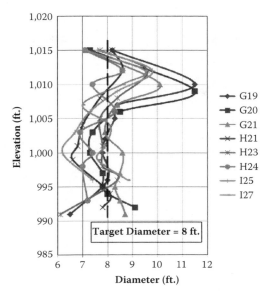

Figure 2.25 Diameter of the double-fluid columns with Es = 130–180 MJ/m. (From Mauro, M., and F. Santillan, "Large Scale Jet Grouting and Deep Mixing Test Program at Tuttle Creek Dam." *Proceedings of the 33rd Annual and 11th International Conference on Deep Foundations*, 2008. With permission.)

The new curtain was 450 feet long, tied into the rock ridge ("septum") serving as the boundary between the dam embankment and the right abutment. The other (eastern) end terminated in landslide materials, where it was calculated that any seepage around the end would be drawn to the existing downstream drainage tunnel and its drains, and would exit the abutment far from the "short path" seepage area. The curtain toed into

Photo 2.15 Effect of an obstruction in a double-fluid jet grout column, Tuttle Creek Dam, Kansas. (From Mauro, M., and F. Santillan, "Large Scale Jet Grouting and Deep Mixing Test Program at Tuttle Creek Dam." *Proceedings of the 33rd Annual and 11th International Conference on Deep Foundations*, 2008. With permission.)

Photo 2.16 Saw-cut of three triple-fluid jet grout columns, Tuttle Creek Dam, Kansas. (From Mauro, M., and F. Santillan, "Large Scale Jet Grouting and Deep Mixing Test Program at Tuttle Creek Dam." *Proceedings of the 33rd Annual and 11th International Conference on Deep Foundations*, 2008. With permission.)

rock, or the silt aquitard and its upper elevation of 1,206 feet corresponded to that of an existing work platform.

The downstream grout row (A) was largely completed first before the upstream row, and its thickness was designed to significantly lower the hydraulic gradient across the curtain and increase the likelihood that existing piping or internal erosion features would be intercepted. From station 1+00 to 4+50 in the A row and for all holes in the B row the holes, after the tertiary phase, were spaced at 5-foot centers. For the first 100 feet of the A row, in the heart of the short path area, primary-tertiary spacings were reduced to 2 feet. This also acted against grout migration into the embankment core gravel, by permitting stage grout volumes to be limited. As a consequence, 122 holes were foreseen in Row A and 91 in Row B, with the possibility of local quaternaries depending on actual conditions.

Simple end of casing injection methods were used with a packer at the end of the casing to inject a suite of four balanced cement-based grouts of superior bleed and pressure filtration characteristics. This method was acceptable since the purpose of the program was to quickly locate and fill preferential seepage paths in the heterogeneous abutment. Drilling was limited to rotary sonic or rotary duplex for dam safety, grout amenability and progress reasons. Holes were generally grouted in 15-foot-long upstages to a stage apparent Lugeon criterion of 6-inch overburden and 3-inch bedrock. A volumetric criterion was also applied, being 80 gallons per foot generally, but 40 gallons per foot in the A row "short path" holes.

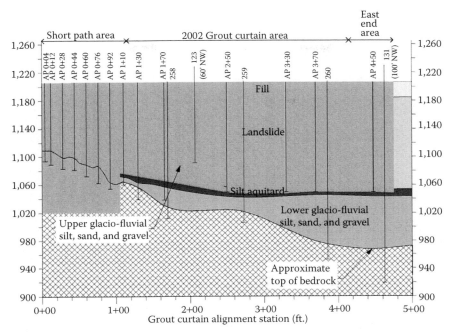

Figure 2.26 Section looking downstream through 2009 interim grout curtain (gray shaded area) Howard Hanson Dam, Washington. (From Danielson, T. J., A. F. Ebnet, R. E. Smith, M. I. Bookshier, and R. P. Sullivan, "Howard Hanson Dam Right Abutment Seepage: 2009 Interim Grout Curtain and Drainage Tunnel Improvements." *ASDSO Dam Safety Conference*, 2010. With permission.)

Other details included:

- The use of sonic drilling (6-inch diameter).
- Real-time use of computer monitoring.
- Analysis of results allowed the curtain to be evaluated in three discrete sections, namely the first 100 feet (short path), the next 295 feet (2002 curtain), and the remaining 55 feet (East end). Figure 2.27, Table 2.5, and Table 2.6 illustrate the results, with the reduced takes of the 2002 curtain area confirming the value of that work and so justifying its incorporation into the 2009 work.
- Seventy-four further holes were added to specifically treat the rock septum to EL 1,020 feet (i.e., coincident with the base of the embankment), based on results of dye tests conducted during the 2009 conservation pool raise.
- The curtain was also extended 25 feet further into the abutment as a cost-effective expedient.
- Quantities of work included 40,681 lineal feet of overburden drilling, 1,356 feet of rock drilling, 74 borehole deviation surveys, and

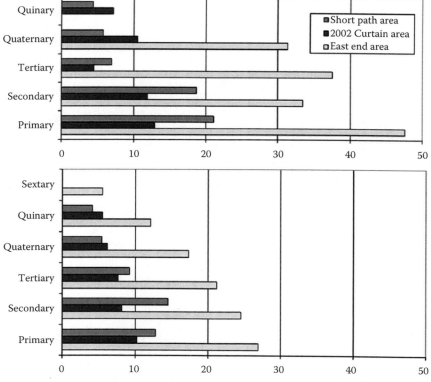

Figure 2.27 Average grout takes per linear foot of hole by area: A Row (top) and B Row (bottom) Howard Hanson Dam, Washington. (From Danielson, T. J., A. F. Ebnet, R. E. Smith, M. I. Bookshier, and R. P. Sullivan, "Howard Hanson Dam Right Abutment Seepage: 2009 Interim Grout Curtain and Drainage Tunnel Improvements." *ASDSO Dam Safety Conference*, 2010. With permission.)

2,121 grout pump hours. Over 2,930 feet of verification drilling was conducted using HMG (Marsh cone value 35 seconds). This confirmed very good closure in Row A, and excellent closure in Row B.
- Modification and extension of the downstream drainage system was conducted simultaneously.

In addition to the previous dam instrumentation, two transects of piezometers were installed across the curtain, each comprising one piezometer upstream, within, and downstream of the curtain. Each instrument has two vibrating wire piezometers in rock, two in the lower aquifer, and two in the upper aquifer. Comparison of readings at the same pool

Table 2.5 Percentage of holes reaching closure by area, Howard Hansen Dam, Washington

Row	Hole series	Number of holes per row, series, and area		
		Short-path area	2002 Curtain area	East end area
A	Primary	13	15	3
A	Secondary	13	14	3
A	Tertiary	25	31	5
A	Quaternary	20	17	11
A	Quinary	6	4	N/A
B	Primary	6	14	3
B	Secondary	5	15	3
B	Tertiary	10	31	5
B	Quaternary	14	25	11
B	Quinary	2	5	9
B	Senary	N/A	N/A	8

Source: Danielson, T. J., A. F. Ebnet, R. E. Smith, M. I. Bookshier, and R. P. Sullivan, "Howard Hanson Dam Right Abutment Seepage: 2009 Interim Grout Curtain and Drainage Tunnel Improvements." *ASDSO Dam Safety Conference*, 2010. With permission.

elevation (1,167) feet before and after the remediation, confirms "significantly lower heads" in the majority of the upper aquifer piezometers in 2010 downstream of the curtain and upstream of the drainage tunnel: in particular, downstream heads in the "short path" area are as much as 24 feet lower.

Table 2.6 Distribution of holes by row, series, and area, Howard Hansen Dam, Washington

Row	Hole series	Percent of stages reaching apparent lugeon closure/criterion		
		Short-path area Sta 0+00 to 1+00	2002 curtain area Sta 1+00 to 3+95	East end area Sta 3+95 to 4+50
A	Primary	80	98	57
A	Secondary	77	95	77
A	Tertiary	96	100	67
A	Quaternary	95	94	74
A	Quinary	100	95	N/A
B	Primary	86	97	66
B	Secondary	81	94	65
B	Tertiary	96	95	73
B	Quaternary	100	96	80
B	Quinary	100	97	85
B	Senary	N/A	N/A	95

Source: Danielson, T. J., A. F. Ebnet, R. E. Smith, M. I. Bookshier, and R. P. Sullivan, "Howard Hanson Dam Right Abutment Seepage: 2009 Interim Grout Curtain and Drainage Tunnel Improvements." *ASDSO Dam Safety Conference*, 2010. With permission.

As a final point, the authors conclude that although these measures, including the grout curtain, have been found to be effective, they are classified as "interim" and further studies and investigations are ongoing to evaluate the possible need for any additional measures.

REFERENCES

Adcock, L., and R. Brimm. 2008. "A History of Karst Seepage—Center Hill Dam Rehab." *International Water Power and Dam Construction*, June 18. http://www.waterpowermagazine.com/story.asp?sc=2049980.

Arora, S., and J. Kinley. 2008. "Chemical Grouting to Eliminate Water Leakage through Concrete Construction Lift Lines at a Dam in Arizona." Paper presented at the ASDSO Dam Safety Conference, Indian Wells, CA, September 7–10, 6 pp.

ASCE. 2005. "Glossary of Grouting Terminology." D. A. Bruce, Committee Chairman, ASCE. *Journal of Geotechnical and Geoenvironmental Engineering* 131 (12): 1534–42.

Baker, W., ed. 1985. "Issues in Dam Grouting." *Proceedings of the Session Sponsored by the Geotechnical Engineering Division of the ASCE in Conjunction with the ASCE Convention*, Denver, CO, April 30.

Baker, W., E. Cording, and H. MacPherson. 1983. "Compaction Grouting to Control Ground Movements during Tunneling." *Underground Space* 7:205–12.

Bandimere, S. W. 1997. "Compaction Grouting State of the Practice 1997." In *Grouting: Compaction, Remediation, and Testing*, edited by C. Vipulanandan, pp. 18–31. ASCE Geotechnical Special Publication 66. Reston, VA: ASCE.

Bliss, M. 2005. "Jet Grouting for Seismic Remediation of Wickiup Dam, Oregon." *Journal of Dam Safety* 3 (1).

Bruce, D. A. 1989. "The Repair of Large Concrete Structures by Epoxy Resin Bonding." *REMR Bulletin* 6 (4): 4–7.

Bruce, D. A. 2003. "The Basics of Drilling for Specialty Geotechnical Construction Processes." *Grouting and Grout Treatment: Proceedings of the Third International Conference*, Geotechnical Special Publication No. 120, edited by L. F. Johnsen, D. A. Bruce, and M. J. Byle, 752–771. Reston, VA: ASCE.

Bruce, D. A., and J. P. Davis. 2005. "Drilling through Embankments: The State of Practice." Paper presented at the USSD Conference, Salt Lake City, UT, June 6–10, 12 pp.

Bruce, D. A., and F. Gallavresi. 1988. "The MPSP System: A New Method of Grouting Difficult Rock Formations." In *Geotechnical Aspects of Karst Terrains*, pp. 97–114. Geotechnical Special Publication 14.

Bruce, D. A., A. Ressi di Cervia, and J. Amos-Venti. 2006. "Seepage Remediation by Positive Cut-Off Walls: A Compendium and Analysis of North American Case Histories." Paper presented at the ASDSO Dam Safety Conference, Boston, MA, September 10–14.

Bruce, D. A., W. G. Smoak, and C. C. Gause. 1997. "Seepage Control: A Review of Grouts for Existing Dams." *Proceedings of the Association of State Dam Safety Officials*, 14th annual conference, Pittsburgh, PA, September 7–10, CD.

Burke, G. 2004. "Jet Grouting Systems: Advantages and Disadvantages." In *GeoSupport 2004: Drilled Shafts, Micropiling, Deep Mixing, Remedial Methods, and Specialty Foundation Systems*, edited by J. P. Turner and P. W. Mayne, pp. 875–86. Geotechnical Special Publication 124. Reston, VA: ASCE.

Bye, G. C. 1999. *Portland Cement: Composition, Production, and Properties*. 2nd ed. London: Thomas Telford.

Byle, M. J. 1997. "Limited Mobility Displacement Grouting: When 'Compaction Grout' Is *Not* Compaction Grout." In *Grouting: Compaction, Remediation, and Testing*, edited by C. Vipulanandan, pp. 32–42. Geotechnical Special Publication 66. Reston, VA: ASCE.

Cadden, A. W., D. A. Bruce, and R. P. Traylor. 2000. "The Value of Limited Mobility Grouts in Dam Remediation." Paper presented at the Annual Meeting of the ASDSO, Providence, RI, September.

Cedergren, H. A. 1989. *Seepage, Drainage, and Flow Nets*. 3rd ed. New York: John Wiley and Sons.

Chuaqui, M., and D. A. Bruce. 2003. "Mix Design and Quality Control Procedures for High Mobility Cement Based Grouts." In *Grouting and Ground Treatment*, edited by L. F. Johnsen, D. A. Bruce, and M. J. Byle, pp. 1153–68. Geotechnical Special Publication 120. Reston, VA: ASCE.

Danielson, T. J., A. F. Ebnet, R. E. Smith, M. I. Bookshier, and R. P. Sullivan. 2010. "Howard Hanson Dam Right Abutment Seepage: 2009 Interim Grout Curtain and Drainage Tunnel Improvements." Paper presented at the ASDSO Dam Safety Conference, Seattle, WA, September 19–23, 41 pp.

De Paoli, B., B. Bosco, R. Granata, and D. A. Bruce. 1992. "Fundamental Observations on Cement Based Grouts (1) Traditional Materials (2) Microfine Grouts and the Cemill Process." *Proceedings of the ASCE Conference on Grouting, Soil Improvement and Geosynthetics*, Vol. 1, edited by H. Borden, R. Holtz, and I. Juran, 474–499. Reston, VA: ASCE.

Fetzer, C. A. 1988. "Performance of the Concrete Wall at Wolf Creek Dam." Trans. of 16th ICOLD Conference, San Francisco, CA, June, 5:277–82.

Garner, S., J. Warner, M. Jefferies, and N. Morrison. 2000. "A Controlled Approach to Deep Compaction Grouting at WAC Bennett Dam." Paper presented at the 53rd Canadian Geotechnical Conference, Montreal, Quebec, October 15–18.

Glossop, R. 1961. "The Invention and Development of Injection Processes, Part 2, 1850–1960." *Géotechnique* 11 (4): 255–79.

Hamby, J. A., and D. A. Bruce. 2000. "Monitoring and Remediation of Reservoir Rim Leakage at TVA's Tims Ford Dam." Paper presented at the U.S. Commission on Large Dams (USCOLD) 20th Annual Meeting, Seattle, WA, July 10–14.

Harris, M. C., R. E. Van Cleave, D. Howell, and M. Piccagli. 2011. "Seepage Cutoff Wall in Karstic Dolomite, Clearwater Dam, Missouri." Paper presented at the U.S. Army Corps of Engineers Infrastructure Systems Conference, Atlanta, GA, June 13–17.

Henn, K. E., III, S. T. Hornbeck, and L. C. Davis. 2000. "Mississinewa Dam Settlement Investigation and Remediation." Paper presented at the U.S. Commission on Large Dams (USCOLD) 20th Annual Meeting, Seattle, WA, July 10–14.

Hockenberry, A. N., M. C. Harris, R. E. Van Cleave, and M. A. Knight. 2009. "Foundation Treatment in Karst Conditions at Clearwater Dam." *Proceedings from 24th Central Pennsylvania Geotechnical Conference*. Reston, VA: ASCE.

Hornbeck, S. T., T. L. Flaherty, and D. B. Wilson. 2003. "Patoka Lake Seepage Remediation." Paper presented at the ASDSO Dam Safety Conference, Minneapolis, MN, September 7–10.

Houlsby, A. C. 1976. "Routine Interpretation of the Lugeon Water Test." *Quarterly Journal of Engineering Geology* 9 (4): 303–13.

Houlsby, A. C. 1990. *Construction and Design of Cement Grouting*. New York: John Wiley and Sons.

Karol, R. H. 1990. *Chemical Grouting*. Revised ed. New York: Marcel Dekker.

Kauschinger, J. L., E. B. Perry, and R. Hankour. 1992. "Jet Grouting: State-of-the-Practice." In *Grouting, Soil Improvement, and Geosynthetics: Proceeding of the Conference*, edited by R. H. Borden, R. O. Holtz, and I. Juran, 2:169–81. Geotechnical Special Publication 30. Reston, VA: ASCE.

Kellberg, J., and M. Simmons. 1977. "Geology of the Cumberland River Basin and the Wolf Creek Damsite, Kentucky." *Bulletin of the Association of Engineering Geologists* 14 (4): 245–69.

Littlejohn, G. S. 2003. "The Development of Practice in Permeation and Compensation Grouting: A Historical Review (1802–2002). Part 1, Permeation Grouting." In *Grouting and Ground Treatment*, edited by L. F. Johnsen, D. A. Bruce, and M. J. Byle, pp. 50–99. Geotechnical Special Publication 120. Reston, VA: ASCE.

Mauro, M., and F. Santillan. 2008. "Large Scale Jet Grouting and Deep Mixing Test Program at Tuttle Creek Dam." *Proceedings of the 33rd Annual and 11th International Conference on Deep Foundations*. New York, NY, October.

Pacchiosi Drill SpA. 2011. Rock Soil Technology and Equipments. http://www.pacchiosi.com.

Shibazaki, M. 2003. "State of Practice of Jet Grouting," In *Grouting and Ground Treatment*, edited by L. F. Johnsen, D. A. Bruce, and M. J. Byle, pp. 198–217. Geotechnical Special Publication 120. Reston, VA: ASCE.

Smoak, W. G., and F. B. Gularte. 1998. "Remedial Grouting at Dworshak Dam." Paper presented at the ASCE Annual Convention, Boston, MA, October 18–21, 18 pp.

Spencer, W. D. 2006. "Wolf Creek Dam Seepage Analysis and 3-D Modeling." Paper presented at the ASDSO Dam Safety Conference, Boston, MA, September 10–14, 36 pp.

Tornaghi, R. 1989. "Trattamento Colonnare dei Terreni Mediante Gettinazione (Jet Grouting)." *Proceedings of the XVII Convegno Nazionale di Geotecnica*, Taormina, Italy, April 26–28.

U.S. Army Corps of Engineers (USACE). 1984. *Grouting Technology*. CECW-EG Report No. EM 1110-2-3506, January 20, 159 pp.

U.S. Army Corps of Engineers (USACE). 1997. *Engineering and Design Procedures for Drilling in Earth Embankments*. CECW-EG, Report No. 1110-1-1807, September 30.

USACE. 2009. *Building Strong*. Tampa, FL: Faircourt Media Group.

Warner, J. 1982. "Compaction Grouting—The First Thirty Years." *Proceedings of Conference on Grouting in Geotechnical Engineering*, ASCE, New Orleans, 694–707.

Warner, J. 2003. "Fifty Years of Low Mobility Grouting." In *Grouting and Ground Treatment*, edited by L. F. Johnsen, D. A. Bruce, and M. J. Byle, pp. 1–25. Geotechnical Special Publication 120. Reston, VA: ASCE.

Warner, J., N. Schmidt, J. Reed, D. Shepardson, R. Lamb, and S. Wong. 1992. "Recent Advances in Compaction Grouting Technology." In *Grouting, Soil Improvement, and Geosynthetics: Proceeding of the Conference*, edited by R. H. Borden, R. O. Holtz, and I. Juran, 1:252–64. Geotechnical Special Publication 30. Reston, VA: ASCE.

Weaver, K. D., and D. A. Bruce. 2007. *Dam Foundation Grouting*. Revised and expanded ed. New York: ASCE Press.

Wilson, D. B., and T. L. Dreese. 1998. "Grouting Technologies for Dam Foundations." *Proceedings of the 1998 Annual Conference Association of State Dam Safety Officials*, Las Vegas, Nevada, October 11–14. Paper No. 68.

Xanthakos, P. P., L. W. Abramson, and D. A. Bruce. 1994. *Ground Control and Improvement*. New York: John Wiley and Sons.

Zoccola, M. F., T. A. Haskins, and D. M. Jackson. 2006. "The Problem Is the Solution: A History of Seepage, Piping, and Remediation in a Karst Foundation at Wolf Creek Dam." U.S. Society on Dams (USSD) 26th Annual Meeting, San Antonio, TX, May 1–6, 10 pp.

Chapter 3

Mix-in-place technologies

*David S. Yang, Yujin Nishimra, George K. Burke,
Shigeru Katsukura, and Ulli Wiedenmann*

3.1 PERSPECTIVE

As described in Chapter 1, mix-in-place techniques are used to blend mechanically the *in-situ* materials with some type of cementing agent, typically referred to as a "binder." In most applications for dam and levee remediation where seepage cutoffs are required, or where seismic mitigation is the goal, the binder is a fluid, cement-based grout. Where the purpose of the treatment is to improve the bearing capacity of the foundation soil to allow raising of a levee embankment, then the "dry method" has also been used. In the dry method, the binder, now typically slag-cement, is introduced into the ground in powder form and seizes the water necessary for hydration from the moisture in the soil itself.

In 2000 the Federal Highway Administration (FHWA) of the U.S. Department of Transportation published a comprehensive review document providing an introduction to the deep mixing methods (DMM). This document described their historical evolution, construction equipment and procedures, properties of treated soils, and applications. However, one of the most useful contributions of this study, and its two comparison volumes (FHWA 2000a and 2001), was to provide a framework to classify the myriad of different DMM variants that had been found to exist especially in Japan, the Nordic countries, and the United States (Yonekura, Terashi, and Shibazaki 1996). The basis for the classification was an evaluation of the fundamental operational characteristics of each of the DMM techniques which, in 1999, numbered 24:

- The method of introducing the "binder" into the soil: wet (i.e., pumped in slurry or grout form), or blown in pneumatically in dry form. Classification is therefore W or D.
- The method used to penetrate the soil and/or mix the agent: purely by rotary method (R) with the binder at relatively low pressure, or by a rotary method aided by jets of fluid grout at high pressure (J). (*Note:* Conventional jet grouting, which does not rely on any rotational

mechanical mixing to create the treated mass, was beyond the scope of the study.)

- The location, or vertical distance over which mixing occurs in the soil—in some systems, the mixing is conducted only at the distal end of the shaft (or within one column diameter from the end), while in the other systems mixing occurs along all, or a significant portion, of the drill shaft. Classification is therefore E or S.

These characteristics were then combined, as shown in Figure 3.1, to provide a generic classification for each variant, based on a combination of these three designators. In theory, with three bases for differentiation, each with two options, there are eight different classification groups. However, in practice, there are only four generic groups since WJS (wet, jetted, shaft mixing) and DRS (dry, rotary, shaft) do not exist and no jetting with dry binder has been developed, and hence DJS or DJE are not feasible. The four generic methods are, therefore, WRS, WRE, WJE, and DRE. One thing that all these DMM techniques have in common, of course, is that they each feature vertical shaft mixing; regardless of the number of shafts used on each machine, they are mounted vertically as they are introduced and withdrawn from the ground (Photo 3.1). These techniques are the subject of Section 3.2.

In recent years, new concepts of *in-situ* mixing have been developed that are not based on vertical axis mixing. Section 3.3 describes the use of the TRD (trench remixing and cutting deep) method, originally developed in Japan. This is, in very simple terms, a large and powerful chainsaw (Photo 3.2) that progresses laterally through the ground, cutting and blending as it passes to create a continuous soilcrete wall.

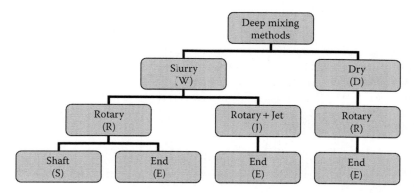

Figure 3.1 Classification of deep mixing methods based on "binder" (wet/dry); penetration/mixing principle (rotary/jet); and location of mixing action (shaft/end). (Modified from Federal Highway Administration, "An Introduction to the Deep Mixing Method as Used in Geotechnical Applications," Prepared by Geosystems, L.P., Document No. FHWA-RD-99-138, 2000a.)

Photo 3.1 Twin Axis DMM machine (WJE system) installing soilcrete panels on a levee in New Orleans, Louisiana. (Courtesy of TREVIICOS Corp.)

In contrast, the CSM (cutter soil mix) method, developed during a German-French cooperation, and the Italian method, CT Jet, use hydro-mill (i.e., cutter) technology previously developed for conventional diaphragm walls (Chapter 4, this volume) to create vertical soilcrete panels, rectangular in plan (Photo 3.3). As is described in Section 3.4, CSM has

Photo 3.2 TRD "cutting post," showing the cutting chain to the right, and the carrier machine to the left. (TRD promotional information.)

Photo 3.3 The CSM cutter head suspended from a Kelly bar. (Courtesy of Bauer Maschinen.)

become very popular throughout the world for constructing cutoffs and earth retaining structures.

3.2 CONVENTIONAL DMM

3.2.1 Introduction

The DMM process increases the strength, decreases the compressibility, and in general reduces the permeability of *in-situ* soils. More than 3,000 projects, both offshore or on land, have been implemented since the first application of the deep mixing method in the mid-1970s. In the United States, the first major application of contemporary DMM techniques was the improvement of the foundation of Jackson Lake Dam, Wyoming, between 1987 and 1989 (Figure 3.2). Since then major levee and dam remedial applications continue to be recorded, with the current focus, at the time of writing, being associated with levee raising in New Orleans, Louisiana. Research also continues apace; the most noticeable work includes the National Deep Mixing Program led by California Department of Transportation and the ongoing development of a deep mixing design manual sponsored by the Federal Highway Administration.

The applications of DMM for dam and levee remediation include the control of water flowing through and under embankments to prevent seepage-induced failure and the reinforcement of embankments for bearing capacity and lateral resistance. For construction of a new embankment, DMM can be used to improve the bearing capacity of soft ground

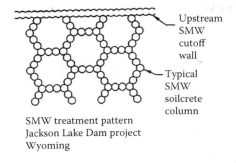

SMW treatment pattern
Jackson Lake Dam project
Wyoming

Upstream
SMW
cutoff
wall

Typical
SMW
soilcrete
column

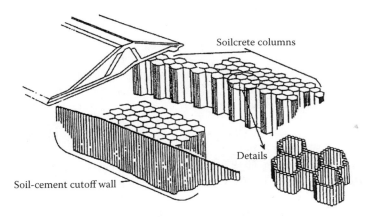

Soilcrete columns

Details

Soil-cement cutoff wall

Figure 3.2 DMM used for liquefaction control and seepage cutoff, Jackson Lake Dam, Wyoming. (From Ryan, C. R., and B. H. Jasperse, *Proceedings of the ASCE 1989 Foundation Engineering Congress, Foundation Engineering: Current Principles and Practices, Vols. 1 and 2*, Evanston, IL, 1989. With permission.)

to provide bearing capacity, to reduce the width of the embankment section, and to maintain the slope stability at the edge of the embankment. For existing embankments, the deep mixed panels, walls, or cells can be installed under the slope and near the toe to perform as shear walls to increase the static and seismic stability of the embankment. Deep mixed walls can be installed along the longitudinal direction of the embankment for the control of water flowing through the embankment section to prevent seepage induced erosion, piping, or instability. For an embankment founded on highly permeable soils, a cutoff wall can be extended through the permeable foundation soils to reach a low permeability stratum, or to a depth sufficient for control of under-seepage to maintain the embankment stability.

3.2.2 Historical background to the deep mixing method

Research and development on deep mixing started with laboratory model tests in 1967 by the Port and Airport Research Institute of the former Japanese Ministry of Transportation for the purpose of stabilizing soft marine soils with lime in harbors or below the sea bed before the construction of harbor facilities. In 1974, the DLM method (deep lime mixing method) became feasible for full-scale construction. Based on technology developed for DLM, the CDM method (cement deep mixing) was initiated using cement grout as the stabilization agent. Starting from laboratory model tests and progressing through on-land full-scale testing and ocean full-scale testing, the CDM method was developed in 1975 for full-scale application. Parallel to the development of the CDM method for large-scale ground treatment, development and research on the SMW (soil mix wall) method was started in 1972 by Seiko Kogyo Co., Ltd., of Osaka, Japan, for the purpose of treating soil on land along a single line to produce a soil-cement wall. In 1976 the SMW method was developed for full-scale application. The soil-cement wall is usually reinforced with steel H-piles when used for excavation support. The Civil Engineering Research Institute of the former Japanese Ministry of Construction started research and development on dry jet mixing (DJM) in 1976 using dry binders, and full-scale application of DJM began in 1981. Independent from the development of the deep mixing methods in Japan, the Swedish lime column method was developed in 1967 for stabilizing soft soil by quicklime. The main application of improvement by lime columns is to reduce foundation settlement under roadway and residential structures. It has also been used to increase the stability of embankments and cut slopes (CDIT 2002).

In the United States, the arrival in 1986 of SMW Seiko, Inc., operating under license from its Japanese parent, stimulated the market. This also had the effect of encouraging U.S. companies to develop their own DMM variants, mainly to address earth retention and environmental barrier applications. A notable exception to this trend was the cutoff built by Geo-Con, Inc., at Lockington Dam, Ohio, in 1993 (Walker 1994).

3.2.3 Product of deep mixing

The deep mixing method mixes *in-situ* soils with binder to produce soilcrete columns using a single-shaft mixing tool, or panel elements using multishaft mixing. On any one base machine the number of mixing shafts can range from one to eight, but for cutoffs, three or four shaft systems predominate. The binder slurry or powder is delivered from ports located in the lower part of the mixing tool. The most frequently used binder in the United States is Portland cement. The column or panel can then be extended to form various configurations (as shown in Figure 3.3) to serve

Layout	Perspective	Applications
Block type 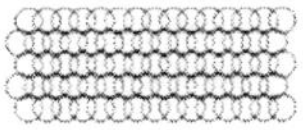	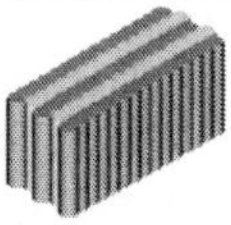	Soil-cement gravity retaining wall; soil-cement seal slab for vertical seepage control
Tangent block type 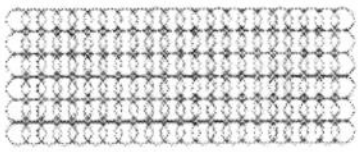	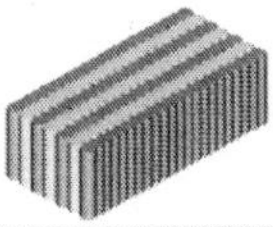	Soil-cement gravity retaining wall; soil-cement dock
Cell type example 1 Vertical load & liquefaction prevention 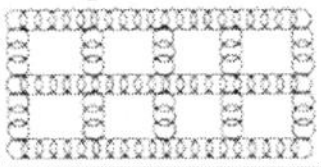	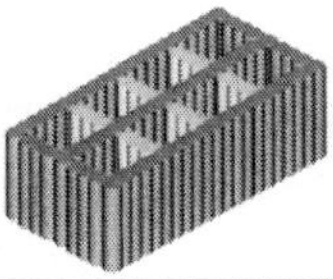	Soil-cement foundation for bearing capacity, settlement control, and liquefaction mitigation
Cell type example 2 Lateral resistance 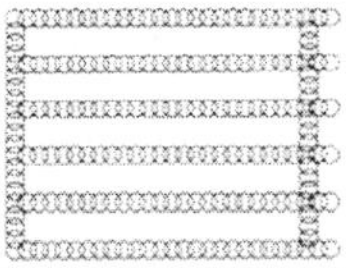	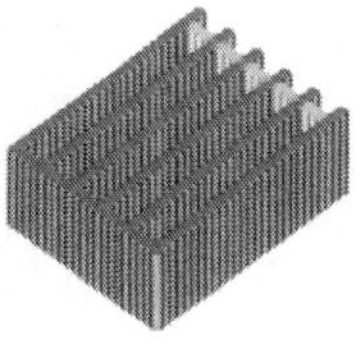	Soil-cement buttress for resisting static and/or seismic lateral forces
Shear wall type 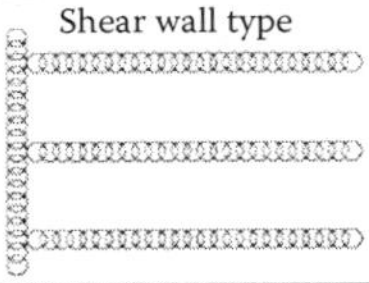		Soil-cement shear walls for resisting static and seismic lateral forces and/or vertical load
Column type 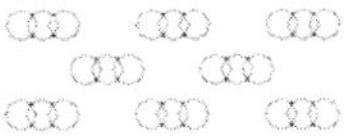		Soil-cement columns for bearing capacity and settlement control
Composite type 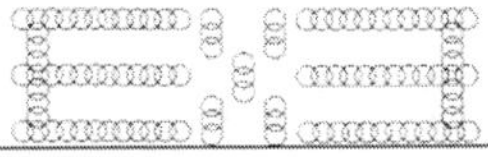		Composite geometric design for bearing capacity at center, and lateral resistance capacity at edge under slope, retaining wall, or MSE wall
Single wall type 		Single wall for seepage control and/or shoring if reinforced with steel member

Figure 3.3 Configuration of DMM structures in the United States.

various design functions, including ground stabilization and seepage control. The arrangement of columns and panels is referred to as layout design, or the geometric design of deep mixing work in contrast to mix design or material design of the soil-cement or soil-binder product. The layout design is generally performed by the owner or engineer, who has the best understanding of the intended function of the DMM work. Preliminary mix design can be performed by the designer or the project owner in order to provide data for design purposes. However, the final mix design for the production work is generally performed by the deep mixing specialty contractors based on their equipment and experience. Grout volume ratios of 30 percent to over 100 percent are used, depending on the ground conditions, desired soilcrete properties, and the particular requirements of each DMM variant. (Grout volume ratio is the volume of grout injected divided by the completed volume of soilcrete.)

3.2.4 Engineering properties of soil-cement

Comprehensive data are provided in the FHWA (2001) document. The following provides a brief summary of major points.

Strength: Unconfined compressive strength, q_u, is the most frequently used strength parameter for the design, construction, and verification of deep mixing work. The unconfined compressive strength of most soil-cement produced by wet mixing ranges from 75 to 450 psi, and the most frequently used design strength for ground stabilization is 150 psi. The test specimens include laboratory samples, field wet-grab samples, and core samples prepared before, during, and after construction, respectively. Triaxial compression tests, direct shear tests, and tensile tests are also performed. However, these tests are only used for research studies or for special projects in which strength parameters other than the unconfined compressive strength are critical for the design and performance of the DMM work. For design purposes, a shear strength ranging from 33 to 50 percent of the unconfined compressive strength is used. The tensile strength of soil-cement is about 15 percent of the unconfined compressive strength and the bending strength varies from 10 to 60 percent of the unconfined compressive strength (CDIT 2002). Without confining pressure, the residual strength of soil-cement is practically zero. However, with even a small confining pressure, the residual strength of treated soil is increased to almost 80 percent of the unconfined compressive strength (CDIT 2002).

Soil type is the most dominant factor that influences the strength of the soil-cement blend. The same mix design and treatment procedure used in different soils would produce soil-cement with a wide variation of strength values. For project sites with complicated subsurface stratigraphy, it is challenging but possible to adjust the mix design and treatment process to cope with the varying subsurface conditions during the deep mixing process. The strength values obtainable in the soil layer that provides the lower range of strength must be used for the design of DMM products, unless layer by layer variation over mixing parameters can be reliably exercised during construction.

Consolidation Yield Pressure: The laboratory consolidation testing results of soil-cement are similar to that of overconsolidated clay, which is characterized by a sharp "bend" at the preconsolidation pressure. For the case of soil-cement, the pressure at the sharp bend is called consolidation yield pressure, p_y. Irrespective of the soil type and binder type, p_y/q_u is approximately 1.3 (CDIT 2002).

Modulus of Elasticity: The modulus of elasticity, E_{50}, is defined as the secant modulus of elasticity in a stress-strain curve at 50 percent of unconfined compressive strength. Based on early studies in Japan, E_{50} of treated soil using wet methods is 350 to 1,000 × q_u (unconfined compressive strength). Most of the data obtained in the United States indicate that the E_{50} ranges from 100 to 150 × q_u, although the difference might be derived from the method of measurement of strain during the strength testing. For treated soil using dry binder, E_{50} is 75 to 200 × q_u when q_u is less than 1.5 MPa and E_{50} is 200 to 1,000 × q_u when q_u exceeds 1.5 MPa (CDIT 2002).

Poisson's Ratio: Although there is a relatively large scatter in the test data, the Poisson's ratio of the treated soil is around 0.25 to 0.45, irrespective of the unconfined compressive strength (CDIT 2002).

Permeability: Based on Japanese test data, the permeability of treated clay is equivalent to or lower than that of untreated soft clays (CDIT 2002). Similar results of treated clays were found in the United States. For seepage control in sandy soils, cement-bentonite slurry with higher water/cement ratio is generally used to produce soil-cement-bentonite cutoff walls. The coefficient of permeability of soil-cement-bentonite generally ranges from 10^{-6} to 10^{-8} cm/s.

Density: The wet density of treated soil is dependent upon the original unit weight of the untreated soil, the amount of binder used, and the water content of the grout. The density change after wet mixing is negligible for soft marine soils due to the low density and the high water content of untreated marine soils. If the wet density of the untreated soil is higher than the unit wet density of the slurry, the wet density of the treated soil will be lower than the untreated soil. For treatment using dry binder, the wet density of treated soil increases by about 3 to 15 percent (CDIT 2002).

3.2.5 Applications and design of deep mixing

Deep mixing creates well-defined configurations of treated soil such as columns, walls, cells (grids), or blocks to provide soil-cement foundations for a wide variety of applications. There are basically three main functions of these deep mixed structures:

Cutoff Walls: Soils are treated panel by panel (element by element) in one row using the procedure shown in Figure 3.4. A full column overlapping of the neighboring panels is essential to ensure the longitudinal continuity of the wall. The existing soils to be treated, in general, are coarse-grained soils with high permeability or interbedded strata of fine- and coarse-grained soils. Cement-bentonite grout is most frequently used for the construction of DMM cutoff walls. Bentonite slurry and clay-bentonite slurry can also be

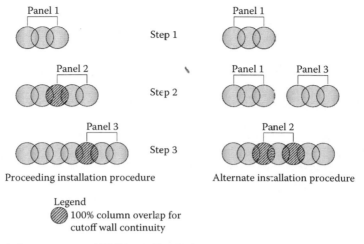

Figure 3.4 Construction of DMM cutoff walls for seepage control.

used to produce soil-bentonite or soil-clay-bentonite cutoff walls. Most soil-cement walls have a strength ranging from 30 to 300 psi and a coefficient of permeability ranging from 10^{-6} to 10^{-7} cm/s. The applications include seepage cutoff within or under levee or dam embankments. An innovative application is to install a soil-cement-bentonite cutoff wall in an aquifer, such as a porous stratum or limestone terrain, in order to contain groundwater and use it as a subsurface dam for water supply or for prevention of salt water intrusion. Steel H-piles or other reinforcement members can be inserted into the cutoff wall before the soil-cement hardens. The reinforced soil-cement wall then becomes a structural wall for excavation support and groundwater control.

Ground Stabilization: Deep mixing increases the strength and reduces the compressibility of the existing subsurface strata to maintain ground stability and to control ground movement under loads induced by new construction. Large-scale civil works in marine environments such as the construction of manmade islands, tunnels, harbors, sea walls, breakwaters, and other harbor facilities were the driving force for the development of DMM. Barges are generally used to support the heavy deep mixing equipment, which can support up to eight shafts for ground stabilization offshore. Deep mixing is also used for the stabilization of soft ground on land for support of highway embankments and levees. Examples of DMM layout design used in the United States are shown in Figure 3.3. These geometries can be installed by any of the DMM variants, although the dry mixing techniques are usually restricted to ground improvement applications under highways and levees, given that they typically have lower strength and modulus than their wet method counterparts. The design of the soil-cement foundation includes external stability analysis and internal stress analysis. External stability includes checks on lateral sliding, overturning, and the bearing capacity of soil-cement foundation and the existing bearing stratum. The internal stress analysis includes checks on compressive stress and shear stress within the soil-cement elements.

Seismic Mitigation: The use of deep mixed walls or cells for liquefaction mitigation includes reinforcement of liquefiable soil and reduction of excessive pore pressure. Reinforcement of liquefiable soils is accomplished by installing soil-cement walls in block, wall, or cell configurations to resist the loads from embankments or other upper structures. The reinforced ground, including the treated soil and untreated soil, would become more rigid and the untreated soil would experience less cyclic strain, which in turn could reduce the generation of excessive pore water pressure and consequently lower the liquefaction potential. Shaking table tests, numerical analyses, and centrifugal studies have been performed to study the effectiveness of various ground treatment configurations in the reduction of liquefaction potential. With the same ground treatment ratio (i.e., ratio of deep mix area to overall site area), the cell type treatment is considered to be the most effective in reducing shear strain and excessive pore water pressure in untreated soil. Such ratios typically very from 25–30 percent.

The effectiveness of the cell type treatment was observed during the 1995 Kobe earthquake when one fourteen-storey hotel//terminal building constructed on a pier at Kobe Harbor survived the moment magnitude 6.9 ground shaking while the adjacent structures suffered severe damage due to ground liquefaction. Post-earthquake studies indicated that there was no structural damage to the building while the sea walls surrounding three sides of the building suffered large vertical and horizontal movements of 2.0 to 6.6 feet, the same as other infrastructures in the area. The structure was supported by drilled piles. To prevent ground liquefaction and the accompanying lateral flow toward the sea, DMM cells were installed through forty feet of liquefiable soils underlying the site and embedded in to competent colluvium fifty-two feet below ground surface. Based on the results of the post-earthquake study, it was clear that no liquefaction or lateral flow had occurred in the foundation soils enclosed by the DMM cells. From these studies, it was concluded that DMM cells are effective in mitigating ground liquefaction and the accompanying lateral flow during major earthquakes (Suzuki et al. 1996; Namikawa, Koseki, and Suzuki 2007).

3.2.6 Case histories of DMM for dam and levee remediation

Six U.S. case examples, comprising three seepage control projects and three ground stabilization projects, are presented in detail to illustrate the typical applicability of DMM for dam and levee remediation. Papers detailing the massive DMM work conducted at LPV 111, New Orleans, from 2010 to 2011 can be found in the proceedings of the Fourth International Conference on Grouting and Deep Mixing, held in New Orleans, February 2012 (DFI 2012).

3.2.6.1 DMM walls for seepage control

3.2.6.1.1 Lewiston, Idaho, levee seepage remediation

3.2.6.1.1.1 BACKGROUND

The West Lewiston Levee System along the Snake River in Idaho was constructed in 1973 by the U.S. Army Corps of Engineers (USACE), Walla Walla District, to protect the city of Lewiston from the reservoir created as a result of the Lower Granite Dam construction (Gibbons and Buechel 2001). The 35-foot-high levees (Figure 3.5) function as dams with normal pool depths ranging from 24 to 28 feet and consist of gravel-fill embankment with a rock riprap shell on the upstream slope and a core at the center over a 6-foot-wide cutoff below the original ground surface.

A drawdown test at the Lower Granite Reservoir downstream of the Lewiston site was performed in 1992 to lower the water from normal pool elevation (EL) at 737 feet to EL 707 feet, the original river elevation. Seepage was first noticed when the reservoir was refilled. Sand boils occurred along the downstream face of the levee when the pool was raised to EL 737 feet in an area that had not experienced seepage since before the construction of the levee system. The seepage and sand boils were considered a threat to the integrity of the levee. A drainage trench, perforated drain, and piezometers were installed for managing and monitoring the seepage. The quantity of seepage increased when the pool was raised higher in the winter.

In July 1998 a seepage berm along the landward toe of the levee was constructed to lower exit gradients and to reduce the potential for internal erosion. The existing drainage pipes were extended to the downstream pond through this berm. Flow in the drainpipe continued at a rate of about 40 gallons per minute (gpm) in December 1998. Seepage was also observed from the toe of the levee upstream and downstream of the berm. In December 1999 the flow in the drain pipe increased to 75 gpm and erupted through the seepage berm. In March 2000 a 120-foot-long trench drain was constructed along the toe of the seepage berm to capture the water and reduce the loss of fines. When the trench drain was completed it was carrying about 90 gpm of water.

Due to the continuing seepage and piping problem and the increased risk of levee failure, the USACE hired a consultant to perform a geotechnical study. The study concluded that a breach of the existing cutoff was the probable cause for the increased seepage and internal erosion or piping observed in the levee.

3.2.6.1.1.2 REMEDIAL ALTERNATIVES AND REMEDIATION DESIGN

The geotechnical study evaluated six potential remedial alternatives and selected a deep mixed cutoff wall for final design. The selection criteria

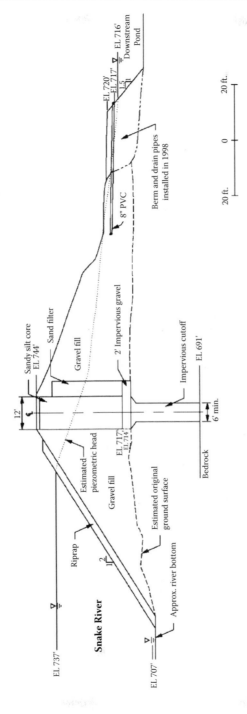

Figure 3.5 Cross section in seepage area, West Lewiston Levee, Idaho. (After Gibbons, Y. R., and G. J. Buechel, "Lewiston Levees DSM Wall Construction," 18th Annual Conference of the Association of State Dam Safety Officials, www.damsafety.org, Snowbird, Utah, 2001; Courtesy of Association of State Dam Safety Officials.)

included potential reliability of the repair, cost, constructability, schedule, and a low potential need for future maintenance. Other options that were eliminated included:

Soil-Bentonite Slurry Cutoff Trenches: A bentonite-filled trench extending along the levee crest centerline would increase the uncertainties associated with the stability of the levee with seepage and piping problems. In addition, it was believed that there was a high potential for spillage or leakage of materials into the river during construction.

Compaction Grouting: Compaction grouting does not result in a reliable seepage cutoff in these conditions and might induce long-term settlement of the very soft existing impervious cutoff located below the existing embankment.

Conventional Grouting: It was believed that conventional (i.e., permeation) grouting could not create a continuous permanent cutoff in the non-homogenous soils. Also its cost would be prohibitive, and grouting had the potential to contaminate the river and ponds.

Jet Grouting: There was significant concern that this method could cause hydrofracture of the embankment soils, which could result in greater leakage through potential piping paths in and under the embankment. Achieving the required column geometry at depth to provide a reliable and continuous cutoff was also a concern.

Trench Drain with Relief Wells: Controlling the seepage at the toe would have been the least expensive option. This option was not viewed as a positive cutoff and furthermore it did not mitigate potential existing piping zones within the embankment. The seepage volume was expected to increase resulting in future maintenance, which was not an acceptable alternative.

A DMM cutoff wall was designed to run along a 500-foot-long section of the levee. It extended from the crest to the bedrock underlying the levee (as shown in Figure 3.6). Based on the seepage analyses, a maximum coefficient of permeability of 1×10^{-6} cm/s was required. However, the project specified a maximum coefficient of permeability of 5×10^{-7} cm/s due to uncertainties in the actual field conditions and differences in the cure conditions between laboratory samples and the *in-situ* wall. A minimum unconfined compressive strength of 20 psi at 28 days was also specified. The specifications required the contractor to develop the mix design using samples of different soil materials encountered along the entire depth of the wall. The contract for construction was awarded in November 2000.

3.2.6.1.1.3 CONSTRUCTION AND VERIFICATION

The DMM rig employed by specialty subcontractor Raito, Inc., consisted of a crawler base machine, a lead to support and guide an electric top drive motor, and triple-shaft mixing tools as shown in Photo 3.4. The

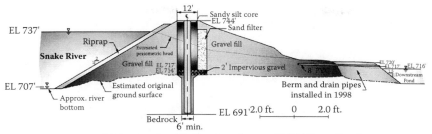

(a) Cross section West Lewiston Levee with DMM cutoff wall

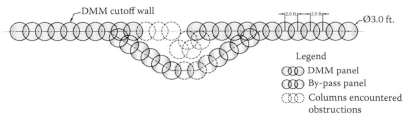

(b) Layout plan of DMM cutoff wall and by-pass panels

Figure 3.6 Cutoff wall section and layout plan, Lewiston Levee, Idaho: (a) cross section
West Lewiston Levee with DMM cutoff wall; (b) layout plan of DMM cutoff
wall and by-pass panels. (After Gibbons, Y. R., and G. J. Buechel, "Lewiston
Levees DSM Wall Construction," 18th Annual Conference of the Association
of State Dam Safety Officials, www.damsafety.org; PowerPoint slides by
Shannon and Wilson, courtesy of Association of State Dam Safety Officials.)

rig was equipped with electronic sensors built into the lead to control
vertical alignment. The rig was also equipped with sensors to monitor
the mixing tool penetration and withdrawal rates, mixing tool rotation
speed, and water-cement-bentonite slurry injection rate. These installation parameters were monitored by a computerized quality-control system on a real-time basis.

The batch plant consisted of two fifty-ton-capacity cement silos and two
slurry mixing systems. Water, cement, and bentonite were measured with
automatic batch scales to accurately determine mix proportions. Due to
limited space on top of the levee, the batch plant shown in Photo 3.5 was
located on the seepage berm at the landside toe of the levee about midway
along the 500-foot alignment.

New piezometers and a data acquisition system were installed in addition to existing observation wells in order to monitor the groundwater
levels prior to, during, and following cutoff wall construction. A metering manhole for recording seepage flows was installed on the drainpipe
that extended through the seepage berm. Vibrating wire piezometers were
installed in all the wells and transducers were also installed in the river and

Photo 3.4 DMM triple axis rig (WRE). (Courtesty of Raito, Inc.)

Photo 3.5 Batch plant on the seepage berm at the landside toe of Lewiston Levee, Idaho. (After Gibbons, Y. R., and G. J. Buechel, "Lewiston Levees DSM Wall Construction," 18th Annual Conference of the Association of State Dam Safety Officials, www.damsafety.org, Snowbird, Utah, 2001; Courtesy of Association of State Dam Safety Officials.)

the pond to monitor water levels. The piezometers located nearest to the suspected internal seepage zone indicated temperatures between 40 to 50 degrees Fahrenheit, which were close to the values measured in the Snake River. The piezometers located in areas that were not experiencing seepage indicated temperatures between 50 and 55 degrees.

A 20-foot-long test section was constructed along the cutoff wall alignment prior to full production to demonstrate that the mix design, equipment, and installation procedure could produce a soil-cement-bentonite wall that would provide adequate mixing for the existing site conditions and achieve the specified material properties and depth.

Several mix designs were used for the test soil-cement wall. The wall section using a mix design containing 220 kg/m³ cement and 80 kg/m³ bentonite met the required permeability and strength requirements. Full construction of the cutoff wall started on February 15, 2001. Production of the wall was slower than expected due to difficult drilling through the dense gravel layer, which resulted in damage to the mixing tool. The auger head of the mixing tool was therefore modified. As the wall installation progressed along the levee, the levels in all of the piezometers immediately dropped from readings reflecting the influence of the river at EL 734 feet to readings indicating the downstream pond at EL 720 feet, as shown in Figure 3.7. The most dramatic results

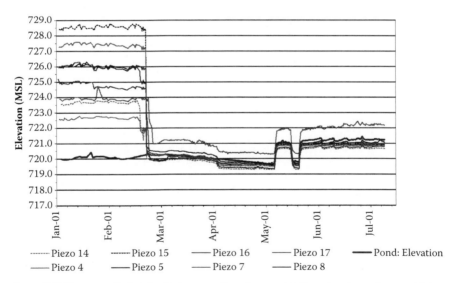

Figure 3.7 Piezometric levels during cutoff wall construction, Lewiston Levee, Idaho. (After Gibbons, Y. R., and G. J. Buechel, "Lewiston Levees DSM Wall Construction," 18th Annual Conference of the Association of State Dam Safety Officials, www.damsafety.org, Snowbird, Utah, 2001; Courtesy of Association of State Dam Safety Officials.)

were obtained from those wells installed in and around the center of the seepage area. All seepage coming out of the collection trench stopped the next morning when wall construction was approximately 20 feet beyond the affected area.

The entire wall was completed by February 28, 2001, except for a section where one of the auger heads was lost when encountering an obstruction. A five-panel bypass wall around the obstruction was installed in an attempt to seal the gap below the obstruction (as shown in Figure 3.6b). However, the modified installation continued to encounter an obstruction at approximately 35 feet below the top of the levee. The USACE finally concluded that the obstruction could not be totally bypassed. This section of the wall had never exhibited a seepage problem. In addition, the USACE felt that the DMM wall had satisfactorily tied into the existing slurry trench.

Piezometer readings continue to indicate that the excessive seepage was effectively addressed. The trench drain remains dry and readings reflect pond fluctuations as expected.

3.2.6.1.2 Sacramento River, California, East Bank levee seepage remediation

3.2.6.1.2.1 BACKGROUND

Geotechnical investigations after the flood of the Sacramento River in 1985 and 1986 concluded that a cutoff wall would be required to control seepage and to prevent sudden levee failure due to piping during future flood conditions (Yang 2008). In early 1990 a DMM cutoff wall was installed along approximately two miles of levee in the Little Pocket Area of Sacramento to protect the residential area from potential flood damage. The cutoff wall was installed through the levee to a depth of approximately 30 feet, a few feet into the alluvium soil below the embankment fills of the levee. Although seepage through the levee embankment was controlled, under-seepage below the cutoff wall continued to occur along a 2,400-foot section of the levee. Further study by the consultant to the Sacramento Area Flood Control Agency (SAFCA) determined that a cutoff wall to depths varying from 90 to 110 feet was needed for the remediation of the under-seepage problem. Following the successful application of a deep mixed wall in the Garden Highway Levee Repair Project in Sacramento, California, SAFCA selected the same method to construct the soil-cement-bentonite cutoff wall. The choice of the deep mixing method was influenced by evaluating the potential of damage to the existing levee, the danger of grout leaking into the river, and minimization of the impact to sensitive residential areas.

The levee embankment fills consisted of sandy silt, silty sand, and sand to depths varying from 15 to 20 feet. The fills were underlain by interbedded layers of clean sand, silty sand, sandy silt, and silty clay to a depth of 150 feet with a higher proportion of silty sand and sandy silt layers at depths below 80 feet. Groundwater fluctuated with river stages at about 30 to 35 feet below the top of the levee during the dry season.

3.2.6.1.2.2 CONSTRUCTION AND VERIFICATION

The construction of the cutoff started in September 2003 following the completion of a 50-foot-long test section. Construction management was performed by the Sacramento District of the U.S. Army Corps of Engineers in conjunction with SAFCA. Two sets of triple-shaft DMM equipment similar to the one used in the Lewiston Levee Seepage Remediation Project were used for cutoff wall installation to a maximum depth of 112 feet. Drilling depth, penetration/ withdrawal speed, shaft rotation, and slurry injection rates were monitored on a real-time basis for accurate mixing control and uniform mixed product. The wall was completed in November 2003 with a total area of 282,300 feet^2.

Soil-cement-bentonite wet samples were retrieved for unconfined compressive strength and permeability testing. Acceptance criteria required a minimum unconfined compressive strength of 45 psi at 7 days and a maximum permeability of 1×10^{-6} cm/s. The strength testing data of all specimens tested ranged from 121 psi to 561 psi and the coefficient of permeability ranged from 3.4×10^{-7} cm/s to 1.6×10^{-8} cm/s at the 28-day curing age. Core samples were also retrieved from the wall for testing and evaluation of uniformity. Representative core samples are shown in Photo 3.6. *In-situ* constant head permeability tests were performed in the same cored holes. The results are presented in Figure 3.8 together with the laboratory permeability testing results of wet samples and core samples. The permeability results on core samples tend to be higher than those from wet samples and *in-situ* bore-hole permeability testing. This trend has also been observed in other soil-cement-bentonite wall projects. The permeability data obtained from core specimens must always be used with caution. The side surface of a core sample tends to be rough and may contain horizontal and/or vertical grooves created during coring by hard particles such as gravels. The coring process may also create micro fissures inside the core sample. The permeability results of core samples may be erratic due to side-wall leakage along the core surface during testing and water permeating through micro fissures or cracks within the core specimen. Wet-grab samples and *in-situ* bore-hole tests generally provide more consistent and reliable permeability testing data for the evaluation of soil-cement-bentonite cutoff walls.

3.2.6.1.3 Lake Cushman, Washington, spillway cutoff

3.2.6.1.3.1 BACKGROUND

In conjunction with the installation of a new radial gated spillway for Lake Cushman near Hoodsport, Washington, two sections of embankment

Photo 3.6 Core samples from Sacramento Levee, California. (After Yang, D. S., *5th International Conference on Landslides, Slope Stability & the Safety of Infrastructures*, CI-Premier, Kuala Lumpur, Malaysia, 2008.)

were constructed abutting the spillway headworks structure as shown in Figure 3.9 (Sehgal, Fischer, and Sabri 1992; Yang and Takeshima 1994). The headworks structure was founded on relatively impermeable bedrock. Soil-cement cutoff walls were installed to bedrock to control water seepage through the embankment fill and the native glacial deposits. The soil-cement cutoff walls were 200 feet long and 180 feet long within the right and left embankments, respectively. The maximum depth was 141 feet. The DMM cutoff wall profile and section are shown in Figure 3.9.

The site was underlain by glacial deposits consisting of recessional outwash, lacustrine deposits, and lodgement till. The glacial deposits were underlain by a submarine-deposited basalt. The recessional outwash materials consisted of dense to very dense fine to coarse sand with trace to little silt, gravelly sand, and sandy gravel with little silt. The lacustrine deposits consisted of stiff to very stiff clayey silt, silt, and medium dense to very dense sand with occasional drop stones. The lodgement till consisted of very dense gravelly sandy silt, silty sand, and gravel with N-values of 50 for less than 6 inches of sampler penetration. The lodgement till contained

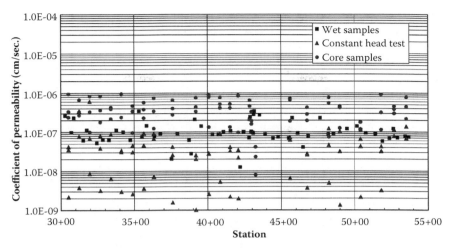

Figure 3.8 Permeability testing results from Sacramento Levee, California. (After Yang, D. S., *5th International Conference on Landslides, Slope Stability & the Safety of Infrastructures*, CI-Premier, Kuala Lumpur, Malaysia, 2008.)

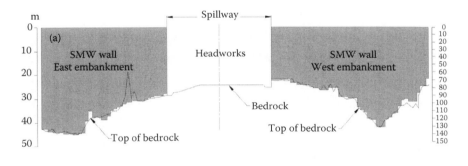

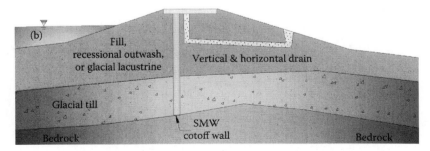

Figure 3.9 Details of DMM cutoff wall, Lake Cushman, Washington: (a) spillway, embankment, and cutoff wall profile; (b) general cross section, spillway embankment. (After Yang, D. S., and S. Takeshima, "Soil Mix Walls in Difficult Ground," American Society of Civil Engineers National Convention, Atlanta, Georgia, 1994. With permission from ASCE.)

cemented zones, cobbles, and boulders, and had been heavily overconsolidated. The basalt bedrock was medium strong to strong with little or no weathered zones. The permeability of the glacial deposits ranged from 10^{-2} to 10^{-5} cm/s based on the results of seepage analysis, slug tests, and packer tests. The permeability of the fresh bedrock ranged between 10^{-6} and 10^{-7} cm/s based on packer tests.

Two short sections of embankment abutting the spillway headworks were constructed on competent basalt bedrock following the excavation of glacial deposits. The embankment fills consist of compacted silt, sand, and gravel.

3.2.6.1.3.2 DESIGN

Due to the absence of low permeability materials at the site for the construction of a clay core in the embankment and the high permeability of the glacial outwash on the left and right abutments, the owner and its consultants designed DMM cutoff walls within the embankment and the glacial outwash to provide for control of water seeping from Lake Cushman. Further investigation revealed that the glacial till contained less fines and was more permeable than expected and so the cutoff wall had to be extended to a maximum depth of 141 feet to reach the bedrock for seepage cutoff.

To prevent leakage along the interface of the headworks structure and the embankment, a joint structure as shown in Figure 3.10 was constructed to connect the concrete spillway headworks and the DMM wall. A layer of bentonite slurry was hand applied to the inside surface of the u-shaped concrete structure and a bentonite-sand mixture was backfilled and compacted within the u-shaped zone during the embankment construction. The cutoff wall was then installed inside the u-shaped zone to form a low permeability

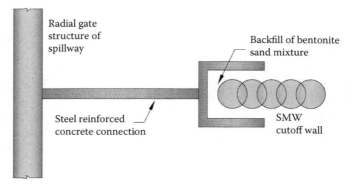

Figure 3.10 Joint detail, Lake Cushman, Washington. (After Yang, D. S. and S. Takeshima, "Soil Mix Walls in Difficult Ground," American Society of Civil Engineers National Convention, Atlanta, Georgia, 1994. With permission from ASCE.)

joint zone and to minimize leakage along the inside perimeter of the U-shaped join structure.

3.2.6.1.3.3 CONSTRUCTION AND VERIFICATION

A 12-foot-long, 15-foot-deep cutoff wall test section was constructed in the highly permeable glacial deposits before full-scale wall installation. Three different mix designs selected from laboratory trial mix tests were used. Unconfined compressive strength values of samples retrieved from the soil-cement columns ranged from 102 to 178 psi. The test wall was excavated for inspection. Based on the strength test results and observed conditions of the exposed test wall, the owner and the design engineers concluded that the mix designs and the installation method would achieve the design requirements for the planned cutoff wall.

Operational challenges included (1) working on top of the narrow embankment, (2) installing the wall in highly permeable embankment and outwash materials, (3) hard drilling in cemented till containing cobbles and boulders, and (4) reaching competent bedrock, which varied in depth. To overcome the hard drilling, predrilling with a single auger was performed to clear the drilling path to the top of the bedrock for triple-shaft augers to follow and produce the cutoff wall. The estimated bedrock depth and drilling resistance encountered by the single predrill auger were used to determine if the drilling had reached the bedrock. In the early stages, borings were drilled along the predrilled section of the cutoff wall to confirm that the predrilling had reached bedrock. In some instances, bedrock fragments could be retrieved from the drill bits after they were withdrawn from a hole during predrilling. Based on this information, field methods were developed to confirm when drilling had reached the top of the irregular bedrock surface.

In several locations, large boulders in the lodgement till prevented even the predrilling auger from reaching bedrock. To avoid leaving a gap or permeable window within the cutoff wall below the large boulders, bypass sections of DMM wall were installed around the boulders to maintain the continuity of the cutoff wall. Both the proceeding installation procedure and alternate installation procedure were used in this project in conjunction with single auger predrilling (as shown in Figure 3.11). The wall was installed element by element, and therefore there was no deep open trench inside the embankment to cause concerns regarding embankment stability.

Mix designs with cement dosages of 350 to 550 kg per cubic meter of *in-situ* soil were used for cutoff wall installation. The west wall consisted of 18,600 ft^2 of soil-cement wall with a 28-day unconfined compressive strength ranging from 85 to 640 psi and a permeability ranging from 2 × 10^{-5} to 6 × 10^{-7} cm/s. The east wall consisted of 23,870 ft^2 of soil-cement wall with strength ranging from 256 to 696 psi and permeability ranging from 1 × 10^{-6} to 7 × 10^{-7} cm/s. Post-construction monitoring of the

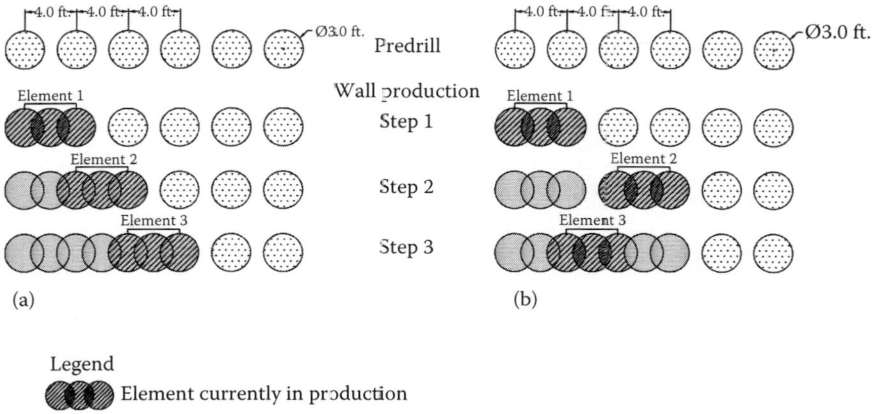

Figure 3.11 Cutoff wall installation procedures, Lake Cushman, Washington: (a) proceeding installation procedure; (b) alternate installation procedure (primary–secondary procedure). (After Yang, D. S., and S. Takeshima, "Soil Mix Walls in Difficult Ground," American Society of Civil Engineers National Convention, Atlanta, Georgia, 1994. With permission from ASCE.)

quantity of seepage at the downstream toe of the spillway embankment has confirmed that the cutoff walls were constructed in compliance with the design specifications.

3.2.6.2 DMM for seismic remediation

3.2.6.2.1 Remediation of Sunset North Basin Dam, California

3.2.6.2.1.1 BACKGROUND

Sunset Reservoir is a lined and covered off-stream reservoir located in San Francisco, California (Barron et al. 2006; Olivia Chen Consultants Report 2004). The 74-foot-high embankment dam, located at the northwest corner of the north basin, was built in 1938 using a combination of cut and fill. The reservoir is owned and operated by the San Francisco Public Utilities Commission (SFPUC). Sunset Reservoir's storage capacity and dam height place it under the jurisdiction of the California Department of Water Resources, Division of Safety of Dams (DSOD). In 1998 SFPUC initiated a field investigation and engineering study to evaluate the seismic performance of the dam and reservoir. In 2000 the SFPUC's consultants and DSOD concluded that strength loss of the foundation soils below the northwest embankment of the North Basin could occur during and after a maximum earthquake on the San Andreas or Hayward Fault, located 5 km and 25 km away from the site, respectively. The controlling seismic event is a moment magnitude (Mw) 8 earthquake on the San Andreas Fault with an estimated bedrock peak ground acceleration of 0.97 g.

The cross section of the embankment dam and subsurface materials at the northwest corner of North Basin are shown in Figure 3.12. The embankment itself has adequate safety factors under both static and seismic conditions. However, the embankment was founded on a layer of dune sand and 10 to 30 feet of silty sand layer over the bedrock. The soils of concern for Sunset North Basin Dam were the saturated loose to medium silty sand (Silty Sand 2) and medium dense to dense silty sand (Silty Sand 3). The study indicated that Silty Sand 2 was susceptible to significant strength loss and Silty Sand 3 might only have strength loss in localized areas. Since the loss of strength would require some time to develop after the beginning of an earthquake and would occur sooner for the loose soils than the denser soils, the consultants performed the stability analyses in three stages (as shown in Figure 3.12). For the upstream slide surfaces, the permanent deformation was estimated to be less than one inch. For the downstream slide surfaces, the analyses revealed that severe deformations were possible as a result of strength loss due to ground shaking. These results were confirmed by the independent analyses performed by DSOD. Based on these studies, it was concluded that foundation improvement of the downstream embankment would be required to maintain seismic stability.

3.2.6.2.1.2 REMEDIAL ALTERNATIVES AND REMEDIATION DESIGN

Six remediation alternatives were evaluated and DMM was selected for the foundation remediation. The other five alternatives were eliminated for the following reasons:

Jet Grouting: This process could not be controlled to an acceptable degree. In addition, this method would be more costly than DMM.

Compaction Grouting: This has limited effectiveness in soils with high fines content: Silty Sand 2 had a fines content between 28 and 35 percent.

Vibro-replacement: This method is also less effective for soils with a high fines content. In addition, it might have had difficulties penetrating embankment soils to effectively treat the target layer.

Permeation Grouting: Silty Sand 2 was not permeable enough to allow uniform penetration of the grout.

Excavation and Recompaction: Excavation and recompaction of the entire embankment was considered and concluded to be prohibitive from cost and schedule perspectives. In addition, the SFPUC wanted the reservoir to remain operational during the foundation remediation.

The selection of the DMM method was also based on previous successful seismic remediation applications in similar projects including Jackson Lake Dam, Wyoming, by the Bureau of Reclamation, Clemson Upper and Lower

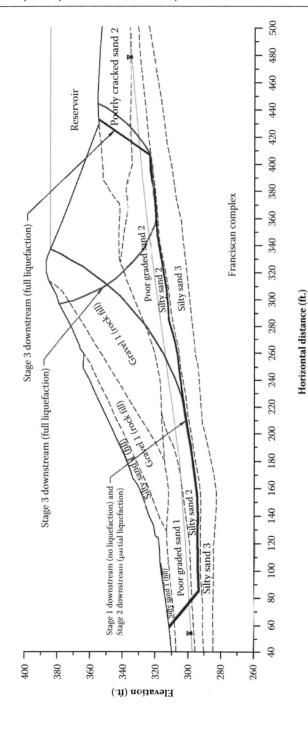

Figure 3.12 Embankment cross section and critical slide surfaces, Sunset Dam, California. (After Barron, R. F., C. Kramer, W. A. Herlache, J. Wright, H. Fung, and C. Liu, *Proceedings of Dam Safety 2006*, www.damsafety.org, 2006; Olivia Chen Consultants Report, *Geotechnical Investigation Embankment Stability Evaluation, Sunset Reservoir North Basin San Francisco Public Utilities Commission,* 2004; Courtesy of Association of State Dam Safety Officials.)

Diversion Dams, South Carolina, for seismic remediation by the U.S. Army Corps of Engineers, and other projects at the Port of Oakland, California.

A series of stability analyses with multiple treatment layout and varying DMM engineering properties were performed by the consultants to develop the final DMM remediation scheme. The final treatment layout consisted of multiple 47-foot-square grids or blocks of DMM columns placed in treatment rows parallel to the longitudinal axis of the embankment (as shown in Figure 3.13). The discrete block layout allowed the regional groundwater to flow between the blocks so minimizing the impact on the hydrogeological conditions at the site. The DMM columns were designed to extend at least 5 feet below the bottom of Silty Sand 2, or to the top of bedrock. In areas where only two rows could be placed, DMM columns were required to key into the bedrock to provide additional lateral resistance.

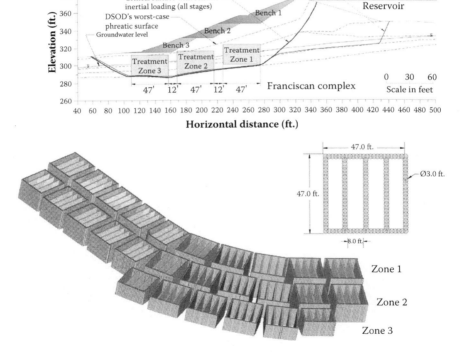

Figure 3.13 DMM treatment plan for Sunset Dam, California. (After Barron, R. F., C. Kramer, W. A. Herlache., J. Wright, H. Fung, and C. Liu, *Proceedings of Dam Safety 2006*, www.damsafety.org, 2006; Olivia Chen Consultants Report, *Geotechnical Investigation Embankment Stability Evaluation*, Sunset Reservoir North Basin San Francisco Public Utilities Commission, 2004; Courtesy of Association of State Dam Safety Officials.)

The treatment layout required that the DMM columns had an average 90-day unconfined compressive strength of at least 400 psi. The potential slide surfaces crossing through the blocks would experience insignificant deformation and the critical slide surfaces would pass below the DMM blocks with static factors of safety greater than 2.8 and a yield acceleration of 0.41g. Under the controlling seismic event, the calculated deformation was less than 6 inches, which met the seismic performance objectives.

Due to the potential for the water table to rise after the DMM treatment, DSOD performed analyses assuming a saturated foundation and a higher phreatic line within the embankment. This conservative approach resulted in a lower factor of safety and greater seismic deformation. However, it was judged that the embankment with DMM treatment would satisfy the embankment performance criterion of not allowing a catastrophic release of water during or after the earthquake. After requesting the DMM treatment be extended upward into the poorly graded Sand 2 layer above the targeted Silty Sand 2, DSOD approved the proposed design.

3.2.6.2.1.3 CONSTRUCTION AND VERIFICATION

Construction of the embankment stabilization began in May 2005 and was substantially completed by December 2006. The ground treatment was performed from the three 60-foot-wide temporary benches cut into the existing embankment slope shown in Figure 3.13. The foundation remediation started from Zone 3 with a DMM rig working on Bench 3 and proceeded to the top of embankment. DMM column installation and restoration of embankment in Bench 3 area were completed before the start of DMM column installation in Zone 2 from Bench 2. The same procedure was followed for DMM column installation in Zone 1 from Bench 1.

A monitoring and testing program was established to assure that DMM would produce the soil-cement columns with geometric requirements and engineering properties required by the design. The program included test sections to verify mix designs, to monitor installation parameters, and to retrieve full-depth core samples for strength testing and evaluation of uniformity of mixing. The monitoring program indicated that the DMM blocks met the geometric requirements for plan location, configurations, depth, inclination, and overlap between adjacent elements. The minimum 28-day unconfined compressive strength of the DMM columns in the treatment zone was 179 psi, which exceeded the specified minimum strength of 120 psi. The average strength of each full-depth core in the treatment zone ranged between 322 and 1,177 psi satisfying the requirement of minimum 300 psi at 28 days or 400 psi at 90 days. The uniformity of mixing was verified from examination of the cores. The average recovery of every full-depth core ranged from 96 to 100 percent within the treatment zone, which exceeded the minimum 85 percent requirement. The percent of unmixed soil was always below the maximum 15 percent allowed.

The particular challenge during the DMM work at this site was evaluating the depth of penetration into the weathered rock in areas where only two rows of DMM blocks could be installed due to site restrictions and the depth to bedrock below the site varied significantly over a short distance. Borings were drilled adjacent to DMM elements to determine the depth to bedrock and to allow correlation with the DMM drilling parameters, including penetration rate, drill energy, and the variation of load on the cable supporting the DMM mixing tool. However, a precise correlation could not be established due to the variation in depth and strength of bedrock across the DMM mixing tool footprint. To evaluate the penetrating capacity, the DMM drilling tool was used to redrill a cured DMM panel. It was found that the equipment was capable of penetrating a cured column with unconfined compressive strength of at least 700 psi. Based on additional analyses, the project team determined that seismic stability would be acceptable when the mixing tool penetrated 5 feet into the weathered bedrock or when it reached bedrock material that reduced the penetration rate to 0.1 foot per minute.

This project marked the first time the California Department of Water Resources, Division of Safety of Dams had approved DMM to remediate a potentially weak foundation and thereby improve the seismic stability of an earth-dam embankment. DMM was a new technology for the SFPUC and DSOD. The successful completion of the embankment dam remediation was based on the clear and effective specifications and sufficient flexibility in the procedures to address variations in the conditions encountered during construction.

3.2.6.2.2 Seismic remediation of Clemson Upper and Lower Diversion Dams, South Carolina

3.2.6.2.2.1 BACKGROUND

Clemson Upper and Lower Diversion Dams were constructed in 1960–1961 to protect lands and facilities at Clemson University in South Carolina (Wooten and Foreman 2005). The Lower and Upper Dams are random earthfill dams with a maximum height of about 80 feet and lengths of approximately 3,000 feet and 2,100 feet, respectively. The foundation soils consist of a loose silty sand/sandy silt alluvial deposit with thickness varying from 7 feet to 28 feet and N-values ranging between 3 and 30 blows per foot. Below the alluvium is a thin layer of gravel underlain by weathered bedrock. Site investigations including borings, laboratory testing, field vane shear testing, and seismic surveys were performed by USACE and its consultants to evaluate the steady-state strengths, the undrained strength available at very large strains, for seismic stability evaluation. The results of analyses indicated that liquefaction slope failure would occur at the downstream section with the level of strains induced in the alluvium by a maximum bedrock acceleration of about 0.08 to 0.1 g. This level of ground shaking is significantly lower than the peak ground acceleration of 0.2 g of the maximum credible earthquake defined by the U.S. Army Waterways Experiment Station (now ERDC) for this site. Additional one-dimensional triggering analyses and two-dimensional finite element triggering analyses were performed to further evaluate post-earthquake stability. These analyses confirmed that accumulated strains during the design event would exceed triggering strains for the downstream slope resulting in estimating safety factors of about 0.6.

3.2.6.2.2.2 REMEDIAL ALTERNATIVES AND REMEDIATION DESIGN

Several remedial alternatives were evaluated to prevent excessive deformations of the downstream sections and liquefaction failures of the dams, including jet grouting, DMM, stone columns, and excavation and replacement. The factors considered in the selection of alternatives included cost, method of design and verification, risk of dam stability during remediation, construction impact, and aesthetics. DMM was selected as the remedial technique, and the chosen design consisted of DMM shear walls beneath the downstream berm of each dam. The DMM walls, oriented perpendicular to the dam axis, function as transverse shear walls to carry the seismic loads and prevent a downstream slope failure during and after the earthquake. A longitudinal wall, oriented parallel to the dam axis, at the upstream end of the transverse walls, prevents the movement of softened soils between the transverse walls during and after the earthquake. These walls were located about 130 feet to 140 feet downstream of the centerline of each dam. The lower portion of the shear walls was keyed into the underlying sand and gravel layer or weathered bedrock and the upper portion of the walls was embedded into the overlying embankment berm to prevent shear failure along the interfaces (Figures 3.14 and 3.15). In order to prevent the buildup of groundwater at the downstream berm and toe, a filtered seepage collection system, consisting of filter gravel and slotted pipes surrounded by a geotextile, was installed upstream of the longitudinal soil-cement wall. Additionally, the longitudinal wall was terminated at the top of the sand and gravel layer.

After selection of the DMM alternative, finite element analyses were performed to evaluate the performance of the modified sections. USACE and its consultants performed the design of seismic remediation with the assistance of others. The results indicated that the modified downstream section of the dams developed negligible deformations under the design earthquake. The dynamic analyses also showed that the DMM reinforcement decreased strains in the alluvium to levels below the triggering strain such that significant strength

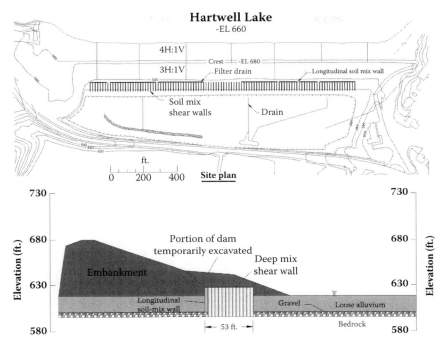

Figure 3.14 Plan and section of DMM layout at Clemson Dam, South Carolina. (After Wooten, R. L., and B. Foreman, *Proceedings of the 25th Annual United States Society on Dams Conference*, 2005; Courtesy of USSD.)

loss in the alluvium was not expected. After the design of the DMM remedial alternative, additional field and laboratory studies were performed to characterize the *in-situ* soils. Bench scale trial mix design testing was conducted on bulk samples from five major soil strata. The results indicated that the DMM

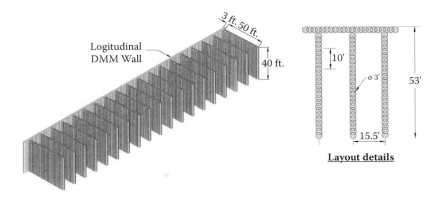

Figure 3.15 DMM treatment layout at Clemson Dam, South Carolina.

treatment could produce soil-cement with an unconfined compressive strength of 400 psi, which was required to maintain the seismic stability with factors of safety between 2 and 3. This target strength (f'_{sc}) was developed using 3-foot-wide DMM shear walls at a 15.5-foot spacing. To allow for flexibility in using different DMM equipment types, the following equation was developed to permit adjustment of the wall width (Wa) and wall spacing (S):

$$f'_{sc} \geq 77.4 \text{ psi} \times (S/Wa) \tag{3.1}$$

However, the wall spacing was restricted to no more than 12.5 feet plus the wall width. This formula also provided a mechanism for revising wall spacing if the strength of soil-cement was lower than the target value due to difficulty during construction in achieving the target strength values. This formula, in fact, defines the relationship between the strength, shear wall width, and wall spacing. The value, 77.4 psi, is a site-specific value that can be developed during the design of a DMM remediation scheme. For this project, it can be considered as the design average unconfined compressive strength of DMM with 100 percent treatment ratio, or a Wa/S of one.

3.2.6.2.2.3 CONSTRUCTION AND VERIFICATION

A portion of the downstream berm was excavated temporarily to create a level work platform for DMM construction. A six-axis DMM rig was used to install the overlapping DMM elements forming the shear walls. Photo 3.7 shows the DMM rig installing a 50-foot-long transverse shear wall. The soil-cement wall had an average width of 2.76 feet. To maintain the wall

Photo 3.7 Six-axis DMM Rig at Clemson Dam, South Carolina. (Courtesy of Ratio, Inc.)

spacing of 15.5 feet, the target unconfined compressive strength (f'_{sc}) was adjusted to 435 psi using the target strength formula (Equation 3.1). Prior to the full DMM wall installation, the specialty contractor, Raito, Inc., performed a bench-scale trial mix study using soil samples from the project site to produce data for selection of mix design and installation procedure. Preproduction test sections, consisting of two test walls at each dam site, were performed to verify that the DMM equipment, installation procedure, and the selected mix design would produce soil-cement walls meeting the geometric configuration and design strength. An electronic real-time quality control system was used to monitor and control the depth, penetration and withdrawal speed, mixing tool rotation rate, and slurry injection rate. The requirements on strength included twelve 28-day unconfined compressive strength tests on wet-grab samples for each transverse wall. The average strength was to exceed 435 psi with only one of the twelve tests allowed to fall below two-thirds of 435 psi, or 290 psi.

For quality assurance, the specifications required strength testing on specimens cast using wet-grab bulk sampling. The core sampling was considered to be slower and more expensive and would be used only in cases where wet-grab sample strengths fell below criteria. However, the contractor tried numerous methods to obtain the wet-grab samples with no success since the soil-slurry mixture was very viscous due to the need to use a mix design with low water:cement ratio in order to achieve the relatively high target strength. As an alternative, full-depth continuous cores were retrieved from the shear walls at about 28-day curing age for quality-control strength testing. The cores are more representative of the wall material than the wet-grab samples and have continuity throughout the entire depth of the wall to allow the selection of test specimens at various depths. USACE selected six specimens from each core for 28-day strength testing and agreed on a frequency of one core per day, or about one core for two shear walls, based on the consistently high strengths of the samples. Representative core samples and test specimens are shown in Photo 3.8. USACE accepted all DMM shear walls installed in the Lower Dam based

Photo 3.8 Core samples and test specimens from Clemson Dam, South Carolina.

on the core testing data. However, 27 of the 105 DMM shear walls in the Upper Dam did not meet the strength criteria. The low strengths might have been caused by localized zones of higher organic content and/or low pH soils within the alluvium layer. Additional shear wall elements were therefore installed adjacent to the low-strength sections to provide a total shear resistance equivalent to that of the original design. The installation of the soil-cement walls began in February 2004 and was completed in February 2005. A total of 45,500 yds^3 of soil-cement was installed to remediate these two dams.

3.2.6.2.3 San Pablo Dam, California, seismic upgrades

3.2.6.2.3.1 BACKGROUND

San Pablo Dam is located in Contra Costa County, California (Geomatrix Consultants 2004, 2005; TNM 2007a, 2007b, 2008). The dam was constructed on San Pablo Creek between 1917 and 1921 using hydraulic fill, and has a clay puddle core. The shells of the dam were constructed using fragments of sandstone and shale that were hydraulically transported from the hills adjacent to the abutments. With the exception of the core trench, the dam is founded on native alluvial and colluvial soils in the San Pablo Creek channel. The alluvial deposits are as much as 100 feet thick. The dam is currently owned and operated by the East Bay Municipal Utility District (EBMUD).

Notable earlier modifications to improve the seismic stability of the dam include a buttress of clayey soils at the downstream toe of the dam in 1967, and a similar buttress on the upstream face of the dam in 1979. The upstream buttress was prompted by the near failure of the Lower San Fernando Dam during the 1971 San Fernando earthquake.

The California Department of Water Resources, Division of Safety of Dams (DSOD), reviewed previous seismic stability reports prepared by several consultants between 1966 and 1978 and concluded that the seismic hazard at the dam site had increased from that estimated in previous studies. Consequently, EBMUD requested their consultant to conduct a reevaluation of the seismic stability of the dam and develop a seismic upgrade scheme. The consultant compared the seismic-induced stresses with cyclic strength of the embankment and foundation soils to estimate the factor of safety against liquefaction and concluded that liquefaction was likely to occur in most of the saturated coarse-grained zones of the embankment shell and foundation alluvium. Using the undrained residual strengths of the liquefied soils, the consultant performed the slope stability analyses and concluded post-earthquake factors of safety of 1.24 and 0.59 for the upstream slope and downstream slope, respectively. The results of further analyses using ground motions from the maximum credible earthquake for the site indicated that the embankment would deform in the downstream direction. The magnitude of deformations would be excessive and could cause an uncontrolled release of the reservoir water at the normal maximum operating level. Based on these studies, the consultant recommended the remediation of the downstream slope of the embankment, if the owner intended to operate the dam at its normal maximum reservoir level.

The objective of remediation was to limit the permanent deformation so the uncontrolled release of reservoir water would not occur and the dam could be repaired and returned to service following the maximum design earthquake. The maximum design earthquake was the maximum credible earthquake with a magnitude M 71/4 on the Hayward-Rodgers Creek fault located two miles from the dam site. The estimated 84th percentile peak horizontal bedrock acceleration of this earthquake was 0.91 g. The maximum allowable deformation at the crest was about 3–5 feet. Several remedial alternatives were evaluated and the two selected for conceptual design in 2005 were (1) removal and replacement of the downstream shell and foundation alluvium with the reservoir completely drained during remediation, and (2) DMM treatment of the foundation soil with the reservoir remaining in operation during remediation at 20 feet below normal maximum operating level. The conceptual design included removal of the existing downstream buttress and replacement with a larger buttress founded on alluvium reinforced by soil-cement cells installed by DMM. This seismic upgrade would reduce the downstream lateral deformation and reduce the movement of the crest of the dam to acceptable levels to prevent the loss of freeboard of the dam during and after the earthquake. The seismic upgrades also included improvement of the alluvium in the vicinity of the portal outlet structure using jet grouting.

3.2.6.2.3.2 DESIGN

The final seismic upgrade scheme of the dam is shown in Figure 3.16 and the DMM treatment layout is shown in Figure 3.17. The designer arranged the DMM elements within the treatment zone as closed-spaced shear walls oriented perpendicular to the axis of the dam to improve the soil properties in the direction of primary loading from the reservoir, and to contain the potentially liquefiable soil between the relatively closely spaced shear walls. The typical length of each shear wall was 150 feet. The designer also developed the following equation that could provide flexibility for the adjustment of field DMM strength obtainable, average wall width, and center to center wall spacing:

$$f'_{sc} = f'_{sc-em} \times (S/W) \tag{3.2}$$

Where f'_{sc} is the minimum average 28-day unconfined compressive strength of DMM shear wall, f'_{sc-em} is the equivalent minimum average 28-day

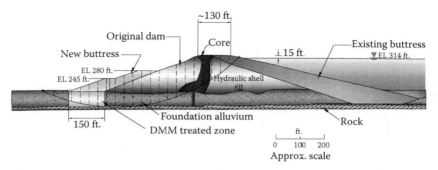

Figure 3.16 San Pablo Dam, California, seismic upgrade details. (After TNM, Results of CDSM Test Section Report for East Bay Municipal Utility District, Design Services for San Pablo Dam Seismic Upgrades, 2007; Courtesy of TNM and EBMUD.)

unconfined compressive strength of DMM treated soil if 100 percent of soil were treated, S is the center to center space of DMM shear walls, and W is the average width of the DMM shear walls.

Based on the analyses of the final design section shown in Figure 3.16, f'_{sc-em} is 225 psi and f'_{sc} is 450 psi, if 50 percent of the soils were treated. W/S, the ratio of average wall width to wall spacing, is equivalent to the DMM treatment ratio. Similar to the one used in the Clemson Upper and

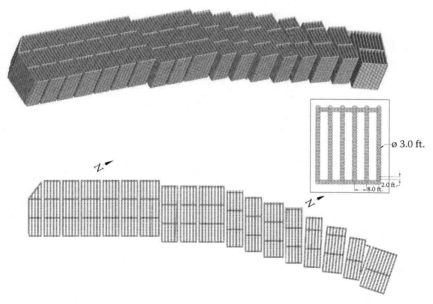

Figure 3.17 DMM treatment layout for San Pablo Dam, California.

Lower Dams project, this equation serves the same function of providing design and construction flexibility. For San Pablo Dam, the center to center wall spacing was required to be less than three times the average wall width to maintain the composite behavior of the shear walls.

A laboratory trial mix testing program was performed to develop strength data on soil-cement mixture using three types of soil samples from the project site. A full-scale test section was also performed in 2007 to evaluate the strength and uniformity of soil-cement that could be produced and the ability of DMM equipment to key the DMM wall into the weathered bedrock. Both laboratory testing and full-scale test section programs were part of the final design scope of work. The triple-shaft DMM equipment was unable to drill three feet into the bedrock at the test section locations as required by the test specification. Therefore, predrilling was needed during the production work to ensure that DMM walls could be extended with sufficient key-in depth into the bedrock to avoid sliding along the interface of the DMM shear walls and bedrock.

As noted above, the main objective of using DMM foundation treatment for the seismic upgrade was to reduce the seismically induced deformation of the embankment to prevent uncontrolled release of reservoir water. Dynamic analyses were performed using various strength values of soil-cement and untreated alluvium/colluvium soils. The DMM treatment ratio was assumed to be 50 percent. The results of analyses indicated that a design shear strength of about 200 psi would be adequate for the control of permanent settlement and downstream deformation and a shear strength of 300 psi would suffice with respect to all appropriate design requirements of the San Pablo Dam. Based on previous studies and experience, the designer assumed the shear strength value to be 33 percent of the unconfined compressive strength and determined that a design unconfined compressive strength of 900 psi for DMM would be required for the seismic upgrade. Based on the review of studies on the long-term strength gain of soil-cement and comparison with data on strength gain with time obtained from tests on core samples from the test section, it was concluded that the long-term strength gain of soil-cement is predictable and that the strength would reach twice the 28-day strength in approximately one year. Considering the insensitivity of dam deformation to shear strength, if greater than 200 psi, and the low probability of occurrence of the design earthquake before the strength reached twice the 28-day strength, a 28-day unconfined compressive strength of 450 psi was selected as the target strength of the DMM foundation with a treatment ratio of 50 percent.

The shear walls were installed at a spacing of 8 feet (Figure 3.17) using 3-foot-diameter overlapping augers, with the auger stems 2 feet apart. The average width of the shear walls was therefore 2.75 feet. Consequently, the required minimum average 28-day unconfined compressive strength of the DMM material, as determined using Equation 3.2, was 655 psi.

3.2.6.2.3.3 CONSTRUCTION AND VERIFICATION

The two main challenges of the DMM work at San Pablo site were toeing the DMM shear wall into bedrock and maintaining the continuity of the deep DMM shear wall. The bedrock varied significantly in depth and strength across the site. The bedrock material within the three-foot key-in depth generally consisted of slightly weathered to severely weathered, very soft to moderately hard siltstone. Based on the results of the 2007 test section, it was concluded that the DMM equipment could not penetrate three feet into the weathered rock as required by the specifications. Therefore during full production in 2009, a single-shaft machine was used to predrill three feet into the weathered rock to allow the triple-shaft DMM mixing tool to reach the key-in depth in bedrock. The predrilling and DMM installation operation are shown in Photo 3.9. The DMM can serve its design function as a shear wall only if it is continuous; if the DMM elements were not aligned, the overall lateral stiffness of the treated zone would be reduced. Verticality of DMM elements can only be maintained to 1:100 (horizontal to vertical). Therefore wall continuity for the deep shear walls needed could not be assured if the neighboring elements were partially overlapped by only one foot. A full column overlapping between neighboring DMM elements was therefore specified to ensure the shear wall continuity for transferring lateral forces. The shear wall construction sequence using predrilling, full column overlapping, and alternate installation procedure (primary and secondary procedure) is the same as the sequence used for the installation of the DMM cutoff wall at Lake Cushman Dam, Washington (Figure 3.11b).

Photo 3.9 Predrilling and DMM treatment at San Pablo Dam, California. (Courtesy of Yujin Nishimura, Raitc, Inc.)

For verification of deep mixing work, both wet-grab samples and continuous core samples were retrieved. The wet-grab samples were taken at the same core locations selected by the engineer. Both molded wet-grab samples and core samples were used for establishing the strength gain of the soil-cement with curing time. The final acceptance of the deep mixing work was based on the uniformity and unconfined compressive strength test results from the core samples. The requirements were as follows: (1) average 28-day unconfined compressive strength tests of seven core samples from each full-depth core should exceed 655 psi and the minimum strength should exceed 327 psi; (2) average core recovery should be greater than 90 percent for every full-depth core and minimum core recovery should be greater than 85 percent for every four-foot core run; and (3) total unmixed soil should be less than 15 percent of every four-foot core run and every occurrence should be less than six inches. The average strength of 655 psi was selected based on the data obtained from the preproduction test section. The preproduction test section, consisting of three 50-foot-long shear walls at three locations of the site, was performed to verify that the equipment, installation procedure, and mix design would produce DMM shear walls meeting the design intent. DMM construction was started in December 2008 and completed in September 2009. A total of 137,300 cubic yards of shear walls with depth varying from 20 to 120 feet were installed. A section of an exposed DMM shear wall is shown in Photo 3.10.

Photo 3.10 Exposed soil-cement shear wall at San Pablo Dam, California.

3.2.6.2.3.4 SPECIAL ACKNOWLEDGMENTS

The reports and papers prepared by the owners and designers are the main sources of information for the preparation of these six project examples. The DMM solution for each case example was created by the innovative thinking process and diligent engineering work of the owner, design team, and contractor. The authors are grateful to the following individuals for sharing valuable project information and/or providing detailed reviews, comments, and editing during the preparation of this section: (1) Gerard Buechel of Shannon and Wilson and Yvonne Gibbons of USACE for the Lewiston Levee Seepage Remediation; (2) Steve Fischer of Tacoma Public Utilities and David Cotton of Kleinfelder for the Lake Cushman Spillway; (3) Andrew Herlache of Fugro West and Howard Fung of City of San Francisco, Rebecca F. Barron of Division of Safety of Dams, California Department of Water Resources, and John Wright of California Department of Water Resources for the Remediation of Sunset North Basin Dam; (4) Ben Foreman of USACE and Lee Wooten of GEI for the Seismic Remediation of Clemson Upper and Lower Diversion Dams; and (5) Atta Yiadom of EBMUD and Robert Kirby of Terra Engineers for the San Pablo Dam Seismic Upgrades. The author is thankful to the American Society of Civil Engineers, the U.S. Society on Dams, the Association of State Dam Safety Officials, CI-Premier, and the authors of the papers for granting the permissions to use the figures or materials in their publications. The authors deeply appreciate the permission and support of Raito, Inc. for the preparation of Section 3.2. Special recognition is extended to Lorena Galvan for the graphic design of certain figures and for her assistance in preparing the manuscript.

3.3 TRD METHOD

3.3.1 Introduction

The TRD method (trench cutting re-mixing deep wall) is a Japanese development conceived in 1993 and tested for the first time in 1994. As of this writing, the TRD method had been used to produce over 28 million square feet of wall in over 500 projects.

The TRD method uses a full-depth vertical cutter-post with a chainsaw-like cutting tool to cut and mix the soil with slurry, which is injected from ports on the post, as the base machine moves along the alignment (Figure 3.18). This method is claimed to offer enhanced production efficiency, and superior quality for the construction of walls, for both groundwater control and excavation support (Aoi, Komoto, and Ashida 1996; Aoi et al. 1998).

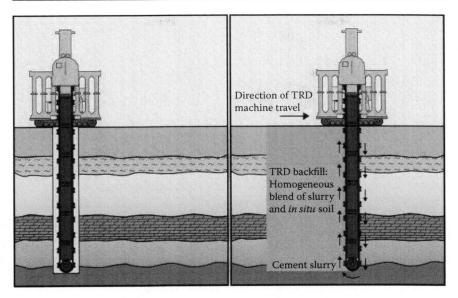

Figure 3.18 TRD method of wall construction. (From Garbin, E. J., J. C. Evans, and J. D. Hussin, *Proceedings of Dam Safety 2009*, Association of State Dam Safety Officials, 2009. With permission.)

The TRD method ensures a given verticality, geometry, and continuity, and assures complete vertical mixing and distribution of all soil strata within the engineered slurry. It is capable of producing a wall thickness from 22 to 33 inches, to depths approaching 180 feet, even in dense soils containing cobbles and hard rock, provided it is "rippable." Most TRD applications have involved the installation of vertical walls, but special equipment has been developed to produce inclined walls.

3.3.2 Means, methods, materials, and properties

3.3.2.1 Excavation and mixing process

The two views of the equipment shown in Figure 3.19 illustrate how the drive mechanism is mounted to the guide frame. In this way, the cutter-post can be adjusted in two planes, and moved independently in the third plane, without having to move the base machine. The cutter consists of a motor to drive the chain, a drive wheel at the top of the post, a post to guide and support the chain, and an idler wheel at the base of the post. The post is a series of connected pieces enabling the adjustment of depth, and facilitating initial insertion into the soil. Inside the post is a series of pipes that permit injection of the engineered slurry at various depths, air if desired, and a pipe for an array of permanent inclinometers that show verticality

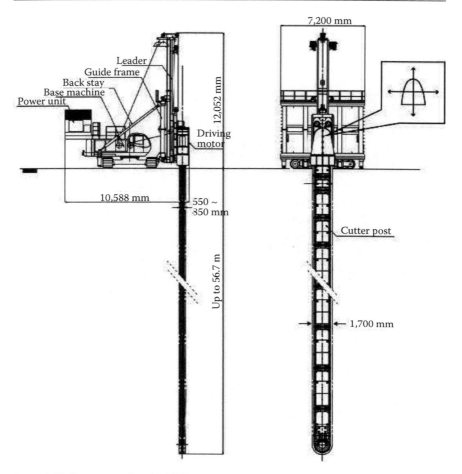

Figure 3.19 Equipment for the TRD method. (Courtesy of TRD Method Association, Japan.)

(in two planes) to the operator in real time in the cabin. Figure 3.20 shows a generic bit configuration, and illustrates how the bits are mounted to the drive chain. A variety of bits are used depending on the soil and/or rock to be cut and mixed.

Figure 3.21 illustrates how the TRD method excavates, disaggregates, and mixes the soil while the slurry is injected. In general, the lateral force on the post is kept small, and the cutter-chain is moved downward at the face of the excavation. With low lateral force, the chain speed can be higher, allowing the bits to fragment the soil/rock cut face into small pieces. The loosened debris is transported with the descending slurry, up the trailing side of the post, gradually passing through the clearance space between the

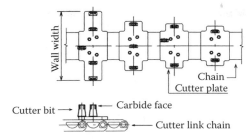

Figure 3.20 Generic bit configuration for the TRD method. (Courtesy of TRD Method Association, Japan.)

excavated width and the post, and then sent forward down the chain on the leading edge of the post. In this way a convection current of soil/rock and slurry is generated that results in a very high degree of uniformity and homogeneity in the wall.

Excellent historic data have been compiled upon which to base production estimates, as illustrated in Figure 3.22.

3.3.2.2 Materials

The TRD method creates a product that consists of *in-situ* materials and an engineered grout slurry. Generally, the *in-situ* materials make up 50–80 percent of the final product. Although the virgin materials predominate, they do not control the engineering properties of the wall. The injected slurry is

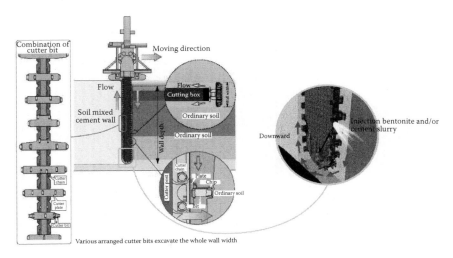

Figure 3.21 Excavation, disaggregation, and mixing processes. (Courtesy of TRD Method Association, Japan.)

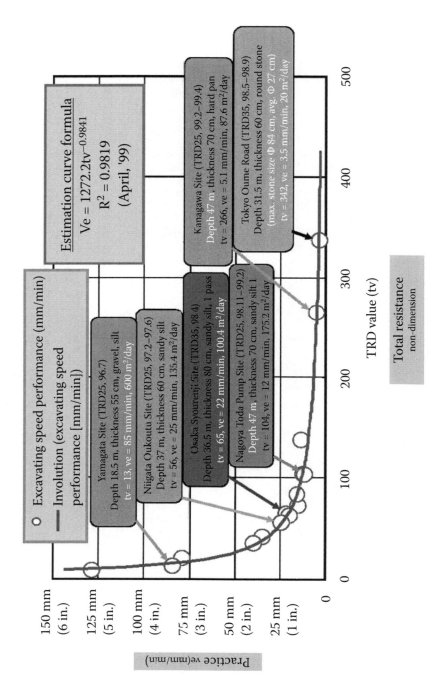

Figure 3.22 Prediction of excavation speed. (Courtesy of TRD Method Association, Japan).

specifically designed to combine with these materials and furnish the properties desired. Most often the slurry consists of water, swelling clays, and cementitious binders. The water should be suitable for the intended product, but does not need to be of potable quality. The clays can be bentonite, attapulgites, or sepiolite, and should be hydrated before blending with binders. Binders can be cement, fly-ash, ground granulated blast furnace slag (GGBFS), lime, or other materials that exhibit cementitious characteristics (Evans 2007). Additives may also be used to enhance fluidity, reduce bleed, or improve some engineering property of the final product. Sometimes air is injected to increase the fluidity and enhance the mobility of the mixed materials.

3.3.2.3 Material properties

The product of the TRD method is closely controlled by managing the composition of the injected slurry and the injection characteristics. Strength and permeability are usually the key properties considered, and it is essential that laboratory testing be performed on all anticipated materials for construction to confirm expectations.

Since the TRD includes full vertical mixing of the soil profile, individual strata are not singled out for testing as is the case with conventional DMM. Rather, the full subsurface profile is mixed with the slurry to produce the final product, in the expected field proportions.

Strength development is somewhat dependent on the type and proportion of the fines in the ground. Sands and gravels have less impact on the engineered strength, as the final strength will be controlled by the grout. Figure 3.23 compares results obtained from conventional multi-axis deep mixing and from the TRD method highlighting the higher degree of vertical homogeneity provided, especially in poorer soils, by the TRD. Most cobbles will be crushed, or can be removed from the mix as they travel up the chain to the surface for conventional excavation. Organic strata should be evaluated closely for their potential influence on the soilcrete material.

3.3.3 Quality control and quality assurance

Quality control and quality assurance are addressed at four stages of each project:

1. Mix design development (preproduction).
2. Mix preparation (preinjection).
3. Wall construction (during production).
4. Verification sampling and testing (postproduction).

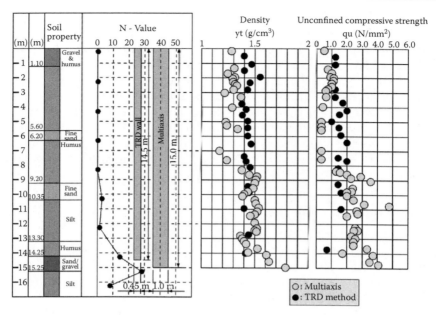

Figure 3.23 Comparison of results from multi-axis deep mixing and the TRD method. (Courtesy of TRD Method Association, Japan.)

3.3.3.1 Preproduction mix design development

This is an essential part of any project, and entails the following tasks:

1. Retrieving representative subsurface materials for the full depth of treatment. This is usually done by drilling with a method that can achieve nearly 100 percent recovery across a wide range of soil types.
2. Preparing a suite of slurry mixes that target the intended product qualities.
3. Mixing the slurries with the "combined" soil profile to achieve the appropriate viscosity. The target range is 150–230 mm, as measured by the flow table test.
4. Performing the laboratory tests that confirm the desired product quality such as initial set, unconfined compressive strength, permeability, and bleed.

3.3.3.2 Preparation of the selected slurry

The slurry, or slurries, defined for use from the laboratory study will have a specific recipe. If swelling clays are used in the mix (e.g., bentonite), they should be hydrated prior to being added to the mix. There is a defined order to the materials added, and there must be an accurate, repeatable means of batching and adding each component. Automatic batching systems exist that can be programmed to do this, but it is always advisable to have a backup control and verification method.

3.3.3.3 Production wall mixing and slurry injection

When mixing the production wall, the rate of wall construction is coordinated with the rate of slurry injection to mirror the selected laboratory test mixes. Routine flow table tests are performed on wet-grab samples from the trench to check for mix viscosity. This ensures adequate flow of the mix past the cutter-post, down to the tip of the cutter-post, and back up the trailing edge of the post to the surface. Depending on the soil encountered and the desired product properties, the volume of slurry will vary from 30 to 60 percent of the wall volume. The depth, inclination, and rate of advance of the post is continuously recorded and displayed in the operator's cabin.

3.3.3.4 Verification sampling and testing

Since the wall is continuous (meaning that the verticality of the post is controlled and the post passes through 100 percent of the soil profile), there is no need to perform testing specifically intended to investigate possible discontinuities, as would be the case between adjacent elements of a secant or panel wall. Due to the efficiency of the vertical mixing process, wet-grab samples (usually from near the surface) can provide most of the samples necessary for verification. These wet samples are then cast into suitable molds for the appropriate tests. However, the wall remains sufficiently fluid for a few hours after installation, thus permitting deep-grab samples to be taken at any desired elevation for this confirmation.

Core samples may also be retrieved. It should be understood that core sampling is a destructive sampling process with inherent difficulties, especially when the engineered slurry is relatively weak in relation to any coarse pieces of gravel or rock in the matrix of the wall. The holes so formed can be subjected to camera inspection and/or *in-situ* permeability testing, depending on the project goals. Core samples can be damaged in the process of retrieval and test results may not be fully representative. However, project guidelines generally require some amount of core retrieval and strength testing. Flexibility should be granted to permit visual observation of the holes by video logging if anomalies in core recovery, or in the cores themselves, are encountered.

3.3.4 Applications

The TRD method can create continuous groundwater barriers along the axis of existing levees. The working platform can be as narrow as twenty-five feet, and the method is especially economical for deep applications through difficult and variable ground conditions. Figures 3.24 and 3.25 show levee and other applications.

3.3.5 Case histories

3.3.5.1 Herbert Hoover Dike, Florida

Lake Okeechobee in southeastern Florida is surrounded by a very heterogeneous levee known as the Herbert Hoover Dike, on which early construction by locals first began around 1915 utilizing mostly sand and topsoil (Garbin, Evans, and Hussin 2009). Portions of the original embankments were overtopped by hurricane-induced surges in 1926 and 1928, resulting

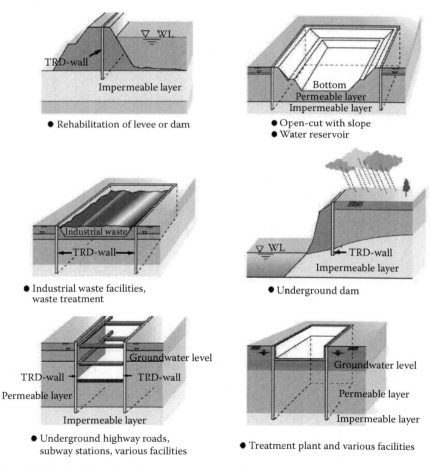

Figure 3.24 TRD cutoff and retaining wall applications. (Courtesy of TRD Method Association, Japan.)

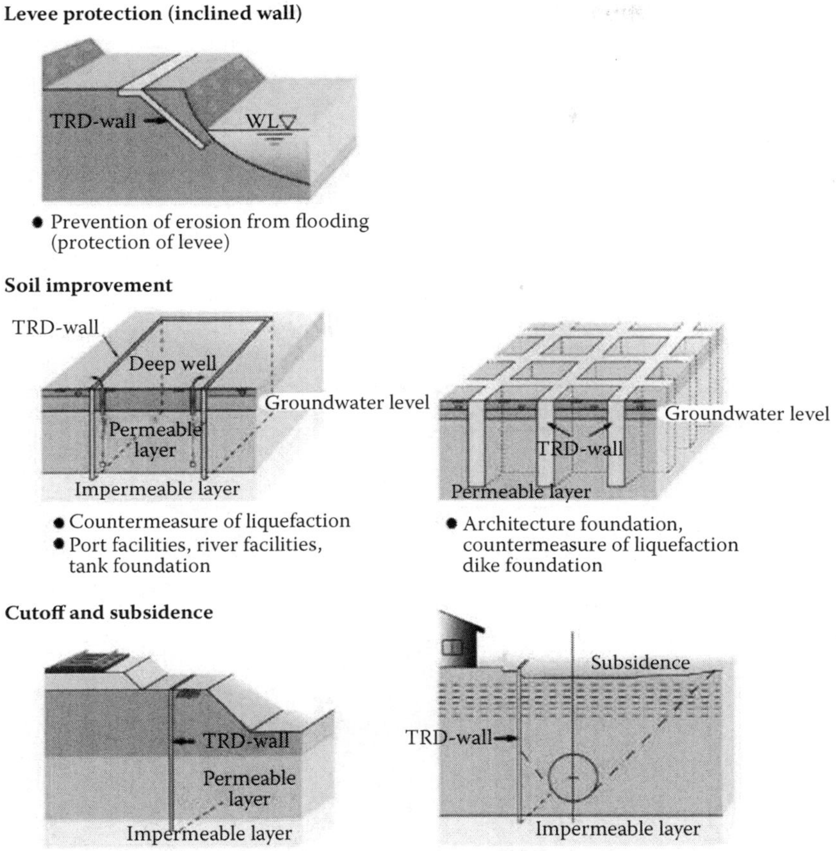
Levee protection (inclined wall)
TRD-wall
WL
• Prevention of erosion from flooding (protection of levee)
Soil improvement
TRD-wall
Deep well
Groundwater level
Permeable layer
Impermeable layer
• Countermeasure of liquefaction
• Port facilities, river facilities, tank foundation
Groundwater level
TRD-wall
Permeable layer
• Architecture foundation, countermeasure of liquefaction dike foundation
Cutoff and subsidence
TRD-wall
Permeable layer
Impermeable layer
Subsidence
TRD-wall
Impermeable layer

Figure 3.25 Miscellaneous applications of TRD walls. (Courtesy of TRD Method Association, Japan.)

in the loss of over 2,500 lives. As a result, about 84 miles of levee were reconstructed by the U.S. Army Corps of Engineers (USACE) between 1932 and 1938. A major hurricane in 1947 emphasized the need for additional flood protection, and the current dike system for Lake Okeechobee was completed in the late 1960s. The dike system now consists of about 140 miles of levee with nineteen culverts, hurricane gates, and other water control structures. Lake Okeechobee has become the third-largest freshwater lake in the continental United States, draining to the ocean through the Everglades. The levee crest is typically at around elevation +36 feet, and the lake is normally at EL +10 to 13 feet. However as constructed, sections of the levee are prone to instability due to seepage and piping, in particular when lake levels increase beyond certain elevations due to surge caused by severe weather (hurricanes) (Davis, Guy, and Nettles 2009). Therefore, the USACE has designed a seepage cutoff wall that is being constructed progressively in prioritized "reaches" (Figure 3.26) to mitigate piping concerns and ensure dike stability during extreme weather events. An illustration of this concept is shown in Figure 3.27. The first 4,000 lineal feet of this wall was completed recently using the TRD method.

The grout used on this project was a blend of hydrated bentonite slurry, Portland cement, and GGBFS. Testing of shallow bulk, deep-grab, and core samples for quality control included unconfined compressive strength (UCS) and permeability at various curing times. Shown on Figure 3.28 are typical shallow bulk and wet-grab strength and permeability test results. As part of a very intense QA/QC program, over 160 samples were tested for strength, a task described by Garbin, Evans, and Hussin (2009) as

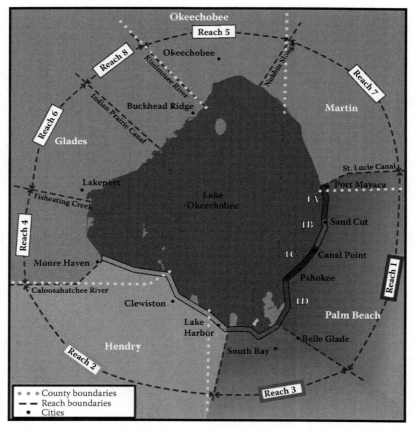

Figure 3.26 General layout of reaches at Herbert Hoover Dike, Florida. The total perimeter length is about 140 miles. (U.S. Army Corps of Engineers, http://www.saj.usace.army.mil/Divisions/Everglades/Branches/HHDProject/HHD.htm, 2007.)

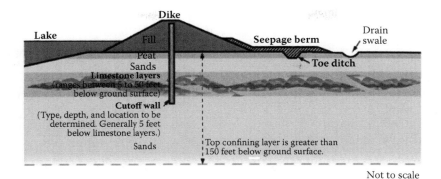

Not to scale

Figure 3.27 Schematic of cutoff wall and seepage berm concept, Herbert Hoover Dike, Florida. (U.S. Army Corps of Engineers, http://www.saj.usace.army.mil/Divisions/Everglades/Branches/HHDProject/HHD.htm, 2007.)

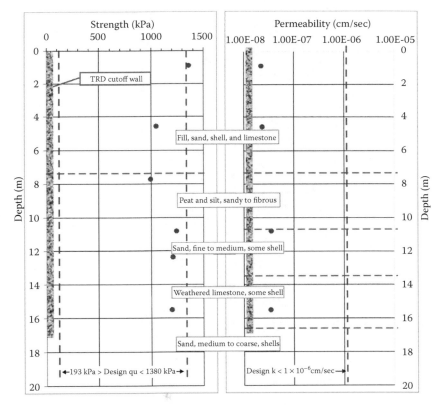

Figure 3.28 Strength and permeability test results, TRD cutoff, Herbert Hoover Dike, Florida. (From Garbin, E. J., J. C. Evans, and J. D. Hussin, *Proceedings of Dam Safety 2009*, Association of State Dam Safety Officials, 2009. With permission.)

"unprecedented." Core strengths averaged 10–20 percent less strength than tests on wet-grab samples extracted from comparable depths.

The excellent homogeneity of the mixture resulting from the TRD method is illustrated by the borehole camera photo (Photo 3.11). This photo, showing the material at a depth of thirty-six feet in one of the verification core holes, is representative and typical of the materials observed in multiple cores throughout the entire length and depth of the wall. It shows a very well-mixed and homogeneous material with a uniform distribution of various particle sizes in a matrix of the hydrated cement-slag blend. The darker areas represent small, discontinuous voids that are likely the result of the coring process dislodging hard aggregate from the soft binder matrix, and/or washing out pieces of organic matter mixed into the matrix. The wider homogeneity of these areas is confirmed by the very low permeability values measured

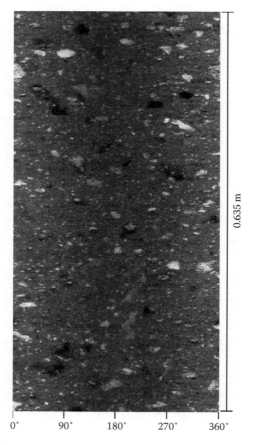

Photo 3.11 Borehole camera photo of *in-situ* cured material created by TRD method, Herbert Hoover Dike, Florida.

in borehole falling permeability tests (Figure 3.28) in those core holes that have not experienced cracking due to drilling and/or shrinkage impacts.

3.3.5.2 Storage dam cutoff wall pilot project, Okinawa

A freshwater storage dam in Okinawa, situated on a foundation that included hard limestone and sandstone with maximum unconfined compressive strength of 5,000 psi and permeability from $<1 \times 10^{-1}$ to 1×10^{-3} cm/s, required a cutoff wall to prevent leakage of the valuable groundwater. To achieve the cutoff, a permeability of $\leq 1 \times 10^{-6}$ cm/s and an unconfined compressive strength of over 75 psi were required. The wall depth was 28 feet, the length 340 feet, and the thickness 20 inches. The grout consisted of a mixture of bentonite, water, and cement. Laboratory tests on field mixed samples confirmed that the design permeability and strength requirements were met. The measured strength and permeability were 1,500 psi (as average value) and $<5 \times 10^{-8}$ cm/s, respectively. The wall verticality was excellent (less than 1:300 deviation).

Photo 3.12 and Figure 3.29 show a cored sample, and the tested permeability and strength data, respectively. Excellent vertical homogeneity and mixing quality were obtained.

3.3.5.3 Alamitos, California, cutoff test section

Two demonstration test cells were to be constructed with *in-situ* walls in the Alamitos Gap between Seal Beach and Long Beach, California (Gularte et al. 2007). Seawater intrusion occurs through a shallow aquifer 100 feet deep and contaminates the freshwater supply aquifers farther inland. The cutoff wall concept called for a deep mix wall constructed into the aquitard that underlies the aquifer and spans the two-mile length of the gap. The

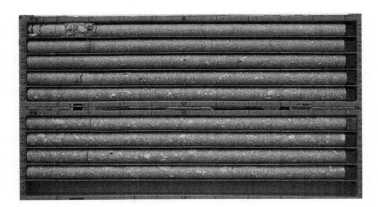

Photo 3.12 TRD core sample from Okinawa, Japan. (Courtesy of Tenox Corporation, Japan.)

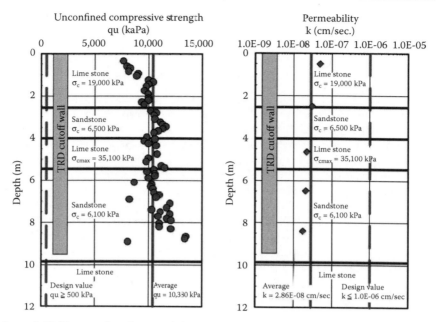

Figure 3.29 Test results of permeability and strength, Okinawa, Japan.

purpose of the proposed 100-foot-deep passive barrier was to optimize the operation of the existing 460-foot-deep injection barrier farther inland.

During design, hydrogeologic evaluation indicated variability in the aquitard beneath the shallow aquifer. Therefore, test cells to depths of 65 and 80 feet were utilized to assess seepage under the wall with respect to wall depth. Preproduction testing of five different soil-mixes indicated a slag-cement-sepiolite-soil mix as optimal for field application that could achieve a permeability of 1×10^{-6} cm/s or less in the saline environment. The pilot test layout in Figure 3.30 shows how the two test cells were constructed by the connection of five individual walls.

The stratigraphy of the site was evaluated using borings and cone penetration tests (CPT). A typical subsurface profile through the site is shown on the interpretive section shown in Figure 3.31. The stratigraphy of the site comprises the merged Recent and I-Zone aquifers, a shallow aquitard, and an underlying deep aquitard.

Field operations consisted of checking for utilities, surface grading to clear and level the area, layout of guide walls, forming guide walls and spoils trench, setting up TRD machine with laser alignment, inserting TRD cutter-post, and production of the TRD walls (Photo 3.13). Operations started with test cutting and mixing to adjust water, cement, clay, and additives and then moved on to production cutting and mixing. Rigorous quality control and assurance measures from depth-of-wall checks through real-time density monitoring of

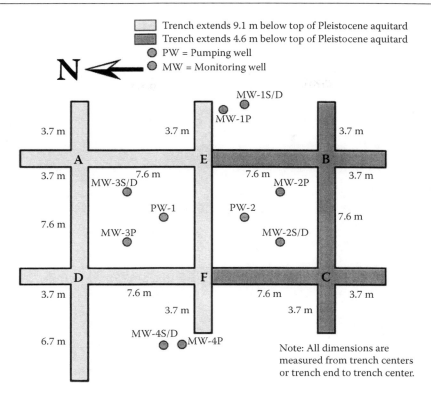

Figure 3.30 TRD pilot test layout, Alamitos Gap, California. (From Gularte, F., E. Fordham, D. Watt, J. Weeks, and T. Johnson, "First Use of TRD Construction Technique for Soil Mix Cutoff Wall Construction in the United States," *Proceedings of GeoDenver*, 2007. With permission.)

the slurry mix ensured compositional consistency. For overnight shutdown and other wait periods, the slurry was modified and/or retarders were added to "shelter" the cutter-post without it becoming cemented in place. Upon completion, the cutter-post was extracted while backfilling with slurry. Spoils produced during the TRD process ranged from 35 to 50 percent of the mixed wall volume.

The following conclusions were developed from the results of the pilot test with respect to the effectiveness of the construction of a passive barrier against sea water intrusion using the TRD method:

1. Based on aquifer testing before cell construction the aquifers were estimated to exhibit a permeability of 2 to 3 × 10^{-2} cm/s and the aquitards to exhibit permeabilities around 10^{-6} cm/s.
2. The TRD provided an extremely effective barrier to water intrusion as evidenced by the difference in drawdown and recovery of water levels for the before and after barrier construction pump tests.

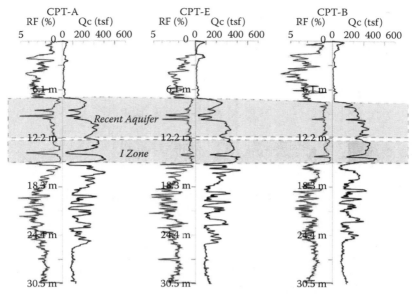

Preliminary CPT Profile "AEB"
Alamitos Barrier Pilot Project

Figure 3.31 Stratigraphy of pilot test site at Alamitos Gap, California. (From Gularte, F., E. Fordham, D. Watt, J. Weeks, and T. Johnson, "First Use of TRD Construction Technique for Soil Mix Cutoff Wall Construction in the United States," *Proceedings of GeoDenver,* 2007. With permission.)

Specifically, the "before" pump test resulted in a 3-foot drawdown when pumped at 34 gpm compared to the 30- to 35-foot drawdown within the walled cell in the same wells when pumping was at a rate of 4 to 7 gpm and no drawdown in the well outside the walled cell.

3. Based on the field recovery rate, during the "after" cell construction pump test, the average permeability of the TRD wall was computed to be 1 to 2×10^{-7} cm/s. The laboratory permeability tests on samples of the wall material yielded the same results indicating that the wall was well mixed and uniform with respect to its permeability.

Photo 3.13 TRD operation at Alamitos Gap, California.

3.3.6 Overview

The biggest advantage of the TRD method is the quality and homogeneity of the wall. It can produce a particularly uniform wall, with certainty of continuity, even in deep, difficult, and heterogeneous ground conditions. The method is also efficient in the rate of production of wall in such conditions, offering economic and schedule advantages, especially when long, straight cutoffs are required. The TRD machine is safe, relatively quiet, with a short, heavy, compact frame with virtually no possibility of tipover.

The greatest efficiency in operating the TRD is achieved by working continuously; a working schedule that allows for day and night work also allows for no stoppages and continuous production. If construction cannot proceed at night, the method requires that the cutter-post produce a 10-foot length of stabilized mixed wall with only a hydrated bentonite slurry. The following day's production starts by retreating two feet into the existing production wall and enriching the bentonite slurry with a cement slurry to achieve the targeted mix in place.

3.3.6.1 Special acknowledgments

The authors acknowledge all of the TRD developers and contractors (the TRD-Method Association) who have offered substantiation for this equipment and its unique method of wet deep mixing. Their dedication to the highest quality construction procedures has been a great benefit to the engineering community.

3.4 CSM (CUTTER SOIL MIXING)

3.4.1 Introduction

Cutter soil mixing (CSM) is an advanced deep mixing method becoming very popular in dam and levee construction, upgrade, and repair. Highly developed equipment and methods allow cementitious materials to be mixed with natural soils in order to construct economic high quality vertical structures for cutoff walls, soil stabilization, and earth retention.

Based on the experience gained with trench cutter technology, and with conventional DMM techniques (Figure 3.32), Bauer Maschinen and Bachy Soletanche commenced development of the CSM method in 2003. The prototype was field-tested in Germany between late 2003 and June 2004, and a patent was granted to Bauer later that year. Thereafter each of the partners proceeded individually with equipment design: Bachy Soletanche use the synonym Geomix. To date, about 150 projects have been completed around the world including Europe, Japan, New Zealand, and North America totaling several million square feet of wall. The first application in North America

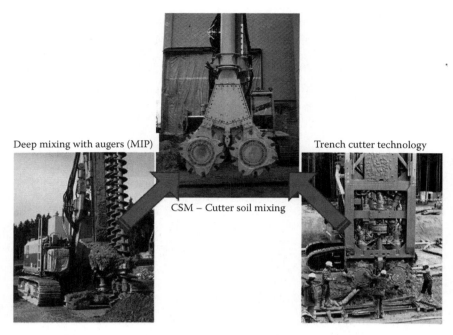

Deep mixing with augers (MIP)

Trench cutter technology

CSM – Cutter soil mixing

Figure 3.32 Origin of the CSM method. (Bauer promotional information.)

was at the Vancouver Island Conference Center in 2006, while the highest-profile current CSM project in the United States is at Herbert Hoover Dike, Florida (see Section 3.4.3).

Compared to traditional deep mixing methods, CSM provides the following advantages:

- It is suitable for most soil types including cobbles and small boulders.
- Harder soil formations can be easily penetrated, broken down, and mixed by using the cutter wheels as both the cutting and mixing axis tool.
- It homogenizes even cohesive soils with self-hardening slurry through horizontal axis mixing.
- Counter-rotating the horizontally aligned cutter wheels achieves a high degree of verticality of wall panels.
- The cutter principle ensures construction of clean and trouble-free panels and joints even between wall panels of different construction age, e.g., after weekend breaks or prolonged stoppages on site.
- No vibrations are induced during construction.
- Construction process is quiet.
- Small base units can generate high daily output and considerable panel depth.

- By producing rectangular panels, the total panel width is effectively used for the structural design and permeability (Figure 3.33).
- Spoils generation is limited, an important factor in contaminated areas.

The limitations of the CSM system are a current maximum depth of approximately 200 feet, while very plastic clay and very hard rock formations are unfavorable soil conditions, although this is a challenge common to all deep mixing methods.

3.4.2 Means, methods, materials, and properties

3.4.2.1 The Method

Traditional deep mixing methods are based on a vertical penetration and vertical mixing process: the CSM method is a vertical cutting and horizontal mixing technique. The following construction procedure is generally adopted:

1. Conventional excavation of a guide trench which is also used for slurry retention and handling.
2. Fluidization of the soil mass during penetration to the target depth. During cutting the natural soil is mixed with cementitious slurry which is induced through a nozzle located between the cutting and mixing wheels (Photo 3.14). Depending on the prevailing conditions, bentonite or cement slurry is used for the mixing and fluidization process during penetration. The rate of slurry usage is determined by the rate of cutter penetration.
3. During withdrawal, the balance of the volume of cement slurry required for producing the final wall product is added.

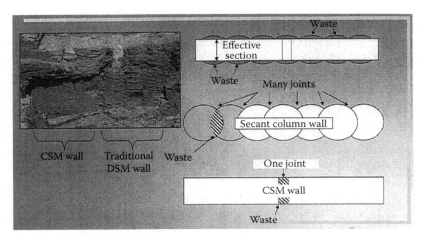

Figure 3.33 Comparison between CSM and conventional DMM products. (Bauer promotional information.)

Photo 3.14 Slurry injected between the cutting and mixing wheels of the CSM head.

3.4.2.2 Construction sequence

A typical construction sequence for the single-panel process is shown in Figure 3.34. This involves up to four steps.

Step 1: Positioning of the CSM head at the required panel location. Construction of a guide wall is not required. For soft surface conditions it is recommended to use steel plates or similar as support for the base machine.

Step 2: Fluidization of the soil mass during penetration to the final depth as an appropriate slurry is simultaneously introduced. This initial slurry volume is typically 50–75 percent of the total volume required. Depending on the prevailing conditions, bentonite, cement slurry, or just water is

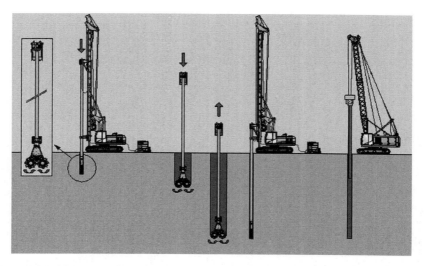

Figure 3.34 CSM working sequence for the single-panel process. (Bauer promotional information.)

added to the mixing and fluidization process during the soil penetration. The rate and volume of slurry injected is determined by the rate of cutter penetration.

Step 3: After reaching the design depth, the CSM head is extracted continuously while the remaining volume of slurry material is pumped to provide the final wall construction material. Efficient blending of the fluid soil and the fresh slurry is ensured by the counter-rotation of the wheels, which is reversed from that during penetration.

Step 4 (optional): To utilize the wall as a structural retaining wall, steel sections may be inserted into the freshly mixed wall panels.

For multipanel approaches, to form a continuous wall, individual panels are installed in an alternating sequence of overlapping primary and secondary panels. Secondary panels can be constructed immediately after completion of the primary panels, i.e., in the "fresh-in-fresh" method. Alternatively, the cutter technology also allows cutting into primary panels that have already hardened, or "fresh-in-hard" (Figure 3.35).

The overcut is influenced by the required mixing depth, soil conditions, and the panel construction sequence. A minimum overcut of approximately twelve inches is recommended for "fresh-in-fresh" sequence, and a four-inch overcut is usually used for the "fresh-in-hard" sequence (Figure 3.36). To utilize the wall as a structural retaining wall, steel columns are inserted into the freshly mixed wall panels and suspended in place until the panel has hardened. It may be observed that the lower water:cement ratio grouts will reduce the performance of delivery pumps and reduce the penetration rate of the CSM.

In applications less than approximately 65 feet deep in soft ground, a one-phase mixing procedure is preferred. In this procedure, the final slurry product consists of cement and water or a cement, bentonite, and water mixture, which is introduced in both directions—downwards and upwards. Advantages of this procedure include the additional mixing of the cement and the soil and the simplicity of only having one slurry mix, no auxiliary desanding equipment is required, and there is a higher speed of extraction. When mixing deeper panels or when penetrating difficult

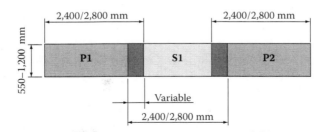

Figure 3.35 Continuous wall construction sequence layout, CSM method. (Bauer promotional information.)

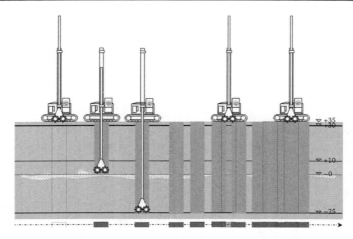

Figure 3.36 Continuous wall construction sequence, CSM method. (Bauer promotional information.)

(slow) to mix soils or rock layers, a two-phase system is used. Instead of using self-hardening slurry from the beginning as in the one-phase system, only bentonite slurry is used on the way downwards. Once the final depth is achieved, the cement slurry is introduced and the soil is mixed during extraction of the CSM tool. This method prevents the mixing head from being trapped in the panel if construction time exceeds the initial set time for the cement slurry. The water:cement ratio should not be less than 0.5. Major advantages of the two-phase system are:

- Increased safety when working at extended depths or when the working process is interrupted.
- Reduced wear and tear on the mixing wheels.
- Economic recycling of spoils (a certain percentage of the slurry can be reused, the remaining spoil can be easily removed as it is a dry material).

The size of individual panels is determined by the type and size of equipment being deployed, and the average productivity is mainly influenced by the soil conditions.

3.4.2.3 Materials

For the construction of CSM panels, cementitious slurries are used containing cement (ordinary Portland or blast furnace cement), bentonite, and water. If required, it is also possible to utilize additives (plasticizer, retarder) or admixtures (such as fly-ash) and while working with bentonite slurries for the two-phase system, polymer additives have shown good results in terms of decreasing viscosity and reducing fluid loss. The mix design should always be determined by suitability tests prior to the start of construction. The following are guideline values for an initial approximate design of the mix proportions and should be used for reference only.

Bentonite suspension (for liquefying the soil in the two-phase system):

Approximately 68 lbs bentonite per cubic yard of slurry.
306 lbs–535 lbs slurry per cubic yard of soil, less the minimum quantity for liquefying the soil.

Typical mix designs for cementitious slurries are shown in Table 3.1.
The mix design and the applicability of the system are highly dependent upon:

The application: For cutoff walls, the main characteristics are permeability, plasticity, and strength (erosion stability). For retaining walls, the main characteristics are strength, permeability, and plasticity of fresh material (as a precondition for installation of reinforcement).
The soil conditions: Particle size distribution, grain size, fines content, organic content, density, SPT values, porosity, groundwater level, and groundwater chemistry are the main influencing factors.

The suitability of different soil conditions for CSM is illustrated in Figure 3.37.

In general, the easier the soils can be fluidized, the higher the homogeneity of the panel and the higher the production. However, increased fluidity is typically dependent on the use of mixes with higher water:cement ratios. Such "wet" grouts are synonymous with lower soilcrete strength, higher permeability, and higher backflow volumes.

3.4.2.4 Wall properties

Typical properties of cured CSM material are shown in Table 3.2.

Table 3.1 Typical grout mix designs for the CSM method

Material	Cutoff walls	Retaining walls
Cement	250–450 kg/m³ slurry (425–765 lbs per cubic yard slurry)	750–1,200 kg/m³ slurry (1,275–2,050 lbs per cubic yard slurry)
Bentonite	15–30 kg/m³ slurry (25–50 lbs per cubic yard slurry)	15–30 kg/m³ slurry (25–50 lbs per cubic yard slurry)
W/C-ratio	2.0–4.0	0.5–1.0

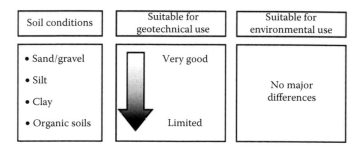

Figure 3.37 Suitability of different soils for CSM construction. (Bauer promotional information.)

3.4.2.5 Equipment and plant

The heart of the CSM technique is the CSM head. It is based on trench-cutter technology and equipped with two standard cutter gear boxes. Both gear boxes are individually driven by hydraulic motors that are located inside a sealed casing that also incorporates the instrumentation. The volume and shape of the casing are designed to enable the unhindered flow of mixed soil around it.

The wheels are designed to cut and loosen the soil matrix and then to mix it with the slurry. The soil type dictates whether more emphasis needs to be put on the wheel's cutting or mixing capabilities. In order to cover varying ground conditions, different types of wheels have been developed (Figure 3.38). Mixing wheels with four tooth holders in one row of teeth were developed for loose to dense noncohesive soils, gravelly soils, cohesive soils, and to provide good mixing efficiency. Mixing wheels with three tooth holders in one row of teeth were developed for dense noncohesive soils, gravelly soils, and hard cohesive soils, and have a good cutting capability. In very abrasive soils, bentonite added to the cement slurry will reduce friction and wear.

The CSM head is either mounted on a guided Kelly bar or on a rope-suspended cutter frame equipped with special steering devices (Photos 3.15 and 3.16). The standard setup is the "Kelly-guided" setup, capable of reaching depths up to 140 feet. "Rope-suspended" systems are particularly suited

Table 3.2 Typical CSM wall properties

Property	Cutoff walls	Retaining walls
Compressive strength	0.5–2 MPa ($\approx$70–290 psi)	5–15 MPa ($\approx$725–2,175 psi)
Permeability	1×10^{-6} cm/s or less	Not applicable
Cement content in treated soil	100–200 kg/m³ soil (170–340 lbs per cubic yard soil)	200–500 kg/m³ soil (340–850 lbs per cubic yard soil)

Figure 3.38 Different types of CSM cutting and mixing wheels. (Bauer promotional information.)

for construction of deep walls. The greatest depth to date at which a rope suspended unit has successfully installed a deep mix wall is 200 feet, using the "QuattroCutter," which mounts cutter wheels both at the base and at the top of the CSM unit (Photo 3.17).

3.4.2.6 The CSM setup

During the penetration process cutting, mixing, fluidifying, and homogenizing are performed while pumping the slurry into the soil. The slurry material is processed in a colloidal mixer and pumped directly to the CSM unit. Adding compressed air is recommended to facilitate the CSM head penetration. Backflow material unsuitable for the process is removed with an excavator. Figure 3.39 provides an illustration of the setup of a one-phase operation.

When encountering a large volume of backflow into the guide trench in a two-phase operation, the backflow material may not be pumpable. It

Photo 3.15 Kelly (round) guided CSM unit for depths up to 65 feet.

Photo 3.16 Kelly (square) guided CSM unit for depths up to 140 feet.

is then recommended to utilize a small sieve as well as a desanding unit combined with a hose pump or a scratching belt for material removal and recycling next to the trench, as illustrated in Figure 3.40.

3.4.3 QA/QC and verification

For each project QA/QC requirements are individual and therefore must be mutually agreed based on the function of the treatment. The minimum QA/QC requirements for CSM technology may be summarized as follows,

Photo 3.17 Rope-suspended CSM unit "QuattroCutter" for depths up to 200 feet.

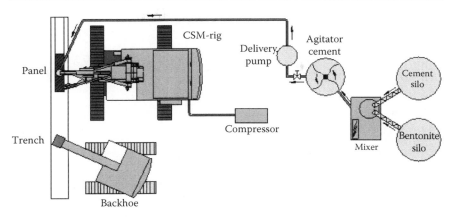

Figure 3.39 Example setup of a CSM one-phase system. (Bauer promotional information.)

and in many respects are little different from those which apply to the other mixing techniques used to construct Category 2 walls.

3.4.3.1 Basics for trial mixes

Before CSM can commence, trial mixes with representative soil samples are executed and tested in a laboratory. For a preliminary mix design, the following soil information shall be made available:

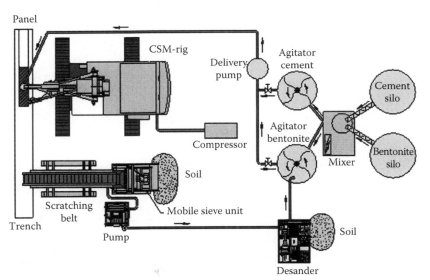

Figure 3.40 Example setup of a CSM two-phase system with scratching belt and mobile desanding unit. (Bauer promotional information.)

- Soil type
- Soil consistency (SPT)
- Density
- Grain size distribution
- Water content
- Atterberg limits
- Soil chemistry

3.4.3.2 Laboratory testing

After a preliminary mix is selected, a number of material trial mixes are prepared, cured, and tested for strength and permeability under laboratory conditions providing some guidance on the soilcrete quality that can be achieved in the field. Based on these data, an initial CSM parameter range can be established. It is important to note that the *in-situ* mix varies with varying soil conditions and therefore CSM parameters may be subject to adjustment during the trial panel phase as well as during the production phase.

3.4.3.3 Trial panels

It is recommended to mix a number of trial panels on site within the established parameter range for flow and production rates. Testing during and after their construction provides reliable quality information on the mixed panels.

3.4.3.4 Testing during construction

During the CSM process, the following tests are recommended:

- Optical survey on individual panel positions.
- Fresh slurry mix testing of samples taken at the storage tank:
  - Density
  - Viscosity (Marsh Cone)
  - Unconfined Compressive Strength
- Soil slurry mix testing of samples taken at different panel elevations:
  - Unconfined compressive strength
  - Wet density
  - Permeability
- Control of production parameters displayed on the CSM operator's monitor:
  - Depth
  - Flow rate and total volume of slurry
  - Slurry pressure

- Slurry-soil pressure in the panel
- Pumped slurry volume vs. time
- Pumped slurry volume vs. depth
- Inclination in "X" and "Y" directions (to a tolerance of 0.2 percent)
- Rotation speed of mixing wheels, and their torque
- General equipment operating data

3.4.3.5 After construction

The following tests can be carried out to monitor the geometric and mechanical characteristics of the CSM panels:

1. After finishing a panel, a double-wall PVC tube can be installed into the fresh mixed soil. After the mix is cured, the inner tube is extracted in order to obtain a continuous vertical sample of hardened soil-slurry mix tested for unconfined compressive strength, permeability and homogeneity.
2. Coring at select locations will provide samples for inspection and testing. These, of course, may be damaged to some extent by the invasive act of coring and may not yield truly representative results. Cored holes can then be subjected to *in-situ* inspection by optical or acoustic televiewers, and permeability testing.
3. If the panels can be exposed, they can be physically inspected and measured, and bulk samples can be taken to test for strength and permeability.
4. Static and dynamic load tests can be carried out on panels that are designed to act as load-bearing elements.
5. Sonic tests can be carried out on the panels to verify continuity of the treatment and to measure the increase in the mechanical characteristics of the treated soil.

3.4.3.6 Documentation

All production parameters are monitored, recorded, and stored inside the rig throughout the construction process and can be printed out in the form of a quality assurance record for each panel, such as shown in Figure 3.41.

3.4.4 Illustrative case history: Herbert Hoover Dike Reach I, Florida

Background information on the evolution of Herbert Hoover Dike is provided in Section 3.3.5.1. One contractor proposed the CSM method for the

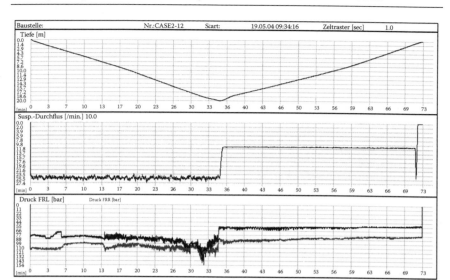

Figure 3.41 Printout of an instrumented CSM panel installation. (Bauer promotional information.)

construction of 19,000 lineal feet of cutoff wall to a depth of about 65 feet as part of the 32-mile-long Reach 1. The obligations and responsibilities placed on the CSM contractor—which were common to all the contractors and their respective methods—were clear. The wall had to satisfy the following requirements:

- To be continuous and homogeneous.
- To be a maximum 36 inches and a minimum of 18 inches wide for the entire depth including at overlaps between panels.
- To have an in-place permeability of less than 1×10^{-6} cm/s.
- To have a minimum unconfined compressive strength of 100 psi and a maximum of 500 psi on core samples.
- To be vertical.
- To be chemically compatible with groundwater and soil conditions.

All these performance parameters had to be proved via a quality control and assurance program featuring:

- Daily bulk sampling and deep-grab sampling from the mixed panels at designated depths.
- Core drilling every 200 feet along the wall alignment, including:

- color photos and high-resolution video logging of the entire depth of the cored hole;
- a recovery of at least 95 percent;
- inclinometer readings to verify verticality;
- falling head *in-situ* permeability testing;
- an assessment of the homogeneity of the wall.

Five-hundred-foot-long demonstration sections were first conducted in order to demonstrate to the government's satisfaction that the contractor's means, methods, and materials were capable of satisfying the project goals. Following experimentation with various mix designs to enhance the stability of the grout under self-weight, the results of the demonstration sections were found acceptable. The production wall is now complete. Table 3.3 summarizes the average composition of three main mixes which were developed.

Predictably, mixes with higher water:cement ratio gave samples with a higher moisture content, 28-day strengths (averaging 380 psi for bulk samples, 315–429 psi for deep grab samples, and 276–478 psi for cores) and more consistent results, reflective of the higher volume ratio. All sample and *in-situ* permeability tests were compliant, often by one or two orders. Due to consolidation of the soilcrete when fluid, it was typical to find moisture content decreasing with depth, and wet density and strength increasing with depth. Long-term strength testing confirmed that these slag-rich mixes continued to gain appreciable strength even over 200 days.

Considering the potential for a negative impact on wall homogeneity and strength, the contractor elected to remove substantial peat layers and replace them with controlled quality granular backfill prior to mixing with the CSM. The cores taken from these horizons, and at all the other elevations in different lithologies, proved to have exceptional quality, homogeneity, and uniformity (Photo 3.18).

Table 3.3 Average mixes (per cubic meter of slurry) used for the CSM work at Herbert Hoover Dike, Florida, Task Order 2

Component	Composition average
Type I cement (kg)	10–60
Slag (kg)	250–62
Bentonite (kg)	20–30
Eucon[a] LR (kg)	0–5
Water (kg)	780–870

[a] Lignosulfonate (i.e., plasticizer).

Photo 3.18 Core recovery and samples obtained during CSM demonstration section coring, Herbert Hoover Dike, Florida.

3.5 OVERVIEW ON CATEGORY 2 WALLS

The rate and diversity of developments in the techniques used to deliver Category 2 walls in the last ten years or so have been astounding. This reflects the skills and inventiveness of motivated, competitive construction companies throughout the world, as much as it emphasizes again the scope and complexity of the market applications. And further, the pace shows no sign of slacking, as is illustrated by the developments made "on the fly" by the DMM contractors currently involved in the huge levee restoration works in New Orleans, unfortunately unpublished at the time of this publication.* One federal project alone—LPV 111—features the round-the-clock deployment of eight simple and twin axis machines using specially developed WJE and WRE methods, and as much as 13,000 tons of binder per calendar week. Construction data are relayed directly from the cabs of these machines to the site's "mission control," as well as to head offices. The quality of the mixed material, even and especially in the fibrous organic and plastic clay horizons is, to the eyes of the most experienced observers, of an unprecedented standard.

Levee remediation work continues apace in other critical projects in California and Florida while more sporadic but still major opportunities for Category 2 walls continue to present themselves throughout the country or dam remediation and even for ash-storage confinement structures. All the more reason, therefore, to remain pragmatic about the relative and systematic advantages, disadvantages, and limitations of the major groups of "soilcrete technologies." For example, given that a major

---

* The reader is referred to the proceedings of the Fourth International Conference on Grouting and Deep Mixing, held in New Orleans, February 2012 (DFI 2012).

component of a Category 2 wall is the virgin soil itself, it is important to recall the fundamental challenges posed by the various basic types of soil. As examples:

Organic Soils: If containing strong, fibrous root structures, they can prove very difficult to mix mechanically, and especially if dry DMM is used. Furthermore, their low pH environment acts against the hydration reaction chemistries, and so will try to defeat strength gain. Organic components, if left unmixed, will typically decompose and/or erode with time, therefore compromising the durability of a mixed wall, especially under sustained hydraulic loading.

Cohesive Soils: Given their high specific surface area, more slurry is needed to efficiently and thoroughly fluidize them during the mixing process. Mixing wheels and blades become clogged with unmixed clumps of soil. Therefore, reduced penetration rate, and so, therefore, a particularly high BRN (blade rotation number) is required to assure dispersion and therefore homogeneity of the final product.

Granular Soils without Fines: Conceptually this should constitute the best aggregate and therefore the highest strength, exactly as is the case with concrete. However, such soils, encompassing as they will, very coarse gravelly beds of very high permeability, may allow slurry loss, and the threat of "washout" of fluid soilcrete under differential head. Even "nice," uniform sand deposits may be problematical occasions: they will be delightful to mix but, due to the reality of the pressure filtration phenomenon (Weaver and Bruce 2007) in unstable grouts, they will provide soilcrete of an *in-situ* strength far in excess of that predicted or perhaps desired. In simple terms, they will draw the moisture out of the fluid soilcrete and, by reducing its total water:cement ratio, will therefore guarantee surprisingly high strengths in the remnant soilcrete. Whereas this may be an added bonus in certain applications, excessive strength (and stiffness) may not be desirable or acceptable in other applications.

Groundwater Characteristics: As noted above, the chemistry has the potential to adversely influence soilcrete behavior and properties during and after placement. Its velocity has the potential to cause washout of uncured soilcrete.

Each of the three major groups of technologies described in this chapter has its own particular strengths and advantages. These are summarized in Table 3.4, which also provides indicative pricing information. Unit rates for walls constructed by Category 2 methods will typically be lower when the walls are between 30 and 100 feet deep; when they are designed in long, straight, or sweeping geometries; when the soil is relatively uniform and consistent; and when the soil is not very dense, cobbly, organic, or very plastic. However, as described in this chapter, none of these factors may be regarded as rendering Category 2 walls *unbuildable*—they simply constitute *unfavorable* conditions.

Table 3.4 Summary of Category 2 techniques characteristics

Method	Principle	Wall dimensions		Typical properties of soilcrete		Costs		Relative benefits/problems	
		Depth	Width	UCS	K	Mob/demob	Unit prod.	Pros	Cons
Category 2—Mix in place									
Conventional DMM	Vertically mounted shafts are rotated into the soil creating panels of soilcrete	Maximum practical about 120'	20–40"	100–1,500 psi	5×10^{-6} to 1×10^{-8} cm/s	Moderate–High	Low in sympathetic soil conditions to Moderate/High in difficult conditions	Low vibrations and noise Experience Several practitioners in U.S. High productivity Good homogeneity	Large equipment needs good access and substantial headroom Depth limitation Very sensitive to obstructions Variable homogeneity with depth due to limited vertical mixing
TRD	Vertical chainsaw providing simultaneous cutting and mixing of soil to produce continuous soilcrete wall	Maximum 170'	18–34"	100–3,000 psi	10^{-6} to 10^{-8} cm/s	Moderate–High	Low–Moderate	Continuity of cutoff is automatically assured (no joints) Homogeneity (especially vertically) Productivity Quality Quick adaptability to wide range of ground conditions Low noise and vibrations Low headroom potential (20') Inclined diaphragms possible Wide range in cutoff properties can be closely engineered Very high degree of real time QC Relatively compact equipment	Difficult wall geometries (sharp turns) Medium-hard rock, and boulder nests (will reduce productivity and increase wear on key components) Currently only one U.S. contractor Requires very specialized equipment Cutting post may become trapped in the wall or may "refuse" on nests of boulders or hard rock

| CSM | Cutting and mixing wheels mounted on horizontal axes create vertical soilcrete panels | Typically 140' with Kelly but maximum 200' with cable suspension | 22–47" with trials to 60" | 70–3,000 psi | 10^{-6} to 10^{-8} cm/s | Low–Moderate | Moderate | Panel continuity/verticality closely controlled
Homogeneity
Productivity
Adaptable to conventional base carriers
Wide range in cutoff properties can be engineered
Can accommodate sharp geometry changes
Applicable in nearly all soil conditions
Cutting teeth can be quickly adapted
Relatively quiet and vibration-free
Fewer joints than conventional DMM
Can change parameters for different soil types
Low headroom potential (15' min) | Rock, boulders, and other obstructions
Requires considerable headroom |

Key to costs (2010 Figures)

Mobi/Demob		Unit costs (i.e., cost per square foot of cutoff)	
<$50,000	Very Low	<$10	Very Low
$50,000–$150,000	Low	$10–$20	Low
$150,000–$300,000	Moderate	$20–$50	Moderate
$300,000–$500,000	High	$50–$100	High
>$500,000	Very High	>$100	Very High

REFERENCES

AK Chemical Co., Ltd. 2006. "TRD Performance Records in Japan and USA." Internal Publication, Prepared for Hayward Baker.

Aoi, M., F. Kinoshita, S. Ashida, H. Kondo, Y. Nakajim, and M. Mizutani. 1998. "Diaphragm Wall Continuous Excavation Method: TRD Method." *Kobelco Technology Review* 21:44–47.

Aoi, M., T. Komoto, and S. Ashida. 1996. "Application of TRD Method to Waste Treatment on the Earth." In *Environmental Geotechnics*, edited by M. Kamon, 1:437–40. Rotterdam, the Netherlands: Balkema.

Barneich, J. 1999. "Alternative Seawater Barrier Feasibility Study." Draft report prepared by URS Greiner Woodward Clyde for the Los Angeles Department of Public Works and the U.S. Bureau of Reclamation 32.

Barron, R. F., C. Kramer, W. A. Herlache, J. Wright, H. Fung, and C. Liu. 2006. "Cement Deep Soil Mixing Remediation of Sunset North Basin Dam." *Proceedings of Dam Safety 2006, the 23rd Annual Conference of the Association of State Dam Safety Officials* (http://www.damsafety.org), Massachusetts, September, pp. 181–199.

Burke, G. K. 1992. "In Situ Containment Barriers and Collection Systems for Environmental Applications." Proceedings from Symposium on Slurry Walls, Canadian Geotechnical Society, Southern Ontario Section, Toronto, Canada, September 24.

Castro, G., T. O. Keller, and S. S. Boynton. 1989. "Re-Evaluation of the Lower San Fernando Dam, Report 1: An Investigation of the February 9, 1971 Slide." Report No. GO-89-2, U.S. Army Corps of Engineers, WES.

Cedergren, Harry R. 1988. *Seepage, Drainage and Flow Nets*. 3rd ed. New York: John Wiley and Sons.

Coastal Development Institute of Technology (CDIT), ed. 2002. *The Deep Mixing Method: Principle, Design and Construction*. Lisse, the Netherlands: Swets & Zeitlinger.

Davis, J. R., E. D. Guy, and R. L. Nettles. 2009. "Preferred Risk Reduction Alternative for Reach 1A of Herbert Hoover Dike." *Proceedings of Dam Safety 2009*. Lexington, KY: Association of State Dam Safety Officials.

Deep Foundations Institute (DFI). 2012. 4th International Conference on Grouting and Deep Mixing. http://www.grout2012.org/

Evans, J. C. 2007. "The TRD Method: Slag-Cement Materials for *in situ* Mixed Vertical Barriers." Conference Proceedings, *GeoDenver 2007*, February.

Federal Highway Administration. 2000a. "An Introduction to the Deep Mixing Method as Used in Geotechnical Applications." Prepared by Geosystems, L.P., Document No. FHWA-RD-99-138, March, 143 pp.

Federal Highway Administration. 2000b. "Supplemental Reference Appendices for an Introduction to the Deep Mixing Method as Used in Geotechnical Applications." Prepared by Geosystems, L.P., Document No. FHWA-RD-99-144, December, 295 pp.

Federal Highway Administration. 2001. "An Introduction to the Deep Mixing Method as Used in Geotechnical Applications: Verification and Properties of Treated Soil." Prepared by Geosystems, L.P., Document No. FHWA-RD-99-167, October, 455 pp.

Fordham, E. 2006. "Flow Modeling as a Tool for Designing and Evaluating a Seawater Barrier Demonstration Project." *MODFLOW and More 2006: Managing Ground-Water Systems Conference Proceedings*, Poeter, Hill & Zheng eds., International Ground Water Modeling Center, pp. 796–800.

Garbin, E. J., J. C. Evans, and J. D. Hussin. 2009. *Trench Cutting Remixing Deep (TRD) Method for Vertically Mixed-in-Place Cutoff Wall Construction at Herbert Hoover Dike*, Proceedings of Dam Safety 2009, Association of State Dam Safety Officials.

Geomatrix Consultants. 2004. "Dynamic Stability Analysis of San Pablo Dam, Contra Costa County, California," vols. 1–2. October 22.

Geomatrix Consultants. 2005. "Refined Conceptual Design Report: Seismic Modifications for San Pablo Dam, Contra Costa County, California." August 11.

Gibbons, Y. R., and G. J. Buechel. 2001. "Lewiston Levees DSM Wall Construction." Paper presented at the 18th Annual Conference of the Association of State Dam Safety Officials (http://www.damsafety.org), Snowbird, UT, 11 pp.

Gularte, F., E. Fordham, D. Watt, J. Weeks, and T. Johnson. 2007. "First Use of TRD Construction Technique for Soil Mix Cutoff Wall Construction in the United States" *Proceedings of GeoDenver 2007*.

Hayward Baker, Psomas and Geopentech. 2006. "Final Report Alamitos Physical Gap Project March 2006." For the U.S. Bureau of Reclamation and the Water Replenishment District. http://www.wrd.org/Project_Alamitos_Deep_Soil_Study.htm.

Namikawa, T., J. Koseki, and Y. Suzuki. 2007. "Finite Element Analysis of Lattice-Shaped Ground Improvement by Cement-Mixing for Liquefaction Mitigation." *Soils and Foundations* 47 (3): 559–76 (in Japanese).

Olivia Chen Consultants Report. 2004. "Geotechnical Investigation Embankment Stability Evaluation." Sunset Reservoir North Basin San Francisco Public Utilities Commission, San Francisco, California, February, 37 pp.

Poulos, S. J. 1981. "The Steady State of Deformation," *Journal of Geotechnical Engineering Division, ASCE* 107 (GT5): 5513–62.

Ryan, C. R., and B. H. Jasperse. 1989. "Deep Soil Mixing at Jackson Lake Dam." *Proceedings of the ASCE 1989 Foundation Engineering Congress, Foundation Engineering: Current Principles and Practices*, Evanston, IL, June 25–29, 1–2:354–67.

Sehgal, C. K., S. H. Fischer, and R. G. Sabri. 1992. "Cushman Spillway Modification." Paper presented at USCOLD Conference, Fort Worth, TX, 25 pp.

Suzuki, Y., S. Saito, S. Onimaru, T. Kimura, A. Uchida, and R. Okumura. 1996. "Grid-shaped Stabilized Ground Improved by Deep Cement Mixing Method against Liquefaction for Building Foundation." *Tsuchi-to-Kiso* 44 (3): 46–48 (in Japanese).

TNM. 2007a. Design Strength of CDSM, Design Memo DM-09 by J. Barneich, Design of San Pablo Dam Seismic Upgrades, East Bay Municipal Utility District. TNM, a Joint Venture of Terra Engineers and Ninyo & Moore, October, 13 pp.

TNM. 2007b. Results of CDSM Test Section Report for East Bay Municipal Utility District. Design Services for San Pablo Dam Seismic Upgrades. TNM, a Joint Venture of Terra Engineers and Ninyo and Moore, August, 59 pp.

TNM. 2008. Design of CDSM, Design Memo DM-11 by Kirby, R., Design of San Pablo Dam Seismic Upgrades, East Bay Municipal Utility District. TNM, a Joint Venture of Terra Engineers and Ninyo and Moore, January, 10 pp.

U.S. Army Corps of Engineers, Jacksonville District. 2007. http://www.saj.usace.army.mil/Divisions/Everglades/Branches/HHDProject/HHD.htm.

Walker, A. D. 1994. "A Deep Soil Mix Cutoff Wall at Lockington Dam in Ohio." *In Situ Deep Soil Improvement.* American Society of Civil Engineers, Geotechnical Special Publication 45, pp. 133–46.

Weaver, K. D., and D. A. Bruce. 2007. *Dam Foundation and Grouting, Revised and Expanded Edition.* New York: ASCE Press.

Wooten, R. L., G. Castro, G. Gregory, and B. Foreman. 2003. "Evaluation and Design of Seismic Remediation for the Clemson Upper and Lower Diversion Dams." *Proceedings of the 23rd Annual U.S. Society on Dams Conference,* pp. 613–622. Charleston, SC, April.

Wooten, R. L., and B. Foreman. 2005. "Deep Soil Mixing for Seismic Remediation of the Clemson Upper and Lower Diversion Dams." *Proceedings of the 25th Annual U.S. Society on Dams Conference,* pp. 385–400. Salt Lake City, UT.

Yang, D. S. 2008. "Application of CDSM for Embankment Stabilization." *Fifth International Conference on Landslides, Slope Stability and the Safety of Infrastructures,* pp. 177–184. Kuala Lumpur, Malaysia: CI-Premier.

Yang, D. S., and S. Takeshima. 1994. "Soil Mix Walls in Difficult Ground." American Society of Civil Engineers National Convention, Atlanta, Georgia, October, 15 pp.

Yonekura, R., M. Terashi, and M. Shibazaki, eds. 1996. "Grouting and Deep Mixing." *Proceedings of the Conference IS-Tokyo Vols. 1 and 2.* Rotterdam, the Netherlands: A.A. Balkema.

Chapter 4

Excavated and backfilled cutoffs (Category 1)

Brian Jasperse, Maurizio Siepi, and Donald A. Bruce

4.1 PERSPECTIVE

Chapter 1 introduced the concept of Category 1 cutoff walls, which are built in the ground by first excavating the native material and then replacing it with an engineered "backfill," typically cement based. During the excavation phase, the trench or panel must usually be stabilized against collapse by employing a bentonite or polymer slurry. Only when the cutoff is being built in rock by the secant pile method is it not necessary to stabilize the excavation with this supporting slurry, although other methods, such as full-length, temporary casing, are required in extreme conditions.

In an earlier review, Bruce, Ressi di Cervia, and Amos-Venti (2006) reported on twenty-two North American dams that had been remediated against foundation seepage between 1975 and 2004 with a Category 1 cutoff. Since then, several other projects have been initiated, associated with the USACE initiatives to remediate embankment dams founded on karstic limestone foundations. Indeed, it may be estimated that the dollar value of Category 1 cutoffs under construction between 2008 and 2013 will exceed the aggregate cost (in current dollars) of all the preceding works.

Table 4.1 provides summary details of the projects for which information has been compiled to date. Figure 4.1 illustrates the progression of construction since the first remedial cutoff at Wolf Creek Dam, Kentucky, in 1975.

Most levee cutoff structures have been built in a continuous and longitudinally progressive fashion using a long-reach backhoe. This is the subject of Section 4.2. In contrast, cutoffs for dams are usually deeper and extend into rock and are mainly constructed by the panel method using clamshells and/or hydromills as the excavating tools. These methods are described in Section 4.3. In certain geological conditions, or when driven by overriding dam safety concerns, dam cutoffs are also constructed using overlapping large-diameter columns, installed in the classic "primary-secondary" fashion or by the "slot" method, typified by the Arapuni Dam, New Zealand, case history (Gillon and Bruce 2003). Such walls are described in Section 4.4. The reader will note that Section 4.2 contains much information that

Table 4.1 Details of remedial dam cutoff projects using Category 1 panel or secant concrete cutoffs, 1975–2005

Dam name and year of remediation	Contractor	Type of wall	Composition of wall	Ground conditions	Purpose of wall	Scope of project				References
						Area	Min. width	Depth	Length	
1. Wolf Creek, KY, 1975–79	ICOS (Phase 1)	24"-dia primary piles, joined by 24"-wide clamshell panels. Two phases of work.	Concrete	Dam FILL, and ALLUVIUM over argillaceous and karstic LIMESTONE with cavities, often clay-filled	To provide a "positive concrete cutoff" through dam and into bedrock to stop seepage, progressively developing in the karst	270,000 sf (Phase 1) plus 261,000 sf (Phase 2)	24"	Max. 280'	2,000' plus 1,250'	ICOS brochures (n.d.) Fetzer 1988
2. W.F. George, AL:										
1981	Soletanche (Phase 1)	26"-thick panels using cable and Kelly-mounted clamshell	Plastic concrete	Random, impervious FILL with silty core over 25–30'	To provide a "positive concrete cutoff" through the dam and alluvials	130,000 sf (Phase 1) plus	26"	Max 138'	Approx. 1,000'	Soletanche brochure (undated)
1983–85	Bencor-Petrifond (Phase 2)	24" panels 15–27' long	3,000-psi concrete	ALLUVIUM over chalky LIMESTONE		951,000 sf (Phase 2)	24"	110–190'	8,000'	Bencor brochure (undated)
3. Addicks and Barker, TX Completed in 1982 (Phase 1 took 5 months)	Soletanche	36"-thick panel wall with clamshell excavation using Kelly.	Soil-bentonite	Dam FILL over CLAY	To prevent seepage and piping through core	450,000 sf (Phase 1) plus 730,000 sf (Phase 2)	36"	Max 66' typically 35 to 52'	8,330' plus 12,900'	Soletanche website

4. St. Stephen, SC, 1984	Soletanche	Concrete and soil-bentonite	24"-thick concrete panel wall, installed by hydromill. Plus upstream joint protection by soil-bentonite panels.	Dam FILL, over sandy marly SHALE	To provide a cutoff through dam	78,600 sf (concrete) plus 28,000 sf (soil-bentonite)	24"	Max 120' including 3' into shale	695'	USACE report 1984, Soletanche (various), Parkinson 1986, Bruce et al. 1989
5. Fontenelle, WY, 1986–88	Soletanche	Concrete and soil-bentonite	24"-thick concrete panel wall installed by hydromill. Minor soil-bentonite panels.	Dam FILL over horizontally bedded SANDSTONE	To prevent piping of core into permeable sandstone abutment	50,000 sf (LA test) plus 100,000 sf (RA test) plus 700,000 sf (production)	24"	Max 180' including 16–160' into rock	Approx. 6,000'	Cyganiewicz 1988, Soletanche (various)
6. Navajo, NM, 1987–88	Soletanche	Concrete	39"-thick panel wall installed by hydromill	Dam FILL over flat-lying SANDSTONE with layers of SILTSTONE and SHALES. Very fractured, weathered and permeable with vertical and horizontal joints.	To prevent piping of core into permeable sandstone abutment	130,000 sf	39"	Max 400' including over 50' into rock	450'	Davidson 1990, Dewey 1988

continued

Table 4.1 Details of remedial dam cutoff projects using Category 1 panel or secant concrete cutoffs, 1975–2005 (Continued)

Dam name and year of remediation	Contractor	Type of wall	Composition of wall	Ground conditions	Purpose of wall	Scope of project				References
						Area	Min. width	Depth	Length	
7. Mud Mountain, WA, 1988–90 (mainly over Dec. 1989–April 1990)	Soletanche	33- and 39'-thick panel wall installed by hydromill. Extensive pregrouting of core.	Concrete	Dam FILL silty and sandy, over very hard, blocky cemented ANDESITE (UCS over 20,000 psi)	To prevent seepage through the core	133,000 sf	33" in abutments, 39" in center	Max 402'	700	Soletanche brochures Eckerlin 1993 ENR 1990 Davidson, Levallois, and Graybeal 1991 Graybeal and Levallois 1991
8. Stewart's Bridge Hydro, NY, 1990	ICANDA-ICOS	30" panel walls installed by clamshell	Plastic concrete	FILL over stratified SAND	To prevent seepage through and under the dam	95,000 sf	30"	Max 110'	About 1,000'	DiCicco (personal communication, 2007)
9. Wister, OK, 1990–91 (6 mos.)	Bauer	24"-thick panel wall installed by hydromill	Plastic concrete	Dam FILL, over 30' of ALLUVIALS overlying SANDSTONE and SHALE	To prevent piping through the embankment	216,000 sf	24"	Approx. 54'	Approx. 4,000'	Erwin 1994 Erwin and Glenn 1991
10. Wells, WA, 1990–91 (7 mos, 208 working shifts)	Bencor-Petrifond	30"-thick panel wall installed by clamshell and joint pipe ends	Concrete	Dam FILL with permeable zones over miscellaneous ALLUVIUM and very dense TILL	Prevent piping through permeable core materials, in gap between original cutoff and rockhead	124,320 sf	30"	80–223'	849'	Kulesza et al. 1994 Roberts and Ho 1991

Project	Contractor	Material	Description	Purpose	Geology	Area	Width	Depth	Length	Reference
11. Beaver, AR, 1992–94 (22 mos)	Rodio-Nicholson	Concrete	24"-thick wall created by 34" secant columns at 24" centers	To prevent seepage through karstic limestone under embankment	Dam FILL over very variable and permeable karstic LIMESTONE with open and clay-filled cavities. Some sandstone.	207,700 sf	24"	80–185'	1,475'	Bruce and Dugnani 1996, Bruce and Stefani 1996
12. Meek's Cabin, WY, 1993	Bauer	Plastic concrete	36"-thick panel wall formed by hydromill	To prevent seepage through glacial outwash deposits	Dam FILL over very variable glacial TILL and OUTWASH comprising sand, gravel, cobbles, and boulders.	125,000 sf	36"	130–170' including minimum 10' into lower glacial till	825'	Pagano and Pache 1995, Gagliardi and Routh 1993
13. McAlpine Locks and Dam, KY, 1994 (6 mos)	ICOS	Concrete	24" panel wall formed by clamshell and chisel. Upper portion pretrenched with backhoe and filled with cement-bentonite.	To prevent seepage through dike and alluvials	Very variable FILL, with rubble, cobbles, and boulders over silty CLAY over SHALE and LIMESTONE	51,000 sf	24"	30–90' plus 5' into bedrock	850'	Murray 1994
14. Twin Buttes, TX, 1996–99	Granite-Bencor-Petrifond	Soil-cement-bentonite	30"-wide wall formed with panel methods (Kelly and cable-suspended grabs, plus chisels.) Hydromill also used.	To prevent seepage through dam foundation causing uplift or blowout	Dam FILL over CLAY and ALLUVIAL gravel often highly cemented (up to 15,000 psi) and SHALEY SAND-STONE bedrock	1,400,000 sf	30"	Max 100' deep including at least 2.5' into rock	21,000'	Dinneen and Sheskier 1997

continued

Table 4.1 Details of remedial dam cutoff projects using Category 1 panel or secant concrete cutoffs, 1975–2005 (Continued)

Dam name and year of remediation	Contractor	Type of wall	Composition of wall	Ground conditions	Purpose of wall	Scope of project				References
						Area	Min. width	Depth	Length	
15. Hodges Village, MA, 1997–99	Bauer of America	31.5"-wide panel wall formed by hydromill plus use of grab for initial excavation and boulder removal.	Concrete	Homogeneous permeable sand and gravel FILL over stratified glacial OUTWASH sands, gravels and cobbles over MICA SCHIST	To prevent seepage through the dam and foundation	185,021 sf (dam) plus 75,486 sf (dike)	31.5"	Max 143' (avg. 89') including 5' into rock (dam) plus max 77' (avg. 57') including 5' into rock (dike) 20–75'	2,078' (dam) plus 1,315' (dike)	Dunbar and Sheahan 1999 USACE 2005
16. Cleveland, BC, 2001–2 (4 mos)	Petrifond and Vancouver Pile Driving	32" panel wall constructed by cable-suspended clamshell	Plastic concrete	Heterogeneous glacial sediments including SILT, SAND, GRAVEL and TILL with hard igneous boulders	To prevent seepage through glacial and interglacial foundation sediments, especially 20-ft sand layer	55,000 sf (est)	32"	20–75'	1,004'	Singh et al. 2005
17. West Hill, MA, 2001–2	Soletanche-McManus JV	31.5" panel wall constructed by hydromill plus clamshell for pre-excavation.	Concrete	Zoned random and impermeable FILL over permeable stratified sand and gravel. Glacial OUTWASH over GRANITITE GNEISS	To prevent seepage through the foundation	143,000 sf	31.5"	Max 120' including 2' into rock (avg. 69')	2,083'	USACE 2004

18.	W.F. George, AL, 2001–3	Treviicos-Rodio	24"-thick secant pile wall (50" dia at 33" centers) plus hydromill through concrete structures	Concrete	Over 90' of water over LIMESTONE with light karst, and very soft horizons (rock strength over 14,000 psi in places)	To prevent seepage through karstified bedrock under concrete dam section	Approx. 300,000 sf including hydromill wall (50,000 sf)	24"	100' of excavation (under 90' water)	2,040'	Ressi 2003 Ressi 2005
19.	Mississinewa, IN, 2001–5 (including shutdown for grouting)	Bencor-Petrifond	18"-thick panel wall, using 30" hydromill, with clamshells through dam	Concrete	Dam FILL over karstic LIMESTONE (to 25,000 psi), very permeable and jointed	To prevent piping into karstic limestone foundation	427,308 sf (330,127 embankment plus 97,181 in rock)	30"	123–230' (including max 148' into limestone) av. 180'	2,600'	Hornbeck and Henn 2001 Henn and Brosi 2005
20.	Waterbury, VT, 2003–5	Raito	18"-thick secant wall using 79" piles at about 66" centers	Concrete in lower 100', permeable backfill above	Zoned EARTHFILL dam over permeable GLACIAL DEPOSITS with detached rock slabs and MICA SCHIST	To prevent seepage and piping in the glacial gorge	20,000 sf designed but only 13 of 22 piles completed	18"	125–180' (avg. 163') with "anomaly" to 222', including 10' into rock	124' designed	USACE 2006 Washington, Rodriguez, and Ogunro 2005

Source: Bruce, D.A., A. Ressi di Cervia, and J. Amos-Venti, "Seepage Remediation by Positive Cutoff Walls: A Compendium and Analysis of North American Case Histories," ASDSO Dam Safety, Boston, MA, 2006.

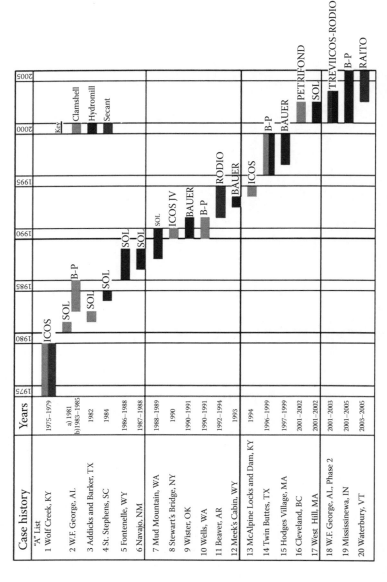

Figure 4.1 Progression of construction since the first remedial cutoff at Wolf Creek Dam, Kentucky, in 1975. BAUER = Bauer Spezialteifbau GmbH; B-P = Bencor-Petrifond; ICOS = TrevilCOS; PETRIFOND = Petrifond; RAITO = Raito, Inc.; RODIO = Rodio; SOL = Soletanche Bachy. (From Bruce, D. A., T. L. Dreese, and D. M. Heenan, "Concrete Walls and Grout Curtains in the Twenty-First Century: The Concept of Composite Cutoffs for Seepage Control," USSD 2008 Conference, 2008. With permission.)

is applicable to the subject matter of Sections 4.3 and 4.4 also. Section 4.2 is therefore presented in much more detail than its subsequent comparable sections, whereas the "art" of secant pile cutoffs is illustrated by describing the details of several case histories.

The intrinsic advantage of Category 1 walls is that the resultant cutoff material (i.e., the backfill) can be engineered to provide an extremely wide range of properties, independent of the native material through which the cutoff is to be excavated. This facility is so fundamental that the actual cutoffs are primarily named after the backfill materials themselves, as opposed to the method of excavation. These materials are:

- conventional concrete walls
- plastic walls
- cement-bentonite walls (CB)
- soil-bentonite walls (SB)
- soil-cement-bentonite walls (SCB)

In all cases except CB walls, excavation is conducted under bentonite (or polymer) slurry, which is thereafter displaced out of the trench or panel by the backfill material of choice. It is generally believed that the concept of excavating under a bentonitic supporting slurry was first developed by Veder, in Austria, in 1938. The relationship between backfill material and excavation method is summarized below:

Type of backfill	Excavation method			
	Backhoe	Clamshell	Hydromill	Secant piles
Conventional concrete	Not feasible	Typical	Typical	Typical
Plastic concrete	Not conducted	Feasible	Feasible	Rare
CB	Common	Feasible	Feasible	Not conducted
SB	Very common	Not conducted	Not conducted	Not conducted
SCB	Common	Very rare	Very rare	Not conducted

Details of the different excavation methods are provided in older, fundamental texts such as Xanthakos (1979), and ASTM (1992), while Bruce, Dreese, and Heenan (2008) summarize case histories of more recent vintage. Much valuable information can also be obtained from the websites of the major specialty geotechnical contractors and the equipment suppliers.

It is often the case that all three techniques may be used on the same project: the backhoe to excavate a "pre-trench," say 20–40 feet deep, the clamshell to excavate through fill or soil, and the hydromill to cut into the underlying or adjacent rock. Furthermore, and as an example, the current cutoff wall being installed at Wolf Creek Dam, Kentucky, features a combination of panel wall (by clamshell and hydromill) and secant pile technologies; such are the challenges posed by the geological conditions and dam safety concerns during construction.

4.2 CUTOFFS CONSTRUCTED BY THE BACKHOE METHOD*

4.2.1 Wall types

4.2.1.1 General background

According to Xanthakos (1979), the first slurry trench cutoff was "probably" built at Terminal Island, near Long Beach, California, in 1948. It was forty-five feet deep and backfilled with soil. Ryan and Day (2003) reported that "thousands" of such walls have been built in the United States since the early 1970s, predominately backfilled with soil-bentonite. The technique is fundamentally very simple: a long-reach bucket excavator (backhoe) is used to dig a long slot in the soil (Figure 4.2) which is temporarily supported by bentonite slurry. Backfilling with SB or SCB is conducted progressively, with reuse of the excavated soil(s) always preferred if at all possible, for simplicity as well as economy. Most often the backfill is prepared by dozers and other earthmoving equipment on the surface adjacent to the trench, or in some type of containment "box," and pushed into the trench where it typically adopts an angle of repose of about 1 vertical to 6 horizontal. On certain projects, a pugmill mixing and blending system is specified, and trucking of the backfill material to the trench may be required, together with tremie placement. Where CB is used, of course, its dual purpose is to support the excavation and then to harden in place as the backfill material. For SC and SCB walls, good technique involves bringing the toe of the backfill close up to the excavated face after completion of the day's work. The following morning, the bottom of the trench is "cleaned" (most effectively by the excavator) and a portion (say 2–5 feet) of the previous day's backfill is dug out of the trench to assure that no highly permeable "stripes" of settled sediment are left *in situ*. It is typical to require a 50- to 150-foot separation between backfill toe and base of excavation slope during routine work, although there seems little engineering logic for this.

Most backhoe cutoffs for dams and levees have been 30–36 inches wide and not more than 60 feet deep. However, recent developments have pushed

---

* Brian Jasperse was assisted in the preparation of this section by Steven M. Artman, Mark E. Kitko, and William A. Buccille, all of GeoCon, a Trade Name of Environmental Barrier Company, LLC.

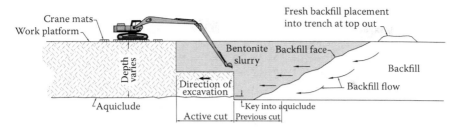

Figure 4.2 Typical installation for soil-bentonite or a soil-cement-bentonite wall.

maximum "comfortable" depths to around 75 feet, while equipment has been developed to excavate to over 100 feet in favorable conditions.

4.2.1.2 Soil-bentonite (SB)

This is the most basic type of cutoff wall normally installed by the slurry-trench method, and consists of the excavation of a vertical trench under a bentonite-water solution (slurry), which acts as hydraulic shoring to support the vertical sidewalls of the trench. The trench is then backfilled incrementally using a mixture of soil, trench slurry, and oftentimes additional dry bentonite.

The soil excavated from the trench is commonly reused for the cutoff wall backfill. This is significant from a project-cost standpoint where off-site soil disposal is a consideration. Depending on the existing soil characteristics, it may be combined with sufficient trench slurry to attain the desired backfill slump and the resultant backfill will attain the target permeability of the backfill. In cases where the existing soil is uniformly sandy or contains very low quantities of plastic fines, additional dry bentonite is added to ensure that target permeability is attained. In rare cases where the existing soils are poorly suited for backfill mixing, imported borrow soil can be substituted for some or all of the existing soil as necessary.

Typically, a hydraulic excavator is used to dig the trench, working from a level work pad, and standard earthmoving equipment is used to mix the backfill on the work pad immediately adjacent to the trench and place the mixed backfill back into the trench. In applications where there is insufficient room along the trench alignment to accommodate standard backfill mixing operations, backfill can still be mixed adjacent to the trench in a "rock-box," or dumpster-style containers or, in the case of extreme space limitations, trench soils can be transported to a remote mixing location and the mixed backfill transported back to the trench location in trucks.

Typical SB backfill material can attain very low permeabilities, in the range of 1×10^{-7} cm/s. If installed by an experienced slurry wall contractor using standard quality control procedures for cutoff wall installation, one can expect an SB wall to attain this level of permeability throughout the extent of the wall.

While the SB backfill mixture will consolidate over time, it will never attain sufficient strength (less than 3 psi) to provide any structural capabilities, and this can be a disadvantage in certain applications. A compacted soil or concrete cap can be installed to accommodate surface traffic over the wall location while also protecting the top of the wall and limiting desiccation.

4.2.1.3 Soil-cement-bentonite (SCB)

In cutoff wall applications where a wall with *no* strength is a concern, a soil-cement-bentonite wall (SCB wall) can be a viable solution. The installation procedures of an SCB wall are identical to those of the SB wall. However, during the mixing of the backfill, dry Portland cement is incorporated into the SCB backfill mixture. Since the backfill mixture needs to remain workable for proper installation purposes, the quantity of Portland cement that can be incorporated into the backfill mixture is limited. Therefore, a modest ultimate strength (typically less than 50 psi) is all that can be attained with this technology. Also, there tends to be a tradeoff between strength and permeability in backfill mixtures. The addition of cement into the backfill mixture will typically have a slightly adverse effect on the permeability of the mixture although a permeability of 5×10^{-7} cm/s can still be attained.

4.2.1.4 Cement-bentonite (CB)

During the 1970s, as cutoff wall technology was being rapidly adapted for new and more challenging applications, the cement-bentonite wall (CB wall) was developed as a means to install a cutoff that attained both a moderate strength and very low and controlled permeability. CB slurry was derived from balanced-stable grouts commonly used at the time for soil and rock grouting. CB is therefore a relatively new technology and less is commonly known about proper installation methods. The implications of this are discussed in more detail in Section 4.2.3.

CB installation methods are a departure from the standard SB or SCB installation methods in that the slurry used to support the trench walls during excavation is self-hardening and becomes the final backfill for the trench. The soil excavated during the installation of a CB wall is not reused for creation of the trench backfill. A specially engineered blend of materials, typically Portland cement, bentonite, and water, creates the self-hardening slurry (SHS). Other materials such as slag cement and alternate types of clays are also frequently used. The recent use of granulated blast-furnace slag has been found very beneficial for strength and permeability development (Bruce, Ressi di Cervia, and Amos-Venti 2006). The SHS typically remains fluid as long as it is agitated during excavation activities.

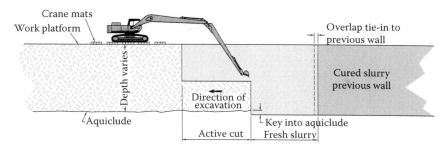

Figure 4.3 "Panel-style" installation for a cement-bentonite wall.

Once excavation is complete for the day, the SHS begins to harden and achieves an initial set overnight. While SB and SCB are installed in a "continuous fashion," a CB wall uses a "panel-style" installation (Figure 4.3). A finite length of trench is excavated each day and allowed to achieve its initial set overnight. The following day, a new panel is started and "keyed" into the previous day's panel to create a longitudinally continuous barrier. Depending on the materials used, SHS has the potential to develop strengths of 50–300 psi and a permeability as low as 1×10^{-10} cm/s.

4.2.2 Construction techniques

The installation of cutoff walls using the backhoe method includes three basic operations: excavation, backfilling, and slurry mixing plant operations. Other ancillary requirements are also discussed in this section.

4.2.2.1 Equipment selection

In general, slurry wall trenches are excavated using a hydraulic track excavator fitted with specialized "long-reach" or "long-front" equipment (Photo 4.1). The long fronts are custom built for slurry wall trenching and

Photo 4.1 Long-reach backhoe, Marysville, California.

are designed to fit on 75–125-metric-ton excavators. Excavation rigs of this size will have the capacity to reach a maximum of 60 to 100 feet below grade. This equipment is therefore adequate to reach the design depths of cutoff walls on smaller dams and levees. When walls are designed for smaller levees such as those around stormwater detention basins, standard excavators can be used to reach depths of 10–30 feet. In any project, the reach should be adequate to dig up to 10 feet deeper than the anticipated depths of the cutoff wall. The buckets used for this type of application are extreme-duty rock trenching buckets, commonly 30–36 inches wide and usually equipped with ripping teeth.

Assembly of these large excavation rigs is completed at the project site. The rigs are broken down and delivered on 3 to 6 trucks. The equipment assembly process can take two to four days and requires the use of a service crane.

Equipment used to blend slurry wall backfill is selected based on the mixing areas available at the project site and in its simplest form can be standard earthmoving equipment. A typical backfill mixing operation will include one or two hydraulic excavators ranging from 15 to 25 metric tons and a 100–150 HP bulldozer. This equipment spread is best suited to blending backfill immediately adjacent to the slurry trench on a cleared, filled working platform. When remote mixing is necessary, off-road haul trucks are used to transport the excavated trench spoils to the mixing area and the prepared backfill back to the slurry trench.

4.2.2.2 Slurry mixing

The slurry used for the excavation of the trench is created at the on-site slurry batching and mixing plant. The slurry is mixed by either tank mixers or Venturi-style mixers, to blend the bentonite and water. The batch plant is established in one fixed location on the project site and the slurry is then pumped to the trench or trenches.

Tank mixers are fitted with high-speed/high-shear paddles and circulation pumps. The mixer is filled with water, and then the powdered bentonite is introduced. The bentonite is added from a bulk silo or by emptying 100-pound bags by hand into the top of the mixer. The slurry is then transferred to a pond or holding tank where it continues to be circulated until it is pumped to the trench.

Venturi mixers blend the bentonite and water utilizing—yes, the Venturi effect. Bulk bags, or "supersacks," of bentonite (ranging from 2,500 to 4,000 pounds each) are positioned over the conical hopper of the mixer. Water passes through the piping attached to the bottom of the hopper. The bentonite slurry is further sheared in a passive mixing chamber and the slurry is ejected from the mixer, usually into a shallow pond.

When mixing a cement-bentonite or "Impermix" preblend (a proprietary mix utilizing clays and pozzolanic binders to achieve high strengths (over 300 psi) and very low permeabilities (less than 1×10^{-8} cm/s) slurry for a self-hardening slurry wall, a more complex batch plant is required. Bentonite-water slurry is typically premixed and stored in tanks. The bentonite slurry is then transferred into additional high-speed/high-shear tank(s) where the cementitious reagent is added. This mix is then delivered directly to the trench without using holding tanks or ponds. Some slurry batch plants incorporate computer-controlled weighing and metering systems that are nearly fully automated and can produce a continuous output of prepared mix.

High-density polyethelene (HDPE) pipe is most often used to deliver slurry to the trench excavation. The pipe is fusion-welded in the field to create long runs. Contemporary, high-capacity pumps at the batch plant can pump the slurry in excess of 10,000 feet at a rate of 200–300 gallons per minute.

4.2.2.3 Excavation

The excavation operation is at the center of the slurry wall installation process. The keys to a successful installation are a properly sized excavator and an experienced, skilled operator. The excavation of the trench is completed while the trench is filled with slurry, so the operator is digging in the "blind" to depths deeper than encountered in other types of construction excavations using similar equipment.

The slurry wall trench is excavated from a prepared working platform. This can be on the crest of a levee or dam, on a degraded levee, or at the toe of a levee. The working platform must have a bearing capacity adequate to safely support the weight of the large excavating rigs without causing settlement.

The slurry wall trench is excavated in a series of approximately 30- to 50-linear-foot-long "cuts." The trench is continually filled with slurry as it is excavated so that the slurry level is maintained within two feet of the working platform surface. Each cut of the slurry wall is excavated to full depth before the machine backs up to begin the next cut. The spoils from the trench are placed on the working platform or directly into haul trucks, depending on the location of the backfill mixing.

The depth of the slurry trench excavation is dictated by the location of the low permeability layer selected as the "toe" of the wall, the engineering design, or a combination of both. A cutoff for a dam or levee is usually designed to cut off a layer of cobbles or other highly permeable soils and, in most cases, the bottom of the wall is "keyed in" up to five feet into a bedrock or aquiclude layer. Some engineering designs do, however, define a predetermined elevation of the bottom of the cutoff wall. Nevertheless, both methods require at minimum a visual verification by the design engineer's representative that adequate depths have been reached. In most cases, the aquiclude is denser or harder than the overlying soils and production is lower in the aquiclude zone. During this time, the trench is more susceptible to collapse due to the extended height of the excavated wall faces and the energy—which can cause vibrations—imparted by the excavator to penetrate the aquiclude. There is, therefore, also more time for the soils suspended in the bentonite to settle out to the bottom of the trench.

Production rates for the installation of the cutoff wall are generally dictated by the rate of excavation. Many factors can affect the rate of production, including soil properties, key-in depths, overall depth of excavation, skill level of the operator, weather, and machine breakdowns. Industrial production rates will typically be between 3,000 and 6,000 vertical square feet of cutoff wall per shift per machine.

When excavating through a slurry-filled trench, a considerable amount of sand can become suspended in the bentonite slurry. The presence of this suspended sand in the slurry creates no detrimental effects to the installation of the cutoff wall. However, if the properties of the slurry become such that it can suspend no additional sand, the sand will begin to drop out of the slurry and create sand lenses. In most cases, this occurs during active excavation and the sand lenses are removed continually as excavation progresses. If sand precipitates from the slurry during off-work hours, it will create a layer on top of the previously placed slurry wall backfill, creating the potential for high-permeability zones in the cutoff wall. This condition is identified by completing depth soundings at regular intervals along the partially backfilled slurry trench. The soundings are conducted at the end of one day's shift and prior to the start of the subsequent shift using a weighted tape measure. If a layer of sand has been deposited along the surface of the backfill, removal is necessary. The long-reach excavator can track out over the slurry filled trench to "clean" the face of the backfill slope. The use of timber crane mats will help to avoid destabilizing the slurry trench when undertaking this task.

If a slurry trench is tied into a previously installed cutoff wall, an overlap of the wall is necessary to create continuity. Slurry cutoff walls can also be tied into other existing structures, such as sheet pile walls or concrete header walls, if necessary.

4.2.2.4 Backfill mixing and placement

Soil-bentonite backfill is most easily created by blending trench spoils, bentonite slurry, and additional dry bentonite, if needed, using earthmoving equipment. The amount of bentonite slurry needed for the mix is determined by the required volume needed to create the desired consistency of the backfill. A concrete slump cone test is conducted on the backfill to ensure it has a flowable consistency, with a range of three to six inches most desirable. If additional dry bentonite is needed to lower the permeability of the soil-bentonite mixture, it can be added in the mixing area. This is usually done by breaking and blending large bags of bentonite directly into the backfill mix.

If the spoils from the trench do not display gradation properties necessary for the backfill, a modification of the soil is necessary. If the cutoff wall profile contains cobbles, these may need to be screened using a grizzly-type screen to remove particles larger than 2–3 inches. Imported soil is often used to adjust the overall gradation of the backfill mixture. Clay, sand, or coarse aggregate can be used, depending on the properties of the native soil. Once the backfill has been properly prepared, it is placed into the slurry trench to form the permanent cutoff wall.

When the backfill is initially placed into a new segment of trench, a technique must be used to allow placement without the potential for backfill segregation, or entrapment of pockets of slurry, which would lead to a "window" in the continuity of cutoff wall. The most efficient way to place initial backfill is through the use of a "lead-in" trench. This is an extension of the cutoff wall beyond the limits of the designed cutoff wall. The lead-in trench is started at surface elevation and is excavated at a consistent slope until the desired depth of the slurry trench is reached at the end of the cutoff wall alignment. The slope of the lead-in trench can be from 1:1 to 4:1 (H:V). If insufficient space is available for a lead-in trench, other methods may be used. These include the use of a tremie to place the backfill from the bottom of the trench up to the surface at the end of the cutoff wall alignment. Also, the backfill may be placed from the bottom up using the long-reach excavator. Free dropping of the backfill through slurry is not permitted due to the potential for entrapping pockets of slurry.

Once the initial backfill has reached the working platform surface at its natural angle of repose, additional placement continues by always placing new backfill on the leading face of previously placed backfill. In this manner, the backfill will displace the slurry, creating a continuous barrier of backfill without voids or entrapped slurry. The most efficient way to place the backfill is to push it directly into the trench using a bulldozer. However, this method is often prohibited by specifications because of concerns that unmixed soil from the underlying earthen working platform will be pushed into the trench. In this case, placement is achieved using the bucket of an excavator.

Backfill in the trench, under the liquid slurry, will form a relatively flat slope of approximately 5:1 to 10:1 (H:V). The toe of the backfill slope is generally kept a distance of 30–100 feet from the toe of excavation to maximize the stability of the trench and also to allow for cleanout of sediments that may accumulate at the leading edge of the backfill slope.

Cementitious binder, such as Portland cement, can be added in a number of ways. The cement can be added by breaking dry "supersacks" in the backfill mix, in a similar manner to how the dry bentonite is incorporated.

Also, the cement can be delivered to the mixing area as cement slurry, created at the on-site batch plant. Some specifications require that the backfill be mixed in a volume-controlled mixing area. This can be accomplished by the use of steel mixing bins alongside the trench or with an earth-lined mixing pit at a remote location: a known volume of soil is first added to the mix area, usually measured using an excavator bucket, followed by a known weight or volume of binder measured by the bag, or by metering slurry from a batch plant.

When a more sophisticated control over the mix proportions is required, a mixer known as a pugmill can be used. This method of mixing backfill is far less efficient and more costly than blending with earthmoving equipment but offers the advantage of better control and greater consistency over the proportioning of backfill components (Photo 4.2). The pugmill plant is set up in a fixed location, to which off-road trucks transport the trench excavation spoils. These excavated trench spoils (and imported soil, if needed) are stored in the aggregate feed bin, from where the soil is fed into the mixing chamber and metered using belt scales. Dry reagents such as bentonite and cement are fed into the mixing chamber from on-board silos. Water, slurried bentonite, and/or slurried cement are added directly into the mixing chamber via "spray bars." Reagent deliveries are controlled by an on-board computer to ensure consistent mix proportions.

4.2.2.5 Top of wall treatment

The top of the wall is usually left within three feet below the temporary working surface or final grade on a dam or levee project. Therefore, it is susceptible to mechanical damage from construction equipment crossing the wall and also vulnerable to desiccation, which could cause the top three to five feet of the wall to be ineffective for its intended use. Hence, treatment of the top of the cutoff wall is necessary to protect the backfill. The

Photo 4.2 Pugmill mixing plant, Mayhew Levee, Sacramento, California.

treatment should be designed to prevent desiccation of the backfill while simultaneously providing a buffer zone to prevent mechanical damage. Top of wall treatment, also known as a temporary curing cap, will generally include a minimum of one foot of uncompacted soil placed over the cutoff wall. A layer of geosynthetic material or visqueen can be used to separate the backfill material from the capping if needed. After a waiting period (3–7 days) to ensure there is no more significant settlement of the backfill, the temporary wall treatment is removed. The remainder of the levee or dam can then be constructed over or around the cutoff wall. Often a permanent clay plug is deposited and compacted in place of the temporary cap.

4.2.2.6 Additional considerations

When constructing a slurry trench cutoff wall on an existing levee or dam, one of the most difficult challenges can be the lack of available space. Existing levees often have narrow crests and do not have sufficient space for excavation spoils or for backfill mixing. These situations often require the use of remote backfill mixing. Slurry cutoff walls installed at the toe of existing levees can be within the floodplain of the adjacent waterway and require special considerations for construction. If there is seasonal flooding, then the construction period may have to be limited to the drier months. If the water table is too close to the surface, then small temporary berms may be needed along the two sides of the excavation to allow higher slurry levels in the trench to offset the *in-situ* hydraulic head of the ground water.

4.2.3 Principles of QA/QC and verification

Knowing and understanding the intent of the cutoff wall design and the final performance expectations of the wall are the essential first step in developing a quality control program.

The primary goals of any cutoff wall installation quality control program are twofold:

1. To ensure the integrity (trench stability)/continuity and minimum thickness of the trench as it is excavated and backfilled.
2. To ensure the cutoff wall backfill properties (e.g., strength, permeability) are within the project requirements.

Accomplishing these two goals ensures the success of the cutoff wall as a structure in meeting the intent of the design.

The QC program should be tailored specifically to the type of cutoff wall being installed. For example, attempting to apply SB specifications to a CB wall installation can be detrimental to the efficiency of the CB wall installation, completion schedule, and installation cost. It is also important to understand the ultimate performance expectations of the completed wall, bearing in mind that the more stringent the quality control standards, the higher the installation costs are likely to be. As discussed in Section 4.2.3.4, a stringent real-time QA/QC construction monitoring program will generally provide the best final product and eliminate costly rework and schedule delays associated with discovering deficiencies in the wall with postconstruction sampling.

4.2.3.1 Mix design

A comprehensive bench-scale mix design is an important first step in the quality control program. Each cutoff project has unique aspects of installation and specific performance criteria and therefore the mix design should be tailored to the individual project. Generally, the goals of the mix design are:

1. To ensure the compatibility of chosen materials with the site soils, contaminants, and groundwater.
2. To determine the binders and addition rates necessary to accomplish the performance goals of the project.

It is important to obtain representative soil samples from all strata for performance of the mix design. Ideally, the cutoff wall contractor should perform the mix design, which will ensure the contractor has the information and data required to perform the project. Split-spoon sampling is one of the best methods of sample collection since it provides blow count data, which are beneficial information for the contractor. Research should be performed to ensure collection of samples of the soil strata and potential contaminants that will represent a "worst case scenario" are collected and used in the mix design.

4.2.3.2 Cutoff wall quality control: Standard tests

The following are typical cutoff wall quality control procedures. Frequency, acceptable parameters, and details are provided in Section 4.2.3.3:

Marsh funnel viscosity: Provides viscosity readings for trench slurry. Performed in accordance with API Standards RP 13B-1.
Mud balance: Provides unit weight of slurry and trench backfill. Performed in accordance with API Standards RP 13B-1.
Filter press: Provides an indication of bentonite quality and ability to hydrate completely as well as the slurry's ability to form a filter cake. Performed in accordance with API Standards RP 13B-1.
Slump cone: Measures the flowability of the backfill mixture. Performed in accordance with ASTM C-143 (2010).

Minus 200 sieve analysis/moisture content: Measures the amount of fines in the backfill that will pass the 200 sieve. Can be correlated to target permeability during mix design and provide a rough field indication of backfill performance. Performed in accordance with ASTM D-1140/D-2216 (2010).

Sand content: Measures the quantity of particles in the trench slurry larger than the 200 sieve. Performed using a sand content kit (available from Cetco or Ofite testing) per the manufacturer's instructions.

Hydraulic conductivity: Measures the permeability of the trench backfill. This is typically performed by a third-party laboratory in accordance with ASTM D-5084 (2010).

Unconfined compressive strength: Measures the strength of an SCB or CB backfill. Typically performed by a third-party laboratory in accordance with ASTM D-4832 (2010).

The following tests are common to all cutoff wall types and should be performed *in addition* to the cutoff wall type-specific QC discussed in Section 4.2.3.3:

Mix water analyses: Water to be used for slurry mixing should be tested for pH, total hardness, and total dissolved solids. These tests will ensure the water being used for slurry mixing is in compliance with the water used in the mix design. It will also provide initial indicators that the mix water could present a problem with slurry quality such as inhibited bentonite hydration.

Material certifications: The material suppliers should provide certifications with each load of materials delivered. Certifications should be reviewed to ensure that materials are in compliance with the mix design and any applicable standards (such as API or ASTM).

Excavation verticality (perpendicular to the trench alignment): The excavator should be checked periodically to ensure that it is generally level. A carpenter's level placed on the tracks is sufficient to demonstrate this. Loss of verticality on the plane perpendicular to the trench alignment could result in excavation difficulty, especially between cuts, and erroneous depth measurements.

Slurry level maintenance: From a stability standpoint it is important that trench slurry levels are maintained within three feet of the work-pad surface. This is even more critical when installing a cutoff trench in very sandy or unstable soils, or in the presence of a high water table. The density of the slurry, before it is introduced into the trench, is typically 64 pounds per cubic foot (pcf). The density of water is 62.4 pcf. Because the slurry is denser, it exerts a net positive thrust against the sides of the trench, creating the lateral support. In order to provide an adequate safety factor, it is necessary to provide the slurry with a larger hydraulic head. This is accomplished by ensuring the slurry level is at least three feet above the water table. In loose sands, the failure mechanism is generally a slip plane. To resist the force of the wedge of soil from a slip plane, it is imperative that the slurry hydraulic head advantage be maintained.

Key verification (if applicable): A design "cut sheet" should be developed based on the site borings along the trench alignment, and followed throughout the cutoff wall excavation. However, variations can always exist between the design depth and the depth at which the key stratum is encountered during actual installation. Therefore, the project team should be familiar with the identification of the design key material. Additionally, key material should be verified for each cut to ensure the cutoff has been installed to the proper depth. Samples of key material from each cut should be saved and archived.

Depth measurements: Depth-measurement records will be one of the largest data sets obtained from a cutoff wall installation. It is important to be able to obtain accurate, repeatable measurements for excavation and AM/PM trench profiles. A simple weighted measuring tape is the best tool for depth measurements. As the work-pad surface is constantly changing throughout excavation and backfill operations, it is important to maintain accurate stationing and repeatable "measuring points" especially for AM/PM profile measurements. Typically, ten-foot stations are sufficient between depth measurements.

Equipment calibration: Typical slurry wall QC apparatus including mud balances, marsh funnels, slump cones, filter press apparatus, and sand content kits should be kept clean and in good working condition at all times. Calibration for items such as marsh funnels and mud balances should be checked and documented on a daily basis in accordance with ASTM standards and manufacturer recommendations.

Reporting: Quality control reports should be generated each day and detail the excavation completed each day with depth measurements, results of all quality control tests, calibrations, samples taken, and an AM/PM sounding (if applicable).

4.2.3.3 Specific quality control aspects

4.2.3.3.1 SB walls

Since the slurry acts as the shoring to hold the trench open during excavation, ensuring quality trench slurry is paramount in accomplishing the primary goal of the quality control program, which is the placement of a homogeneous soil-bentonite backfill that does not contain pockets of slurry and/or excavated soils, to the required lines and dimensions.

4.2.3.3.1.1 FRESH SLURRY

Bentonite slurry for trenching can be mixed in an on-site batch plant and pumped directly to the trench, or stored in ponds or tanks to ensure sufficient supply of fully hydrated slurry is available for trenching activities. Slurry can also be produced using a Venturi-style mixer and stored in an agitated pond or tanks to fully hydrate. Slurry should never be mixed in the trench. A quality high-shear colloidal-type batch plant will produce the most consistent high-quality slurry for trenching purposes. If the slurry is well mixed at the plant, target slurry viscosities are often easily attained without any additional hydration time. Slurry should also be completely mixed and free of "fish eyes," which are small pockets of unhydrated, agglomerated bentonite typically observed in poorly mixed slurry.

Fresh slurry viscosity and unit weight should be monitored during production to ensure only fully hydrated slurry is used for trenching. A 5 percent bentonite/water solution is typically used for trenching. This slurry should have a mud balance unit weight of approximately 64 pcf and should, assuming a quality 90-barrel yield bentonite is used and there are no compatibility issues, attain a Marsh funnel viscosity of 40 seconds or greater. Slurry quality and bentonite hydration should be further confirmed using a filter press apparatus. A pressure of 100 psi is applied to the slurry for a period of 30 minutes. During that time, a pure bentonite filter cake should form on the filter paper and should not allow any more than 25 mL of fluid to be ejected from the apparatus.

The following frequencies will provide a good indication of overall fresh slurry quality:

- Viscosity: four times per work shift.
- Mud balance unit weight: two times per work shift (concurrent with two of the viscosity measurements).
- Filtrate test: one time per work shift (concurrently with one of the viscosity and unit weight measurements).

All tests should be performed at the batch plant or storage pond/tank prior pumping to the trench.

4.2.3.3.1.2 TRENCH SLURRY

Once the fresh slurry is placed in the trench, the properties can begin to vary from those observed during fresh slurry testing. It will mix with soil cuttings during excavation and suspend varying amounts of soil particles; it will be subjected to pressure filtration and deposit bentonite, in the form of a filter cake, on the trench walls, and could be subjected to thinning due to groundwater infiltration.

The properties (unit weight, viscosity, and sand content) of the trench slurry will vary depending on the location from which they are obtained. For example, trench properties will be most desirable (and will more closely mimic fresh slurry properties) at the surface near the point of fresh slurry injection, while slurry at the bottom of the trench near the backfill toe will typically exhibit the least-desirable properties. Therefore, trench samples should be obtained from various locations and depths during excavation to ascertain best- and worst-case scenarios considering that most of the slurry in the trench will fall somewhere in between that range of properties. Trench slurry samples can be obtained by a variety of means. The excavator bucket can be used to obtain a sample from the point of excavation at various depths. A simple slurry sampler can be made using PVC pipe and check valves available from most hardware stores. This apparatus can be weighted and attached to a rope and used to obtain samples from any location and depth.

Generally, the viscosity and unit weight of the slurry in the trench will increase in response to its suspension of soil particles. From a stability standpoint, this is beneficial. However, as the slurry becomes excessively heavy and laden with soil particles, two adverse consequences in particular can result:

1. Sands begin to fall out of suspension and can leave lenses on the backfill face, adversely affecting the overall permeability of the completed wall.
2. The excessive unit weight of the slurry begins to match or exceed the unit weight of the backfill; proper backfill placement becomes problematic and will result in poor backfill consolidation and trapped pockets of slurry. This is referred to as backfill "float."

To preclude the above, the trench slurry viscosity should typically be kept at less than 90 seconds and the unit weight at less than 85 pcf, or 15 pcf less than the corresponding trench backfill weight.

The following frequencies will provide a good indication of overall trench slurry quality:

- Viscosity: two times per work shift.
- Mud balance unit weight: two times per work shift (concurrently with the viscosity measurements).

Trench slurry filtrate tests generally do not provide any meaningful information and are not recommended. A sand-content test kit can be used to monitor the quantity of coarse soil particles suspended in the slurry. While this can be useful information, imposing a strict sand-content specification can actually be *detrimental* to the slurry wall installation from a quality standpoint. Imposing hold points for failed slurry sand-content tests and requiring the use of cumbersome desanding techniques will delay the progress of the slurry wall.

As the wall progresses, four key processes are taking place:

1. Excavation is progressing forward.
2. Fresh slurry is being pumped to the trench to hold the trench open.
3. Old trench slurry is being used for backfill mixing and is being replaced with fresh slurry.
4. Backfill is displacing the slurry and the wall is progressing forward.

The longer a given section of trench remains "open" (unbackfilled), the greater the likelihood of trench failure. If the trench cannot progress forward, backfill cannot be mixed and placed. If backfill cannot be mixed, old soil-laden trench slurry cannot be used and replaced with fresh slurry and the trench slurry will become heavier and thicker, possibly causing sand to settle out, necessitating risky trench cleanouts and creating a cycle of difficulties and quality control problems.

The sand-content information should be used to recognize the potential of sand to fall out of suspension. As long as the wall continues to progress steadily, this should not be a problem. However, extra attention should be paid to the AM/PM backfill profiles, especially during any extended period of inactivity (such as an equipment breakdown or weekend downtime), to look for anomalies in the soundings that could indicate sand settlement.

4.2.3.3.1.3 BACKFILL

The second goal of the QC program is ensuring the trench backfill meets the required project parameters.

To ensure the backfill remains at optimum workability, the slump of the backfill is monitored using a standard slump cone. A range of 3–6 inches is optimum for most applications and should result in a backfill slope between 5:1 and 10:1 (L:H). Unit weight should be in compliance with the completed mix design and, as noted above, should be compared with the trench slurry weight. A representative sample of the backfill mixture should be obtained daily (or more often if specified) prior to placement in the trench. The sample should be split three ways:

1. Used for slump testing, moisture content, and minus 200 sieve testing.
2. Sent to an off-site laboratory for hydraulic conductivity testing.
3. Archived and stored on site.

Fines content alone does not indicate that a sample will meet target permeability: plastic and nonplastic fines are indistinguishable from each other in a field minus 200-sieve test, but nonplastic fines will not significantly contribute to a lower permeability. A correlation should be developed during the mix design testing or during the startup of the project between the target permeability and the fines content of the backfill. This will then serve as a field index that the backfill will meet target permeability during laboratory hydraulic conductivity testing.

If dry bentonite addition is determined by the mix design to be necessary to meet the target permeability, this should be demonstrated empirically as trench installation progresses. While there are laboratory methods for the determination of bentonite content in a backfill mixture (such as methylene blue methods), that level of accuracy is neither appropriate nor cost-effective for a slurry wall installation. Typically, dry bentonite is supplied in 2,000-lb (or larger) jumbo sacks. These are placed along the work pad using a simple calculation to determine the spacing of bags based on the required dry addition, weight of the bentonite bag, and volume of soil per lineal foot of trench excavation. Observing that the backfill equipment operators are distributing the bentonite evenly throughout backfill mixture is generally all that is required to assure an acceptable level of homogeneity.

4.2.3.3.1.4 TRENCH EXCAVATION AND BACKFILL PLACEMENT

It is imperative that the trench is accurately stationed prior to any slurry wall activities taking place and this stationing be maintained throughout the project. Due to the nature of slurry wall installation, this can prove challenging. Appropriate offsets should be established so that field personnel can easily re-establish stationing during operations. All depth measurements, testing locations, and AM/PM profiling measurements should be related to this stationing.

The trench should be completed in "cuts" defined by the safe reach of the chosen excavator. Each cut should be completed to depth, verified, the key documented (if applicable), and cleaned by reaching the length of the cut and scraping the bottom from the front to the back of the cut before the next cut is started. Once the next cut is completed, the excavator should be able to reach into the previous cut during cleaning. This will ensure the cuts are continuous and clean as excavation progresses.

The backfill slope should be kept as close as practical to the active excavation. Backfill should never be placed in freezing weather. Backfill should be uniformly mixed. An experienced contractor and backfill equipment operator will be able to rapidly mix the backfill on the work-pad surface with trench slurry to a paste-like consistency meeting the desired slump and accurately incorporating the desired bentonite dry addition (if specified). Backfill should be placed initially with a tremie pipe, or a "lead-in" slope not exceeding 1:1. Once the backfill establishes its own angle of repose, and a "top out" is attained (the top of the natural backfill slope reaches the work-pad surface), subsequent freshly mixed backfill should only be placed on this top out, which will result in the entire backfill slope "slumping" forward. This method will avoid trapping slurry pockets within the backfill or any separation of the backfill resulting from free falling of the backfill through the slurry.

The trench should be inspected each morning, and periodically throughout the day for signs of trench instability or cave-ins along the alignment. Additionally AM/PM backfill profile measurements should be reviewed closely for any anomalies that could indicate a cave-in or trench wall sloughing. In the case of a cave-in or cleanout required beyond the reach of the excavator, timber mats should be used to support the excavator as it tracks into position to clean out the area in question. This will ensure the safety of the excavator operator and help preclude additional instability or damage to the trench walls as a result of the cleanout.

Upon completion, the trench should, as a minimum protection, receive an antidesiccation cap. This can be as simple as mounding up any extra backfill on the top of the wall to protect the wall during consolidation and prevent it from drying and cracking.

4.2.3.3.2 SCB walls

Since the installation procedures are nearly identical to the SB wall, the quality control is also nearly identical. One notable difference is associated with the addition of cement: this addition should be empirically demonstrated in a similar fashion as dry bentonite addition.

Samples obtained for laboratory testing should be collected and molded in three-by-six-inch plastic cylinders, which can be used for both UCS and permeability testing.

Oftentimes, the success or failure of a project rides solely on the results of the backfill QC samples. Even if the project is executed perfectly by all parties involved, sample mishandling and/or poor curing disciplines of samples can cause an otherwise successful cutoff wall project to end in controversy and, possibly, costly rework. Such samples will have extremely low strengths and must therefore be handled and tested with extreme care. By no means should they be cured and handled in a similar manner as standard concrete cylinders. Low-strength samples that will be tested for both strength and permeability will be extremely susceptible to micro fracturing if disturbed or mishandled, which can cause poor permeability results. The following procedures should be followed:

- Samples should be stored on site for a minimum of three days or until initial set is attained. Care should be taken not to disturb the samples during the critical time between the "gelling" of the sample and the achievement of initial set.
- They should be stored in an undisturbed location in a temperature/humidity-controlled environment (which can be as simple as a sealed plastic container in a heated office trailer). Under no circumstances should they be allowed to freeze.
- They should be transported, by the contractor or engineer, to the laboratory in a careful manner (securely in the front seat or cab of truck or car, not in the trunk or pickup truck bed). Under no circumstance should a sample be shipped or mailed.
- They should be tested in an experienced and accredited laboratory that is familiar with handling and breaking low-strength samples.

4.2.3.3.3 CB walls

As noted above, CB walls are a departure from the traditional slurry wall concept. Thus, it is important to understand the fundamental differences between the two siblings and adapt QC requirements accordingly. It should also be borne in mind that a "CB wall" is a misleading name, since it implies that the mix is simply cement and bentonite. Modern-day self-hardening slurry (SHS) mixes are often multicomponent and can include blends of Portland and slag cements, bentonite, attapulgite or other clays, accelerants, and other admixtures. It is often overlooked that the mix *in situ* may well contain 10 percent or much more of the native soil incorporated during the excavation process.

4.2.3.3.3.1 FRESH SELF-HARDENING SLURRY (SHS)

Since the SHS, which also becomes the backfill, is manufactured in the batch plant, this mixing process should be a major focus of the QC program. Close attention should be paid to ensuring the batch plant can produce batches of SHS that are consistently in compliance with the completed mix design. Components are often delivered in bulk and stored in a silo and metered into the batch plant from the silo. The most efficient and accurate way to accomplish this is the use of batch scales. A typical mud balance can usually only measure the slurry weight to an accuracy of 1 pcf. When these variances are extrapolated to a full batch, the amount of material in the batch can vary (up to 8 percent) enough to affect the characteristics of the slurry. Batch scale measurements in concert with automated batching will result in a slurry within a range of 1 percent of the required material usage.

Since the SHS is primarily cement and clay with a high water content, bleed can be a significant issue. The clay in the mix plays a part in combatting this bleed and therefore needs to be fully hydrated during SHS production. Batch plants should use high-speed colloidal-type mixers and utilize high-shear mixing pumps. Clay slurry can also be premixed to ensure full hydration prior to being used in SHS slurry. Unit weight and viscosity (if applicable) should be confirmed at the batch plant at least twice per shift to be in compliance with the completed mix design.

Since SHS primarily relies on weight rather than the production of a filter cake to aid in maintaining trench stability, filtrate testing does not provide useful information and need not be performed: the air pressure is applied to the slurry before any significant hydration has occurred and, as there is minimal bonding between the cement and water, most of the water will squeeze out in the sample in a few minutes and air-breakthrough will then occur.

4.2.3.3.3.2 TRENCH SHS

Once the SHS is being worked in the trench, it will suspend soil particles just as bentonite slurry does. Since SHS relies on its weight for trench stability, an increase in unit weight is considered beneficial and oftentimes relied upon to ensure trench stability. The suspended soil particles are encapsulated by the SHS particles, acting similar to an "aggregate" in concrete, and do not adversely affect strength or permeability. Further, SHS does not form a traditional filter cake on the trench wall and so it can be affected by groundwater infiltration.

SHS viscosity and unit weight in the trench should be monitored at least twice per shift. Samples should be collected from various locations and depths within the trench. In addition to unit weight and viscosity testing, samples should be molded into three-by-six-inch sample cylinders to be tested for UCS and permeability in accordance with the specifications. Sample curing, handling, and testing procedures should be as for SCB materials. Since some SHS formulations (especially those containing slag cement) are prone to desiccation and cracking if allowed to dry out, special care should be taken to cap cylinders or keep the tops moist at all times.

4.2.3.3.3.3 EXCAVATION

Because the excavator arm is constantly moving through the slurry in the trench, SHS will stay fluid long enough to complete the excavation as it is a thixotropic product and relies on hydration (which is initially retarded by agitation) to begin the set process. Production should be limited to a single shift to allow the completed panels to consolidate and hydrate. At the end of each shift, the trench should be completely filled to capacity with SHS. During the excavation of the first panel of the day, particular care should be taken to ensure it is "tied in" to the last panel excavated the previous day. This tie-in should extend a minimum of three feet into the previous panel. Accurate trench stationing and depth measurements are necessary to demonstrate this has been accomplished. Since the SHS tends to bleed and may shrink considerably, it is important to monitor this even during weekends or shutdowns and constantly "top off" the completed trench. Certain formulations of SHS, particularly those containing slag cement, tend to desiccate and crack if left exposed to the elements. Therefore, any completed sections that will not be receiving any more SHS "top-offs" should be treated with an antidesiccation cap.

4.2.3.4 In-situ *sampling for verification*

Strict adherence to a well-designed quality control program during construction will result in a cutoff wall of acceptable quality that will perform as designed. However, many specifications continue to trend toward requiring more tangible proof of successful installation, and *in-situ* sampling is often specified. *In-situ* sampling requirements can range from Shelby tube sampling in a SB wall to coring in a CB wall. These requirements should be regarded with extreme caution. Much as poor sample handling and testing can result in the rejection of what should be acceptable work, *in-situ* sampling, imposed without realistic expectations and engineering common sense, can cause the *in-situ* quality of an acceptable cutoff wall installation to be called into question. First and foremost, cores or split-spoon samples from an SCB or SB wall cannot be used for verification testing for UCS and permeability. The nature of coring and split spooning will inevitably cause damage, whether visible or not, to any low-strength sample. Oftentimes coring in low-strength material will result in the material "spinning" in the core barrel, breaking up, and washing out, resulting in an "empty" core run, especially when the backfill contains coarse aggregate.

Verticality can also be problematic especially in a deep and narrow cutoff wall. Even though excavator verticality is monitored, due to the nature of slurry wall installation, it is difficult to maintain it within a strict tolerance: it is within acceptable industry practice for the excavator to be several degrees from vertical during installation, the concept and reality being that while the wall may "ripple" a little, it is still longitudinally continuous. Even the most accurate drilling methods are susceptible to deviation of up to several degrees during drilling. These two factors combined can easily lead to the sampling effort "drifting out" of the trench and into the trench sidewall, leading to erroneous conclusions regarding discontinuities within the cutoff wall.

4.2.4 Cost

Typical pricing for slurry walls installed with hydraulic excavators is summarized below:

Wall type	Mobilization	Unit cost per vertical square foot
Soil-bentonite	$75,000–$125,000	$3.00–$6.00
Cement-bentonite	$90,000–$130,000	$7.00–$10.00
Soil-cement-bentonite	$100,000–$150,000	$8.00–$12.00

4.2.5 Case histories

The following brief case history summaries have been selected to illustrate various factors influencing the design and construction of the three different wall types. While there are commonalities, such as for example the use of the long-reach excavator, each project faced different sets of challenges.

4.2.5.1 Soil-bentonite cutoff wall, Canton Dam, Oklahoma

The U.S. Army Corps of Engineers (USACE) let a contract in 2007 to install a soil-bentonite cutoff wall that ties into the existing Canton Dam. The cutoff wall is part of an active retention structure between Canton Lake and a new auxiliary spillway system being installed for Canton Lake. It extends though several permeable zones into the Dog Creek Shale key-in layer, creating an impermeable barrier. The contractor installed 83,998 vertical square feet of trench to an average depth of 54 feet with a maximum depth of 81 feet below ground surface.

Prior to the beginning of the project, a mix design was performed with the following results:

Mix number	Dry bentonite addition in sample[a]	Bentonite in sample from slurry[a]	Permeability (cm/s)
1	6.0%	0.5%	6.0×10^{-9}
2	4.0%	0.5%	1.6×10^{-8}

[a] By weight of soil

The specifications required the soil-bentonite backfill to be mixed with a pugmill mixing plant. Trench spoils and backfill were transported to and from the pugmill with off-road dump trucks. All backfill samples satisfied the permeability requirement of 1×10^{-8} cm/s.

The scope of work included exploratory borings along the alignment prior to installing the slurry wall in order to determine the profile of the Dog Creek Shale layer. The borings showed that the shale layer was deeper than anticipated. This led to a change to the contract to extend the slurry wall on the east and west ends. In addition, a change to the east end of the wall was necessary and was completed with a vertical tie-in to the existing dam structure. This section of wall was backfilled using a tremie pipe held with a crane to avoid freefall and voids within the slurry wall. A permanent cap on the slurry wall using geosynthetics and compacted soil was also installed.

4.2.5.2 Soil-bentonite cutoff wall, Feather River Setback Levee, Marysville, California

The Feather River Setback Levee is a multiphase project intended to increase the flood capacity of the Feather River near Marysville, California. The project is sponsored by the State of California, with oversight by USACE, and is administered by local flood-control authorities.

Approximately 475,000 square feet of slurry wall on Phase IV of this project were required to be installed. The slurry wall serves as a seepage barrier under the newly constructed levee to reduce the risk of flooding to the nearby towns of Olivehurst, Linda, Marysville, and Yuba City. The soil-bentonite walls were built before the new levee was raised (Photo 4.1). The depth of the trench varied from 54 to 71 feet.

Slurry wall operations included the incorporation of sand and gravel into the slurry wall backfill in order to meet the specified gradation requirements, the incorporation of 1.5 percent dry bentonite addition into the backfill using "side mixing," and the desanding of the trench slurry to meet the 15 percent maximum sand content requirement prior to backfill placement.

4.2.5.3 Soil-cement-bentonite cutoff wall, Sacramento, California

The project entailed construction of 2,000,000 square feet of soil-cement-bentonite cutoff wall to a maximum depth of 79 feet. The cutoff wall, with a maximum specified permeability of 5×10^{-7} cm/s and an average unconfined compressive strength of 65 psi, protects the stability of an existing levee by cutting off river water seeping into the levee and its foundation soils. The cutoff wall was constructed during two seasons from September 2000 to October 2000, and from April 2001 to September 2001. Five large excavators capable of excavating to a maximum depth of 82 feet were utilized working six days per week to complete the 25,316-foot-long cutoff wall before the mandatory completion date of September 28, 2001. In order to meet the tight schedule and performance requirements, the contractor designed a backfill mix that met the specified 28-day permeability requirement after only 14 days. The soil-cement-bentonite backfill material was prepared in a 30-cubic-yard steel bin adjacent to the trench. This innovative construction method minimized equipment traffic on the levee and areas of disturbance.

The project also included the completion of emergency work that consisted of excavation and replacement of discrete 200-foot, 400-foot, and 2,000-foot sections of the levee that contained fractures and/or unsuitable material.

The site, with severe space limitations, presented a significant safety challenge. Over thirty pieces of equipment working six days a week in addition to material delivery, required extensive coordination and scheduling of site activities. The project included construction of the cutoff wall around existing underground utilities and under overhead power lines. Several underground utilities including fiber optic, sewer, drainage and electrical existed at depths of 5 to 20 feet below the levee crest and were protected in place during construction activities. Extensive coordination with the utility owners was required to ensure in place protection during construction activities and minimize service disruption in the event of an emergency.

Other challenges encountered during construction activities included real estate procurement, environmental protection, availability of resources, design changes, severe space limitations, and meeting the construction schedule, most of which entailed significant public relations efforts. Real estate procurement issues included supply of staging areas and safe access to the site. The contractor negotiated with a local property owner to supply a staging area to support construction of a three-mile portion adjacent to an industrial area, and negotiated with a local church to provide access to the project site (portion adjacent to residential neighborhood). Access through the residential streets was prohibited to minimize disturbance to the local community. Environmental protection included measures to protect the endangered elderberry bushes, local recreation facilities, and the American River.

This project introduced other unique challenges related to working in close proximity to the American River recreational areas and residential homes within the city of Sacramento:

- Maintaining access to the park and other recreational areas.
- Maintaining access to and working around the bike trail.
- Maintaining communication with the public via articles in the newspaper; Internet; meetings on changes in the bike trail; USACE spokesmen updates on progress of the project; and weekly meetings with the USACE discussing any public concerns.
- Minimizing dust near residential homes by implementing dust-control measures, primarily using water trucks coupled with vehicle/heavy equipment speed monitoring to maintain designated speed limits.
- Continuously monitoring noise and vibration levels. Results were checked throughout the day and reported daily. All readings met the specifications provided by the USACE.

4.2.5.4 Soil-cement-bentonite cutoff wall, Mayhew Levee and Drain, Sacramento, California

The Mayhew Levee is located along the American River in Sacramento, and protects the neighborhood to the south. The Mayhew Levee-Raising and

Drain project was designed to improve and increase the capacity of the levee and to replace the Mayhew Drain structure at the west end of this reach of levee. The Mayhew Levee portion of the project included the installation of a slurry wall within the existing levee as well as the raising and widening of the levee. The Mayhew Drain portion of the project consisted of installing a slurry wall under a slough that drains storm water into the American River. The new slurry wall tied into an existing slurry wall previously installed on the west side of the drain, and to the levee slurry wall installed under this contract. The Mayhew Drain section was approximately 143 feet long and comprised 8,200 square feet. The Mayhew Levee section was approximately 4,300 feet long and 220,000 square feet, to depths up to 70 feet below the work platform.

The cutoff wall was installed to predetermined depths designed by the USACE. The contractor was required to have a geologist on site at all times to log the material and verify the low permeability layer at the design depths. The soil layers included the well-compacted levee material, poorly graded sand layers, a cobble layer up to 40 feet thick, and a sandy clay/clayey sand material, which was the key layer.

The backfill for the project was created using a pugmill mixing plant. This project is believed to be the first slurry wall constructed using a pugmill for mixing soil-cement-bentonite backfill. The backfill was mixed at a fixed plant location along the waterside toe of levee. The pugmill plant included an aggregate feed belt with a belt scale, a dry bentonite silo and feed belt with a scale and variable speed drive, and a slurried cement mix tank with a mass-flow meter and variable speed pump, all of which fed the material into the pugmill mix chamber. The material feed systems were computer controlled at the pugmill operator's control trailer. The excavated material was trucked to the plant, and the mixed backfill was trucked back to the slurry trench using off-road dump trucks.

The project specifications required the slurry wall backfill to have a permeability of 5×10^{-7} cm/s or lower, and have strengths of 50–300 psi. The backfill was mixed using primarily the soil excavated from the trench and some imported material. The trench spoils were screened at the pugmill plant to smaller than three-inch material prior to placing in the pugmill aggregate hopper. The project specifications also called for a maximum sand content in the trench slurry of 15 percent. A full-time desanding effort was required to meet this specification requirement. A complex slurry desanding rig was used on the site which included a hydraulic submersible pump, a primary scalping screen, multiple hydocyclones, and a secondary scalping screen. The desanding rig was self-contained and moved on the levee crest behind the trench excavation equipment.

A key challenge on the project was access. The centerline of the trench was approximately 10 feet from the water side of the levee, leaving approximately 15 feet on the land side for delivery of the SCB backfill. The levee was also the only access for material and equipment to the plant area.

4.2.5.5 Slag-cement-bentonite cutoff wall, Horsethief Reservoir, Jetmore, Kansas

The contract called for the construction of two cement-bentonite slurry cutoffs at the site of the future embankment dam. The two walls were 30 inches wide and keyed into shale at depths ranging from 5 to 28 feet. The slurry used for excavation was a mixture of ground granulated blast-furnace slag, bentonite, and Portland cement. This self-hardening slurry provided a permanent, low-permeability trench backfill. The work platform was created by benching the alignment down into a deep, narrow "canyon." Because the long-reach excavator used to dig the slurry wall was working from the narrow platform, the spoils from the trench were loaded directly into haul trucks for disposal (Photo 4.3).

The barrier walls were constructed using a Caterpillar 375 excavator with a long-reach boom and stick combination and a 30-inch trenching bucket. The specifications required a permeability of 1×10^{-6} cm/s and minimum 25 psi unconfined compressive strength after 28 days. The achieved permeability was in the range of 1×10^{-7} to 1×10^{-8} cm/s and actual unconfined compressive strengths were in the range of 100 to 200 psi.

4.2.5.6 Slag-cement-bentonite cutoff wall, Winyah Generating Station Levees, Georgetown, South Carolina

A series of cement-bentonite slurry walls were required at the Winyah Generating Station. The slurry walls were designed as seepage barriers for levees around the three ponds used for storing water to cool the power plant.

Photo 4.3 Excavation for CB cutoff, Horsethief Dam, Kansas.

Each of the ponds had a 30-inch corrugated metal pipe (CMP) approximately 32 feet from ground surface; the CMPs had all deteriorated to varying degrees allowing water to seep through the levees that housed them.

The three cement-bentonite walls were each 30 inches wide and were excavated into the shelly limestone toe material at depths between 45 and 47 feet. The self-hardening slurry used for excavation was a mixture of ground granulated blast-furnace slag, bentonite, and Portland cement. Because the long-reach excavator used to dig the slurry wall was working from the narrow crest of the levee, the spoils from the trench were loaded directly into haul trucks for disposal.

One CMP was leaking worse than the others because both ends were underwater during the project to the extent that the owner felt that levee failure was imminent. The result would be a disastrous environmental release of contaminated water into a nearby waterway as well as disruption to the owner's operation. The contractor's personnel were mobilized to site in less than 48 hours, working out technical details of the emergency repair with the owner. Construction began within two further days. At this location, the contractor grouted the CMP to stop water flow and to mitigate potential slurry loss before starting the slurry wall excavation. The grouting took place over several days, using a thicker mix each day until the inside of the CMP was fully grouted and the water was displaced. The owner drained the downstream holding pond to verify that flow through the pipe had been arrested.

Following the verification of the grouting, the first wall segment was excavated across this CMP over a 225-foot-long portion of the levee. The contractor constructed the second wall segment without grouting the CMP first since the ends of this pipe were plugged with flyash. The CB slurry was allowed to fill the inside of the CMP when the excavation crossed it. A 6-foot-deep, 30-foot-long "trough" was constructed at the surface in front of this slurry wall to hold a similar volume of slurry to that which filled the CMP upon breaching it. The owner filled the third CMP with flowable fill before the contractor excavated this 60-foot-long segment.

The specifications required a permeability of 1×10^{-6} cm/s and 20 psi unconfined compressive strength after 28 days. The achieved permeability was in the range of 1×10^{-7} to 1×10^{-8} cm/s and unconfined compressive strengths in the range of 80 to 130 psi were recorded.

4.2.5.7 Slag-cement-bentonite cutoff wall, Taylorsville Dam, Ohio

In 1999 the Miami Conservancy District (MCD) instituted the Dam Safety Initiative (DSI) out of concerns for their aging flood protection system (Fisher, Andromalos, and Johnson 2005). As a result of this program, it was determined that the Taylorsville Dam would need repairs and upgrades in order to ensure its integrity in a flood event. The Taylorsville Dam is 2,980 feet long, 67 feet high, and 397 feet wide at its base. It was constructed of hydraulic fill between 1918 and 1922. In order to be sufficiently improved, the dam required the installation of additional relief wells, an increase in the size of the toe berm, and the raising of the dam cutoff from the existing clay core.

The methodology chosen to raise the cutoff was a self-hardening slurry wall (SHSW) using a backfill mixture of ground granulated blast-furnace slag (GBFS) and bentonite to create the extended impermeable core. The required wall depth averaged 34 feet and was installed using a Cat 330 excavator with an extended boom and stick. The wall is 2,400 feet in length and took 22 days to install. The required properties of the SHSW were a minimum UCS of 100 psi in 28 days and a maximum permeability of 1×10^{-6} cm/s. As verified in testing of field samples, the average 28-day compressive strength was 155 psi and the average permeability was 2.5×10^{-7} cm/s.

4.2.5.8 Cement-bentonite slurry cutoff wall, A.V. Watkins Dam, Utah

The A.V. Watkins Dam is a ∪-shaped zoned earthfill structure 36 feet high at its maximum section and more than 14.5 miles long (Barrett and Bliss 2008 and Demars et al. 2009). It is located between the Great Salt Lake on its west and I-15 on its east. On November 13, 2006, the dam came perilously close to a disastrous foundation piping failure due to internal seepage and erosion. Emergency measures were taken at the time by personnel of the Bureau of Reclamation and the Weber Basin Water Conservancy District. After the dam's situation was stabilized, a two-phase remediation program was implemented: Phase I allowed some interim storage prior to the implementation of the Phase II permanent modifications.

Phase I corrective actions included an upstream ring dike embankment centered on the incident area, a 200-foot-long interceptor trench located at the upstream toe of the dam, complete replacement of the downstream toe drain for a distance of 700 feet, and restricting the reservoir to EL 4,217 feet, approximately 9 feet below the active conservation reservoir water-surface elevation.

Phase II was a modification to the dam with the major feature being the construction of a 5-mile-long CB wall through the dam and up to 40 feet into the foundation materials. The wall design required a minimum UCS of 15 psi and a maximum permeability of 10 feet/year.

This option was evaluated against other remedial alternatives such as a new toe drain on the downstream side, a filter zone, an interceptor trench, and various combinations of the previous alternatives. Only a cutoff wall could address all of the failure modes and allowed for the use of several different construction techniques. Three types of walls were then considered.

A soil-cement-bentonite wall was rejected because it was considered too expensive and was not ductile enough to withstand the movement associated with a "floating" earthen dam. A soil-bentonite wall was considered and rejected due to the narrow crest of the dam which did not allow room for a backfill mixing area, and also its inability to resist internal erosion. A CB wall was judged ductile enough to flex with the embankment, had sufficient strength to resist erosion, and did not require widening the crest for a temporary working area. Additionally, the construction process of the CB wall would have the ability to fill voids and defects encountered in the trench since the CB slurry would flow into the voids during the trench exaction, thereby removing the need to fill these voids separately with structural material.

The wall was constructed using large hydraulic excavators with a "long stick" excavation arm that could dig up to 65 feet deep. The work took place from the fall of 2008 to spring of 2009 and is one of the largest cement-bentonite walls ever constructed in the United States.

4.2.5.9 Soil-bentonite slurry wall for seepage control at L-8 Reservoir, Florida

The Loxahatchee Reservoir, also termed L-8 Reservoir, is part of the Comprehensive Everglades Restoration Plan, a federal project to improve water quality and the distribution of fresh water in south Florida (Christman et al. 2009). The reservoir was created by constructing earth embankments to EL +23 feet. It was planned to increase the reservoir storage by dredging the bottom to EL −42 feet. However, during the initial design stage, some of the borings around the south storage area revealed zones of higher hydraulic conductivity than found elsewhere at the site. Because of the deepening of the reservoir by the dredging, the zones of higher hydraulic conductivity could potentially increase the seepage inflow during drawdown conditions.

As part of the performance criteria, the reservoir had to pass a strict seepage test before acceptance by the South Florida Water Management District. In order to comply with the performance criteria, it was decided to install a SB wall around the west, south, and east sides of an area of the reservoir designated Cell 6. The wall would have to penetrate to EL −50 to be effective. During excavation, five different strata were encountered including silts, clays, silty and clayey sands, limestone and weathered limestone, with the key-in stratum being dense cemented sand to dense slightly silty sand to dense slightly clayey sand. Although this bottom stratum would not be considered an aquiclude, *in-situ* permeability tests combined with seepage analyses showed that a SB slurry wall installed to this layer would reduce the potential seepage to levels that would meet the required criteria.

By analyzing composite samples from the test borings, it was determined that there were only about 8 percent fines, which would be too low to make the proper SB backfill, which generally needs fine contents greater than 20 percent to achieve the required permeability of 1.0×10^{-7} cm/s. It was decided to augment the excavated soils with soils from a nearby process pond that contained over 25 percent fines. Additionally, 2 percent bentonite by dry weight was added to ensure that the very low permeability could be achieved.

The wall was installed along the 10,982-foot alignment from an elevation of +18 feet (the berms were cut down 5 feet to widen the working area) to –50 feet, producing therefore 763,800 vertical square feet of wall. A large hydraulic excavator with a specially designed long boom and stick performed the excavation. Due to the cemented soils and the limestone, some unusual measures were taken to facilitate the excavation. In particular, a subcontractor was engaged to perform preblasting on 8-foot centers along the alignment in the limestone zone to create fractures that would allow the excavator to penetrate and remove the limestone.

4.3 CUTOFFS CONSTRUCTED BY THE PANEL METHOD

4.3.1 Clamshell excavation

The technology was first practiced by ICOS (under patent protection) on a project on the Venatro River in Campania, Italy, in 1950 and quickly spread throughout Europe as a very adaptable method for constructing deep foundation systems. The first Canadian application was in 1957 and the first use in the United States was in 1962. The first example for dam remediation appears to have been the seminal project at Wolf Creek Dam, Kentucky, between 1975 and 1979, although this was, in fact, a combination of rotary drilling and clamshell excavation techniques.

Clamshells (excavating buckets) can be cable-suspended or Kelly-mounted, mechanically or hydraulically activated (Photos 4.4 and 4.5). They are used to excavate panels 16 to 66 inches wide, to maximum depths of about 250 feet depending on the choice of crane. Most clamshell excavations are 24 to 36 inches wide and less than 150 feet deep since control over panel verticality becomes more difficult at greater depths. One "bite" is typically 6 to 10 feet long, and primary panels may consist of one to three bites. The exact length of a primary panel is reflective of dam safety concerns. In critical areas of the project, the length is typically restricted to one bite, since this minimizes the volume of excavation to be supported by the bentonite or polymer slurry prior to concreting. In less-critical areas, where the geological conditions are more favorable and where the consequences of the loss of slurry are not dire from a dam safety perspective, then the longer, multibite panels may be acceptable. The intervening secondary

Photo 4.4 Cable-suspended and activated clamshell. (Courtesy of TREVIICOS Corp.)

panel is most typically installed in one bite, with special attention required to assure the cleanliness and integrity of the interpanel joints.

4.3.2 Hydromill ("cutter") excavation

Hydromills, or "cutters," evolved from earlier Japanese and European reverse circulation excavating equipment in the late 1970s. Developed principally by Bauer, Casagrande, Rodio, Soletanche, and SoilMech, these machines basically consist of a large rigid frame housing two pairs of cutting wheels set below a high-capacity reverse-circulation suction pump (Figure 4.4). Such machines are best suited for excavating deep walls toed considerable distances into bedrock, for cutting through especially resistant horizons, and for assuring efficient tie-in into very steep valley sections or existing concrete structures. Due to their relatively high cost of operation (equipment depreciation charges, maintenance, and often significant downtime contribute mainly to this reality), their use is typically not competitive in the conditions prevalent on most levee repairs. As an exception,

Photo 4.5 Semi Kelly-mounted, hydraulically activated clamshell. (Courtesy of TREVIICOS Corp.)

the particular geological conditions and the length of the project at Herbert Hoover Dike, Florida, have rendered the use of a hydromill cost-effective, even though the cutoff is barely 80 feet deep.

As detailed in Bruce et al. (2006), hydromills had been used on nine major U.S. dam remediations between 1984 (St. Stephens Dam, SC) and 2005 (Mississinewa, IN) for a combined total area of almost 2.4 million square feet, while current work is ongoing at Clearwater Dam, Missouri, Wolf Creek Dam, Kentucky, and Center Hill Dam, Tennessee, which encompasses the same order of combined wall area. Wall thicknesses range from 24 to 72 inches with most being in the range of 33 to 39 inches. The maximum depth of just over 400 feet was recorded at Mud Mountain Dam, Washington, in 1990. Short, one-bite secondary panels (6–10 feet long) are typically used to mate at least 4 inches into the larger, three-bite (18–26 feet long) primaries (Figure 4.5). The same caveats on primary and secondary panel lengths applicable for clamshells are valid for hydromill operations also. Recent developments allow the hydromill to be guided in real time to

Figure 4.4 Hydromill being extracted from trench, at Herbert Hoover Dike, Florida. (Courtesy of TREVIICOS South.)

assure deviations from verticality considerably less than 1 percent of depth during excavation, and this is the standard of performance being exercised in the recent USACE seepage remediations (Photos 4.6 and 4.7).

4.3.3 Backfill materials and properties

Most panel wall cutoffs are created using a concrete that can be regarded as, more or less, "conventional." It must be stable, in the sense that it will not bleed when placed, it must be pumpable or flowable, when tremied, it must not segregate or compact significantly during or immediately after placement, and it must achieve reasonable and specified permeability and strength characteristics when it hardens. Most mixes incorporate some amount of pozzolanic substitution for the Portland cement portion of the mix, to benefit rheological, heat of hydration, and hardened properties. This is illustrated in the following two examples of recent mix designs.

The project specifications for the Mississinewa Dam, Idaho, remedial cutoff wall required 3,000 psi concrete for the 2,600-foot-long, 18-inch-wide,

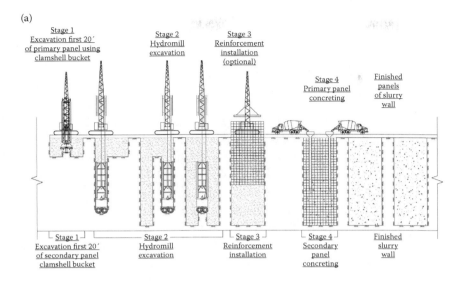

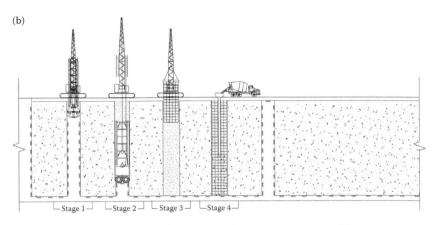

Figure 4.5 Typical sequence of work for a hydromill panel well: (a) primary panels;
(b) secondary panels. (Courtesy of Bencor, Inc.)

427,358-square-foot cutoff to depths of 147-230 feet (Section 4.3.5.7). The
initial mix used in the test section yielded a strength of nearly 5,000 psi,
which led to a concern that the concrete panels would become brittle and
crack over time. The mix design changes (Table 4.2) resulted in a 3,200-psi
strength. Slump was maintained in the range 6–9 inches with an average of
8 inches and an air entrainment of 4 percent. It is noted that 6 percent was
originally specified but, due to the high content of the flyash, an abnormally

Photo 4.6 Hydromill being moved along the work platform at Mississinewa Dam, Indiana. (Courtesy of Bencor-Petrifond JV and USACE.)

Photo 4.7 Hydromill operation at Wolf Creek Dam, Kentucky. (Courtesy of TREVIICOS Soletanche JV USACE.)

high amount of air entrainment additive was needed to achieve the 6 percent figure.

Rate of gain of strength data were as follows:

	UCS (psi)	
Age (days)	Test section	Production section
7	2,290	828
28	4,100	2,887
90	4,875	3,230

At Hodges Village Dam, Massachusetts, the 248,425 sf remedial cutoff was built in 213 panels, each 31.5 inches wide. The concrete mix produced a 28-day strength of 4,000 psi and comprised (per cubic yard of mix):

Cement	500 lbs
Flyash	124 lbs
Coarse aggregate	1,629 lbs
Fine aggregate	1,276 lbs
Water	33.7 gallons (280 lbs)
Air entrainer	0.6 oz
Retarder	18.7 oz

Some panel walls, as described in Section 4.2, have been constructed using a cement-bentonite mix, also known as a self-hardening slurry (SHS). There is a very wide range in the composition of these mixes reflecting different contractors' preferences but, in general, they can be expected to include 3–5 percent bentonite and 15–30 percent cement. It is common to include a retarder, while it is often overlooked that the mix *in situ* may well contain at least 10 percent of the native soil that has not been removed from the slurry during routine recirculation and cleaning. An example of a mix used by Trevi as a "plastic" cutoff for a dam in North Africa comprised (per cubic meter of mix):

Bentonite: 45–50 kg
Cement: 200–230 kg
Water: 900–950 kg

This provided:

- A permeability of less than 10^{-6} cm/s, decreasing further to 10^{-7} cm/s and less with time
- UCS ≥ 100 psi
- Strain at failure: 1–2 percent

Table 4.2 Cutoff wall mix designs, Mississinewa Dam, Indiana

Materials	Test section mix	Production section mix 60:40	70:30
Flyash:cement ratio	24:76	60:40	70:30
Portland cement (lbs) (Type I)	338	150	200
Flyash class C (lbs)	106	–	–
Flyash class F (lbs)	–	350	300
No. 8 limestone aggregate (lbs)	1,679	1,400	1,400
No. 23 sand (lbs)	1,503	1,491	1,505
Water (lbs)	222	285	285
Air-entraining admixture "A" (oz/100 lbs)	1 [8.9 oz./cy]	–	–
Air-entraining admixture "B" (oz)	–	21.5	21.5
Water-reducing admixture (oz/100 lbs)	3 [13.3 oz/cy]	–	–
Water-reducing admixture (oz)	–	10	10

Excellent and detailed background on specific U.S. projects has been provided by Khoury, Harris, and Dutko (1989), Hillis and Van Aller (1992) and Fisher, Andromalos, and Johnson (2005). Blast-furnace slag is proving to be a popular substitution for significant weights of Portland cement, especially where relatively low strength and long setting times are required.

There are special circumstances that demand the use of a "plastic" concrete backfill mix: in the United States the fear of a significant seismic event causing rupture to a stiff, hard concrete wall is often expressed. Some examples of mixes that have been used on recent dam remediation projects in such circumstances include the following:

Mix "A" (per cubic meter of mix)
Water: 400 kg
Bentonite: 30 kg
Cement: 150 kg
Sand and gravel: 1,300 kg

This provided $k = 10^{-7}$ cm/s; UCS = 60–120 psi, and E = 1,400–10,000 psi.

Mix "B" (per cubic meter of mix)
Water: 400 kg
Bentonite: 100 kg
Cement: 100 kg
Sand and gravel: 1,150 kg

This provided $k = 10^{-6}$ to 10^{-7} cm/s; UCS < 60 psi and failure strains of up to 5 percent.

To repeat, excellent general guidance on mix designs was provided in Xanthakos (1979), while the standard of care in the design and testing of such mixes was set by Davidson, Dennis, et al. (1992). The plastic concrete mix developed for their project comprised (per cubic meter of mix):

Water: 400 kg
Bentonite: 32 kg
Cement: 143 kg
Fine aggregate: 798 kg
Coarse aggregate: 798 kg

This provided $k = 4 \times 10^{-6}$ to 10^{-7} cm/s, UCS = 220 psi and an unconfined tangent modulus of 90,000 psi. A "jet erosion" test was also performed on trial mixes to attempt to quantify the mix's resistance to piping under service conditions.

A similar suite of tests was run by Anastasopoulos et al. (2011) on their plastic concrete mix, which comprised (per cubic meter of mix):

Cement: 150 kg
Dry bentonite: 35 kg
Water for bentonite: 350 liters
Free water: 35 liters
Sand: 675 kg
Gravel: 675 kg

This mix provided a wet density of about 120 pcf, a dry density of about 110 pcf, moisture content around 20 percent, a 28-day unconfined strength averaging 150 psi, and a secant deformation modulus at 5 percent strain of less than 29,000 psi. The permeability (at gradients from 50 to 300:1) was significantly less than 10^{-6} cm/s. Pinhole erosion tests also provided "not susceptible to erosion" results. CPT testing confirmed that in fact the material was acting *in-situ* like a very stiff sandy silty clay.

In similar vein, Dinneen and Sheskier (1997) detailed soil-cement-bentonite (SCB) mix used as backfill for the 1,400,000 sf of cutoff wall installed by panel methods at Twin Buttes Dam, Texas (Section 4.3.5.5). They noted that, despite previous SCB utilization in the Sacramento Levees, and at Sam Rayburn Dam, Texas, there was "limited experience" with this material upon commencing their project. Their mix design featured 4–10 percent cement (and/or pozzolan) by dry mass of soil ("aggregate") and 4–5 percent bentonite slurry (i.e., about 1 percent by dry weight).

The aggregate was reasonably well graded with a maximum size of 1½ inches and 10–20 percent fines. The mix needed a continuous-type plant capable of accurate batching and homogeneous mixing. Trucks were used for tremie placement. The mix had a 7-to-10-inch slump, a

28-day UCS of around 100 psi (or twice the potential 120 feet of head differential in service), and a target permeability of 1×10^{-6} cm/s. On this project, the slurry had to have a density less than 1.20, a sand content of less than 5 percent, and a Marsh cone value of less than 45 seconds, prior to SCB placement.

4.3.4 Principles of QA/QC and verification

In Section 4.2.3, the specific protocols for controlling and evaluating the quality of Category 1 walls were detailed, and the majority of these are valid for walls constructed with the panel method and, in large measure, also for cutoffs built with the secant pile method (Section 4.4). Readers may therefore find it convenient to consider at this point a more generic overarching appraisal of the subject.

As a general statement, quality control, assurance, and verification procedures must demonstrate that the cutoff wall is and has been installed in accordance with the requirements of the specifications and/or design intent of the project. Although the parameters and requirements often vary from project to project, the following performance requirements need to be demonstrated for any cutoff wall:

Geometry of the cutoff:
    location in plan
    depth
    width (including minimum overlap width at joints)
    length
    continuity
    verticality
Homogeneity and integrity
Material properties:
    strength, deformability, unit weight (wet and dry)
    permeability
    chemical compatibility (within the backfill components and with the
        ambient environment)

It is, of course, the case that quality control (QC) refers to measures implemented by the contractor during the execution of his work, and that quality assurance (QA) refers to measures taken by the owner, or his agent (either directly or via a third party) during and/or after the work is installed. The exact scope of each respective quality program is defined on a project-specific basis. This section identifies the various tests and measurements. It does not dictate which test or measurement must be conducted by each party, although it is natural that some are conducted by both parties, within a relatively short time frame, and often simultaneously. Further, additional or alternative tests or controls may be required for the particular type of construction method that has been adopted.

In contrast, the verification of the effectiveness of the cutoff, as a durable seepage barrier, is normally a longer-term project being conducted or reaching fruition long after the construction has been completed and the contractor has demobilized from the site: many climatic seasons or hydraulic cycles may be necessary before the intended contribution of the cutoff can be challenged and evaluated within the framework of its intended purpose. Bearing in mind a dearth of data on such long-term performance characteristics, the recent works of Rice (2009), and Rice and Duncan (2010a,b) are insightful and timely.

Another basic precept of QA/QC is that, to the extent practical, possible, and reasonable, each parameter should be capable of being verified by at least two independent means and methods and further, that all data and results, whether measured or recorded by contractor or owner, should be shared to the maximum extent contractually permissible for the overall benefit of the project.

4.3.4.1 Geometry of the cutoff

The plan location of panel or secant walls is most simply and effectively controlled by the use of guide walls (Photo 4.8). These are reinforced concrete structures, firmly and very accurately prepositioned in the working platform, so that the *starting* position of the cutoff is verified. The traditional "land surveyor" techniques of former years have been complemented in more recent years with the use of GPS techniques of astonishing

Photo 4.8 Guide wall installed prior to excavating a remedial cutoff wall. (Courtesy of Bencor.)

precision. Furthermore, even cutoffs installed with the backhoe method (Section 4.2), which does not incorporate guide walls in its process, can now be verified with similar means as being installed in the designed location to within inches of accuracy.

In short, there should be no reasonable or defensible argument that any contemporary cutoff wall has not been built in the requisite location, and to acceptable and anticipated tolerances.

4.3.4.2 Depth

Every cutoff wall must reach the minimum depth specified and, depending on the nature of the construction technique and the contractor's proposal, may have to extend some finite distance lower, to insure that the design intent is met. Contemporary excavation equipment of most types of cutoff walls is characterized by on-board instrumentation that provides the machine operator (and remote observers) with a real-time display of the depth of the excavation tool below ground surface as well as other information on tool verticality and other mechanical characteristics. These data are generated, very simply, from a sensor that records the movement of a steel cord attached to the excavation tool or the drill head (corrected for distance above ground level), or a sensor reading the drum revolutions. A good QA/QC program will allow for frequent, periodic calibration of such systems.

Following the excavation phase, the depth of the excavated element (panel or pile) is measured manually with a weighted tape, or some other simple mechanical device. Certain instruments used in Category 1 excavations to measure the shape or verticality of an element, for example, the Koden ultrasonic sensor, also have very accurate depth-recording capability. However, if required, following construction the depth of the wall can be further verified by full-depth coring and from the depth information provided from down-the-hole (DTH) logging devices such as the optical televiewer.

Again, it must be concluded that the installed depth of a cutoff wall should not be an issue for debate given contemporary construction standards.

4.3.4.3 Width

Cutoff width is dictated by the width of the individual elements, and by their overlap. Category 2 walls have a lateral dimension equivalent to the diameter or width of the mixing tool: provided there is no interruption to the injection of the grout during cutting and mixing, or there are no extraordinary hydrogeological conditions, then there is no real possibility that the thickness of the as-built elements can be doubted. Even then, a test or measurement is often specified at no more than 100-foot centers, and at 10-foot vertical intervals. Category 1 walls may occasionally be suspected of having a final *in-situ* width smaller than their excavated width, as a result of trench instability. In such cases, however, the real potential for such instability must be rationalized; this is in fact a very remote occurrence in dam and levee remediation. The simplest way for element width to be proved is to demonstrate the free movement of the excavation tool for the entire depth and length of each element so constructed. Further assurance is provided once the excavation tool is extracted, by one or more of the following assessments:

> An ultrasonic scanner (e.g., Koden) or sonic caliper can be used to provide a three-dimensional representation of the trench.
> The volume of backfill placed should be carefully logged against the rise in the backfill level in the trench, so allowing calculation of "overbreak" (typically 10–25 percent), and therefore confirmation of no excavation collapse, which would be reflected in an "underbreak."

In addition, for circular elements (i.e., as used to build a secant pile cutoff) that are water filled before concreting, an inverted plumb-bob can provide a quite surprising degree of accuracy (fractions of an inch) for such an old and simple tool.

For the geophysical methods in particular, the slurry in Category 1 trenches must have a low unit weight (e.g., ≤75 pcf), and a small amount of suspended solids (e.g., ≤5 percent) for accurate and effective results. Regarding the issue of overlap of adjacent elements, recent advances in on-board instrumentation for the excavation tools afford the contractor a surprisingly high degree of verticality measurement and control. As a consequence, the as-built geometry and location of each element—panel or pile—is provided in real time by inclinometers in the hydromill, clamshell, or rotary rig. Such data are then double-checked by one or more of three post-excavation methods outlined above. Further processing with CAD can then be done to illustrate the inter-element overlap at any depth, thus proving that the minimum wall thickness has been assured at joints.

It is now not unusual for wall verticality to be measured to an accuracy of 0.25 percent depth, although it must be emphasized that the key issue is wall *continuity*, not necessary *verticality*.

4.3.4.4 Structural continuity

Most Category 1 walls, and certainly those deeper than 100 feet, are constructed in discrete elements, such as panels or large-diameter piles. They therefore have inherent discontinuities (i.e., joints) at regular intervals. A continuity acceptance criterion should address the quality of inter-element joints, and assure that they are properly constructed, with full contact and without defects such as entrapped slurry or unmixed material, open seams, or open cracks. The key to creating good joints is, of course, appropriate quality control measures such as proper forming and cleaning of joints, thorough desanding of the slurry, adequate "bite in" and overlap between adjacent elements, and rigorous control over the quality and placement of the backfill material.

In general, walls with an unconfined compressive strength of about 100 psi or more can be cored, provided the appropriate coring equipment and methods are used: there is no question that the coring of these walls is a specialized form of drilling, and one wherein penetration rate must be sacrificed for good recovery and verticality. Special attention must also be paid to the selection of the drill flush characteristics and parameters. Otherwise there is the potential for core recovery to be poor, and/or for borehole walls to be cracked. Special standards of care must be imposed during the selection of an acceptably qualified driller. Cores can be taken of the interior of the element itself (i.e., vertical), of inter-element joints (i.e., vertical, but difficult due to hole deviation tendencies), or both (i.e., by holes inclined across joints, in the longitudinal plane of the wall). Interpretation of the core samples is facilitated when the concrete used in the primary and secondary panels has been colored with distinctive dyes. Whereas it is not atypical to find minor smearing of joints in Category 1 walls formed with high-strength concrete, it is equally common to find excellent contact in "softer" walls, for example, the self-hardening slurry wall joint shown in Photo 4.9, or in plastic concrete walls. Concerns over the quality of extracted cores can be resolved by conducting optical or televiewer surveys of these holes, which permit the actual *in-situ* conditions to be clearly exposed. Permeability tests (falling head or constant low head or rising head) can be a run on such joints

Photo 4.9 Compression test on a cored inter-element sample that shows different colors of self-hardening slurry. (Courtesy of TREVIICOS Corp.)

to further demonstrate continuity, although care must be exercised not to cause hydrofracture during water pressure testing.

4.3.4.5 Cutoff wall homogeneity

The definition of homogeneity varies from project to project and is different for Category 1 and Category 2 walls. It is not unreasonable to expect Category 1 walls—with the exception of deeper backhoe walls—to comprise backfill with no foreign debris inclusions, and minimal bleed or segregation. In other words, the *in-situ* material should not be sensibly different in uniformity, composition, appearance, and in other properties from the material as batched on the surface. Due to their relatively simple method of construction, backhoe walls can equally reasonably be anticipated to be somewhat less homogeneous, while still remaining fit for purpose, as described in Section 4.2.3.

For walls stronger than about 100 psi, coring is the standard method of *in-situ* evaluation of homogeneity. Cores should be inspected and logged by a professional and the drilling parameters of each hole (penetration rate, drillability, flush returns, etc.) carefully logged. Recovery targets should be pragmatically set—95 percent or more is not unreasonable to specify in "hard" walls, whereas 85 percent may be a more realistic criterion in soft and/or Category 2 walls, provided always that the lost 15 percent can be rationalized as not being truly representative of a void, soft inclusion, or segregation and honeycombing. Likewise, high rock quality designation (RQD) targets (≥80 percent) should be set. Core should be not less than 2¼ inches in diameter and retrieved in runs not more than 10 feet long. Alignment checks need to be conducted to verify that hole deviation is within acceptable limits (e.g., within a drill depth of 100 feet, a maximum deviation of the order of 0.5 percent can be achieved with adequate care and technique).

In-hole permeability testing or logging with an optical or acoustic televiewer run at a relatively modest rate, say not more than 3 feet per minute, also illustrates material homogeneity. Such tests have particular relevance when the drilling has been targeted at sampling specific interpanel or intercolumn joints. In this regard, it is especially difficult to "chase" a specific vertical joint with a vertical drill hole, due to deviation tendencies, whereas it is common to find inclined holes being drilled (within the vertical plane of the cutoff) that can intersect numerous joints at successively greater depths.

4.3.4.6 Material properties

Strength is not typically a fundamental design property of a cutoff wall since structural stresses induced in service are not significant. However, strength is linked to durability and to the resistance of the wall to piping-induced erosion, under service conditions.

Conceptually, a cutoff's deformability characteristic should be compatible with that of the surrounding embankment material at the time of the installation. This, of course, is a critical consideration when constructing a cutoff wall through deep alluvium under a new dam. This drove the recent decision to install a plastic concrete diaphragm at, for example, Papadia Dam in Greece (Anastasopoulos et al. 2011) wherein the 28-day average strength was restricted to 150 psi to assure a correspondingly low degree of stiffness (29,000 psi). (Incidentally, inclined cores taken from the Papadia cutoff were unable to differentiate inter-element joints, so intimate were their contacts.)

For Category 1 walls, strength and deformability tests are routinely conducted from samples of the backfill materials as delivered to the excavation, in addition to measurement of slump and bleed before placement.

In-situ sampling of Category 1 walls is commonly conducted by coring, except in "soft" walls where some other type of sampling (e.g., piston or spoon sampler) is used, if indeed any *in-situ* sampling is requested. Samples are subject to the broad battery of tests, usually at 28 days of curing, although there can be great benefit from conducting similar tests at 7, 14, 56, and 112 days (and more). When assessing the results of such tests, it is important to closely rationalize exceptional or unexpected data. For example, anomalously low strength can result from drill-damaged cores, or from the presence of relatively large inclusions. It is also important to seek out trends and, in this regard, a running 10-point average is a responsive way to proceed.

Cutoff walls are built to arrest seepage. Therefore, the assessment of permeability is of prime importance. Permeability is typically (but not necessarily) measured at 28 days after backfill placement, and it does tend to decrease with age as the backfill continues to hydrate. Samples taken of the backfill before placement, during placement, or after placement can indeed be tested—most accurately in a triaxial cell. Such tests invariably give uniformly low values (10^{-6} to 10^{-8} cm/s), which, of course, reflect the concrete- or grout-like nature of the backfill material. However, such tests will not reflect any potentially disruptive effects created by the construction and placement methods on the permeability of the cutoff as a structure. So, when the wall can be cored, the most representative test is to conduct an *in-situ* borehole permeability test, typically by rising head or falling head methods so as not to overpressurize the wall and cause fracturing. Also, bentonite should not be used as a core drilling fluid and completed holes should be flushed to ensure that the actual *in-situ* permeability is not being masked by mud smearing any fissures or joints. It must be noted, however, that the results of such coring must be viewed with care and understanding, for several reasons:

- In the case of lower-strength materials, coring may damage the wall, causing or triggering fissures to develop that would artificially increase the measured permeability. It is in such cases that borehole logging with the optical or acoustic televiewer is so useful, combined with a close examination of the cores themselves.

- Especially in the case of deeper walls, natural tendencies for boreholes to deviate can lead to perforation of the side of the cutoff, or the phenomenon of having only a very thin "skin" of backfill on one side of the hole, and so a susceptibility to coring- or testing-induced cracking. In certain cases, special directionally controlled drilling systems may be necessary, although examples of this in practice are very rare.
- The interpretation of the actual field-test data is not always straight-forward because of the cutoff wall geometry in relation to the bore-hole diameter; simplified equations (e.g., Hvorslev 1951) to calculate *in-situ* permeability do not take into account the complexities created by boundary conditions. The use of more rigorous numerical meth-ods can be advocated instead (e.g., Choi and Daniels 2006) to calcu-late more accurately borehole permeability results. Equally, when the focus of the water test is a specific interpanel joint, regular Lugeon tests can be run, but at modest excess pressures, of course. For walls that are too weak to be cored without creating artificially induced permeabilities, *in-situ* permeability must be verified with other types of testing such as a piezocone.

On the large scale, the hydraulic effectiveness of a cutoff is most accu-rately and responsively demonstrated by its effect on piezometric levels upstream and downstream of it, its effect on seepage volumes, and its elimination of suspended sediments or dissolved minerals in the seep-age outlets. Effectiveness can be verified by large-scale pumping tests on discrete stretches, or "cells," although to be meaningful, these must be conducted with extraordinary levels of engineering common sense (but frequently are not), and tend to be very costly. Alternatively, one must wait for the cutoff to be naturally tested, by a significant amount of reservoir raising. The benefit and accuracy of such testing is directly proportional to the extent of the historical "baseline" information available.

Chemical compatibility among the backfill materials themselves, and later between the backfill mix and the surrounding dam and foundation materi-als, are questions that are often raised, but infrequently addressed specifi-cally. Rather, during preconstruction lab testing, the mix is verified as having acceptable, repeatable, and controllable rheological properties, while rate of gain of strength data, especially if extended well beyond 28 days, tacitly confirm that no structural deterioration of the mix will occur with time. Regarding *in-situ* compatibility, on-site batching plants invariably use the local water supply (often just the lake or river water itself) and so imbal-ances based on water chemistry are not feasible.

There have been no published accounts of walls deteriorating with time, other than observations of desiccation in the tops of softer walls (of high water content) left exposed to the elements.

4.3.5 Case histories

As detailed in Table 4.1, there were at least twenty major dam remediations conducted on North American dams using panel or secant pile concrete cutoff walls in the period 1975–2005. Seven are now discussed further in this section to illustrate the development of the technology of panel walls with time, the range of backfill mixes used, and how site-specific problems were addressed. These seven projects are listed in Table 4.3.

4.3.5.1 Wolf Creek Dam, Kentucky

This USACE structure, near Jamestown, Kentucky, was built between 1941 and 1951, including a three-year interruption during World War II (Fetzer 1988; ICOS 1980; Ressi di Cervia, personal communication, 2011). As described in Section 2.3.4.2, the 3,940-foot-long, homogeneous, low-plasticity clay embankment has a 1,796-foot-long gated overflow concrete section forming the left abutment, which rises a maximum of 258 feet above its karstic limestone foundation. It impounds Lake Cumberland, which is the ninth-largest constructed reservoir in the United States. The original design relied on an upstream clay-filled trench to intercept major interconnected solution features that strike across it and extends to over 75 feet beneath top of rock.

The appearance in October 1967 of muddy flows in the tailrace, and in March and April 1968 of major sinkholes near the switchyard, prompted the massive emergency grouting operation of 1968–73. This arguably saved

Table 4.3 Listing of case histories described further in Section 4.3.5

Dam	Date	Backfill	Approximate area (SF)	Specialty contractor
Wolf Creek, KY	1975–1979	Concrete	531,000	ICOS
St. Stephen, SC	1984	Concrete and soil-bentonite	107,000	Soletanche[a]
Navajo, NM	1987–1988	Concrete	130,000	Soletanche[a]
Meeks Cabin, WY	1993	Plastic concrete	125,000	Bauer
Twin Buttes, TX	1996–1999	Soil-cement-bentonite	1,400,000	Bencor-Petrifond JV
West Hill, MA	2002	Concrete	143,000	Soletanche[a]
Mississinewa, IN	2001–2005	Concrete	427,000	Bencor-Petrifond JV

[a] in various business associations

the dam from a piping-induced failure through the critical area adjacent to the concrete section. It was concluded at the time that design issues existed with the depth and configuration of the core trench, and that there were deficiencies in the original single-line grout curtain installed from January 1942 to August 1943. A specially convened international Board of Consultants agreed that a "permanent" solution be implemented, since the presence of large amounts of potentially erodible clay in the karstic features would defeat the durability of the emergency grouting effort in the long term. They favored the installation of a continuous concrete cutoff wall starting at the concrete section and extending 2,237 feet along the dam crest, to a maximum depth of 280 feet, of which almost 100 feet would be in rock. They also recommended a shallower cutoff into rock around the switchyard. Such a project had not previously been undertaken on a major existing dam, a task further complicated by the fact that the lake could be lowered only by a small amount.

USACE elected to procure the work under what was, at the time, a very innovative contracting procedure. Specialty foundation contractors were solicited to provide unpriced technical proposals as a first step. Five of the seven schemes were rejected, and the remaining two qualified contractors were invited to price their own schemes as the second step. Furthermore, USACE defined a Phase 1 comprising 1,000 lft of wall, plus the switchyard, to reduce the amount of bonding required of the bidders. The ICOS Corporation of America was awarded the contract in 1975, having had excellent experience with a deep diaphragm wall at Manicouagan 3 Dam, Quebec—a new structure, however.

Their proposal was to construct the wall of minimum thickness 24 inches by first installing circular primary elements (piles), which would then be connected by biconcave secondary elements (panels). The self-imposed tolerance on pile installation was a maximum deviation of 6 inches at 280-foot depth. The concept is illustrated in Figures 4.6 through 4.8 and involved, in practice, sixty "painstaking steps."

The procedure for construction of each of these interlocking piles began with the excavation of the dry by clamshell of a 51-inch outside-diameter casing approximately 75 feet into the compacted clay of the embankment. After this hole was filled with bentonite slurry, a temporary one-piece, casing 47-inch outside diameter and 80 feet long, was inserted into a hydraulic casing driver and positioned. Then, with additional casing linked by special mechanical joints, this 47-inch casing was driven while a clamshell continued internal excavation to a depth of 140 feet. Throughout this operation, verticality was regularly checked by a direct plumb-bob method.

Temporary casing 140 feet long and 41¼-inch diameter was then placed inside the 47-inch casing and advanced downward to continue the excavation past the alluvium (about 150 feet depth) to the bedrock (about 200 feet depth). The oscillation imparted by the casing driver along with the weight of the steel casing sealed its notched shoe into the rock. A rotary drill with

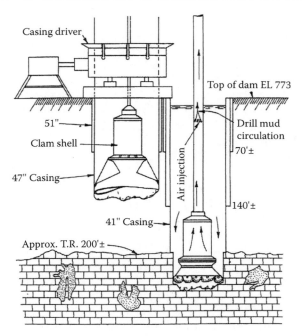

Figure 4.6 Primary element excavation, Wolf Creek Dam, Kentucky. (From A. Ressi, personal communication, 2011.)

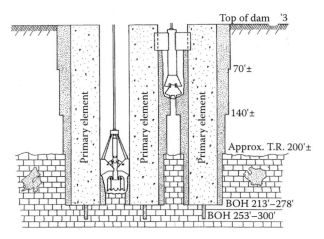

Figure 4.7 Secondary element excavation, Wolf Creek Dam, Kentucky. (From A. Ressi, personal communication, 2011.)

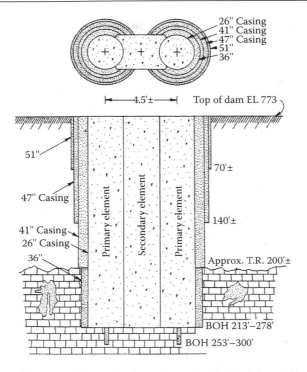

Figure 4.8 Typical section of completed diaphragm wall, Wolf Creek Dam, Kentucky. (From A. Ressi, personal communication, 2011.)

reverse circulation then excavated a 36-inch outside-diameter hole through the rock. To keep within the tolerance limits for the permanent casing, verticality was checked every 10 or 20 feet.

When the bottom elevation of the wall was reached, exploratory drilling was carried out in order to test the underlying rock. Once the rock was determined sound and tight by water-pressure testing, the exploratory hole was grouted and the bottom of the 36-inch hole cleaned to remove grout and rock cuttings. The 47-inch casing was withdrawn and the 41¼-inch casing freed. A 26-inch outside diameter permanent casing was weighted with ballast and lowered into position. With the permanent casing in place, a bentonite-cement grout was tremied into the annular space as the 41¼-inch casing was withdrawn.

After a 24-hour wait for the grout to strengthen, the ballast was lifted out of the permanent casing and a tremie pipe inserted. Concrete (3,000 psi) was tremied into the permanent casing. Once two permanent casings (primary elements) were completed, the embankment and overburden between them was excavated under a thin bentonite slurry by a "Wolf Creek" rig and chisel bucket (Photo 4.10). The chisel bucket was a specially designed

Photo 4.10 Cable-suspended grab in operation off a specially developed tracked rig, Wolf Creek Dam, 1975–79. (Courtesy of Arturo Ressi.)

clamshell that had a small set of jaws and biconcave chisels that rode the outsides of the two permanent casings. Once excavation reached top of the rock, a star chisel broke out the rock remaining between the two permanent casings. After the cuttings were removed by alternate use of a special clamshell and bailer, the entire excavation for the secondary element was filled with 3,000 psi concrete using the tremie method.

In July 1977, with the first phase of construction nearing completion, the second phase, comprising an additional contiguous 1,250 lft of wall, was also awarded to ICOS. Basically the same method was used but, due to improvements in technology, the verticality of the piles in Phase 2 was superior:

Deviation	Phase 1 (221 primaries)		Phase 2 (278 primaries)	
	Number of piles	Total	Number of piles	Total
0–3 inches	82	37%	122	44%
0–4	144	65	200	72
0–5	184	83	257	92
0–6	213	96	277	99

Note: 8 piles over 6 inches, 4 for specific reasons.   Note: Only 1 pile over 6 inches.

Another challenge faced by the project was the minimum 2-foot tie-in of the cutoff into the sloped (1 in 10) face of the concrete structure: this was accomplished by using a rotary drill rig to create a series of descending "steps" into the concrete.

Upon completion of the wall, piezometers on the downstream side dropped up to 60 feet (although some critical instruments remained high near the junction with the concrete structure), and there were slight (but erratic) downstream and vertical crest movements. Total flow from the measuring weir was about 1 gpm. Thus, at the conclusion of the project, there were strong indications that the wall was acting as a successful seepage barrier. However, by 2002, wet areas downstream of the dam—in similar and different locations to those prior to the 1975–79 wall installation—became more prominent, some critical piezometers had risen by about 10 feet, and other "distress indicators" were noted, prompting further phases of embankment exploration. This led to the USACE's decision to build a longer and deeper diaphragm wall, upstream of the first: this contract was awarded to a Joint Venture of TreviICOS-Soletanche in late 2008, for completion in 2014.

4.3.5.2 St. Stephen Dam, South Carolina

The dam is located on the Cooper River, and consists of a central concrete power station flanked by two earth-filled embankments (Soletanche Bachy 1999). The maximum height of the embankment above river bed is approximately 120 feet (Figure 4.9). This USACE structure rests on horizontally bedded sediments comprising, from the top down, interlayered sand, silt and clay, clay stratified with sand, sand with some clay, shale, and the limestone on which the powerhouse sits.

Various attempts to stop unacceptable seepages under the embankments, including blanketing, and sheet pile installation were unsuccessful, and so in 1984 St. Stephen Dam became the first example in the United States of the use of a hydromill to create a cutoff through an existing dam. USACE had just designed two such walls as new structures at the Clemson

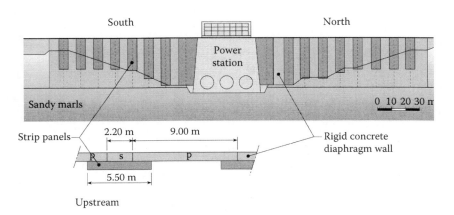

Figure 4.9 Arrangement of "complementary" panels, St. Stephen Dam, South Carolina. (Courtesy of Soletanche Bachy.)

Diversion Dams, and so had familiarity with, and confidence in, "slurry wall" technology.

A 24-inch-wide wall was excavated from the crest, through the abutments, toeing into the shale about 3 feet. The depth was about 120 feet, and the wall was built in 30-foot (primary) and 7-foot (secondary) panels, comprising conventional 3,000 psi concrete. Special care had to be taken to ensure that the panels adjacent to the sloping faces (1 in 10) of the powerhouse were properly keyed into the existing concrete structure: this was perceived as another advantage of using a hydromill. Given the high seismicity of the area, USACE also required that each interpanel joint be further protected and so upstream full-depth panels, 18 feet long, were installed and backfilled with soil bentonite. Each protective panel protected two cutoff wall joints.

Core drilling of the cutoff panels revealed in two cases some slurry trapped at their base, a situation remediated by grouting. Such experiences were put to good use in later hydromill walls where specific attention was focused on continuous trench desanding, which also facilitated the easy vertical travel of the mill through the trench.

This landmark project was in fact completed in about 180 days, and involved barely 78,600 sf of concrete cutoff wall, and 28,000 sf of soil-bentonite panels. The cost was less than $3 million.

4.3.5.3 Navajo Dam, New Mexico

The embankment was built on the San Juan River, 38 miles east of Farmington, New Mexico, by the U.S. Bureau of Reclamation ("Reclamation") between 1958 and 1963 (Dewey 1988; Davidson 1990). It is a zoned earth-fill of maximum height 402 feet and a crest length of 2,648 feet. The bed-rock is flat-bedded Eocene poorly to moderately cemented sandstones, with interbedded siltstones and shales. The sandstones are moderately to highly permeable and are weathered to a 200-foot depth in both abutments. This was particularly severe on the right abutment where the more intense weathering had removed the cementation, thereby increasing permeability. Deep-cutting river erosion had also created joints and cracking in the abutments, parallel to the very steep canyon walls.

Later evaluations of the original grout curtain concluded that "following the technical specifications of that period [it] was actually too light" (Dewey 1988). Seepage was observed within one year of initial reservoir filling, and increased thereafter to a rate of about 1,800 gpm by 1987. The left abutment contributed 600 gpm, and this flow had saturated the adjacent embankment materials for a distance of 50 feet from the contact. The embankment itself was found to be impermeable although the core material was potentially erodible. Historical and current data were evaluated, which concluded a high probability of seepage flowing along the embankment-abutment contact with the potential to erode core material into untreated joints and fissures, that is, the Teton Dam failure mode.

Reclamation therefore commissioned the construction of a concrete cutoff wall, which was built between May 1987 and April 1988. It extended for 436 lft at the left abutment, was 40 inches wide, and reached a maximum depth of 399 feet—at the time a world record. The contractor was procured under a "request for proposal" bidding system that weighted costs and technical approach. As shown in Figure 4.10, the wall had to be "cut in" to the steeply dipping contact, and this was achieved by a hydromill, the largest built to that time. The mill was 90 feet high, weighed 30 tons, and featured cutterheads that could swivel 2 degrees laterally and longitudinally for verticality control. Inclinometers within the frame provided real-time data to the operator. Panels were also surveyed ultrasonically prior to concreting, and the typical deviation was found to be around 0.1 percent of depth. This new machine was previously tested in December 1986 during a full-scale trial in France, where several test panels 30 inches wide, 7 feet long, and 400 feet deep were constructed and instrumented.

The wall was built in 18.9-foot primary panels with 6.7-foot secondaries, the primaries being installed in three bites. The overlap between adjacent

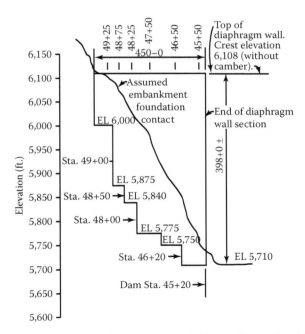

Figure 4.10 Profile of left-abutment diaphragm wall, Navajo Dam, New Mexico. (From Dewey, R., "Installation of a Deep Cutoff Wall at Navajo Dam," *Transactions of 16th ICOLD Conference*, Volume 5, San Francisco, CA, 1988. With permission.)

panels was 3 to 6 inches. The 3,000-psi nominal strength concrete was typically composed as follows (per cubic yard of mix):

Water	207 lb
Cement	363 lb
Pozzolan	155 lb
Sand	1,261 lb
Gravel (max 1½")	1,915 lb

This mix in fact provided 28-day strengths in excess of 5,000 psi.

The lake was drawn down over 60 feet to reduce the potential for damage during construction of the wall and the contractor developed an emergency preparedness plan bearing in mind the unprecedented depths involved and the potential for an uncontrolled loss of slurry. This in fact happened on five occasions, the worst being when 500 cubic yards of slurry plus 100 cubic yards of sand and gravel were suddenly lost 340 feet down: none of this was observed to exit the dam. The area was grouted, leading to successful wall completion. Cutting through the sandstone was slower than foreseen, necessitating the use of diamond teeth on the cutting wheels. An old grout cap and steel grout standpipes were also encountered during milling operations.

Dewey (1988) reported that plots of reservoir elevation and seepage versus time indicated that the flow through the left abutment had significantly dropped following wall completion, while Davidson (1990) provided further information on the performance of the wall. The response of the embankment was closely monitored during and after construction with 20 piezometers, 2 abutment weirs, and crest monitoring. The wall itself had 17 core holes, and 4 inclinometers. The left-abutment piezometers (in rock) indicated that "wall construction caused a decrease in the water level within the abutment but not as much drop as was anticipated" (Davidson 1990). Certain piezometers in the embankment, downstream of the wall, showed up to a 30-foot drop following the installation of the wall, while others there showed flows being forced around the end of the wall "or residual effects from the construction process" (Davidson 1990). Generally, upstream piezometric levels increased. Seepage flows were reduced by 57 percent. No inclinometer movements were recorded. Minor cracking ("shrinkage") was noted in the concrete core holes and up to ¼-inch-thick bentonite seams were found in certain interpanel joints, "the bulk" of which were "judged satisfactory."

In summary, the wall's performance was considered satisfactory, and it had apparently stopped all near surface flow at the embankment-abutment contact; the "remaining surface seepage is constant with reservoir fluctuation and presents no problems" (Davidson 1990).

4.3.5.4 Meek's Cabin, Wyoming

This zoned earthfill dam was built from 1966 to 1971 for the U.S. Bureau of Reclamation ("Reclamation") (Gagliardi and Routh 1993, Pagano and Pashe 1995). It has a maximum height of 174 feet, a crest length of 3,200 feet, and is located on the upper reaches of the Black Ford River, about 22 miles southwest of Ft. Bridger, Wyoming. The left abutment was constructed upon morainic materials comprising layers of sand, gravel, cobbles, and boulders (of very hard quartzite), subclassified as impermeable glacial till and permeable outwash deposits. These deposits overlie shale bedrock. The original design featured a cutoff trench, backfilled with core material, about 100 feet wide at its base and about 20 feet deep, toeing into the shale or impermeable till.

From first filling in 1970, seepage had emerged from the left abutment and had caused slope stability issues, remediated in 1971 by horizontal drains. The combined flow from these was about 500 gpm with the lake at EL 8,679 feet, compared to a dam crest elevation of 8,705 feet and a conservation pool elevation of 8,685.7 feet. By 1984 the seepage had migrated closer to the embankment-abutment contact, prompting the installation of a second set of drains. These intercepted at least 600 gpm and collected about 3 cubic yards of fine sand over the subsequent eight years. Since 1970 small sinkholes had also appeared at the upstream toe (at EL 8,665 feet) at Sta 22+50 (i.e., near the contact). Dye testing confirmed flow in these areas at a rate dependent on reservoir level. Further investigatory drilling in this area showed that there were three cohesive till deposits separated by two coarse granular outwash deposits, each of which was in contact with the embankment core material (Figure 4.11). It was therefore logical to conclude that the potential for internal erosion of the core was very high, and indeed such piping had already initiated. The fear of uncontrolled seepage occurring on the left abutment led Reclamation to select and design a concrete cutoff in this area.

This was designed to penetrate at least 10 feet below the gravels and into the glacial till (as identified by holes drilled at 145-foot centers), and so was 124 to 166 feet deep from the working platform at EL 8,701 feet. Fifty-nine percent of the 840-foot-long cutoff area was in embankment, and 41 percent in the foundation soils, for a total of 125,000 sf. The wall width was selected as 3 feet, based on considerations for panel deviation and erosion resistance in full-service conditions, bearing in mind that a plastic concrete wall was specified given seismicity concerns, and that ongoing settlements and deflections in the embankment were occurring due to consolidation and cyclic reservoir loading.

The performance requirements for the backfill mix were:

Permeability: less than 10^{-7} cm/s.
Strength: at least 200 psi at 28 days.
Ductility: 5 percent axial strain at failure

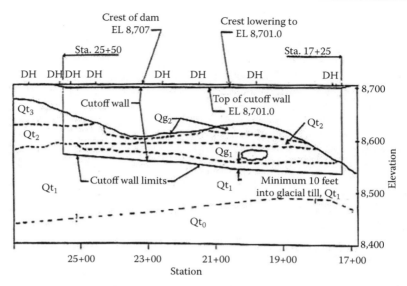

Figure 4.11 Geologic profile of left abutment, Meeks Cabin Dam, Wyoming. (From Gagliardi, J., and R. Routh, "Geotechnical Modifications at Meeks Cabin Dam," *ASCE Specialty Conference on Geotechnical Practice in Dam Rehabilitation*, North Carolina State University, Raleigh, NC, 1993. With permission.)

Fluidity: minimum 8-inch slump
Elastic modulus: 100 ksi (i.e., 4 to 10 times that of the dam's core at 170 feet)
Consolidation: <1 percent
Erodibility: <0.5 percent by weight

In addition, the wall had to (a) perform for the remnant life of the dam, (b) comprise materials available locally, and (c) resist sulfates in the groundwater. The mix design that was developed had the following composition (per cubic yard of mix):

Cement: 255 lb
Bentonite: 45 lb
Sand: 1,350 lb (approximately)
3/8-inch agg: 255 lb
1-inch agg: 1,100 lb
Water:Cement ratio: 1.8

This in fact provided a 28-day strength of about 400 psi, and a measured laboratory permeability of 2.4×10^{-8} cm/s.

Most of the excavation was foreseen to be conducted with a Bauer BC30 cutter, given the depth, quality, environmental and schedule implications, and the benefit of eliminating the need for end stops with primary panels.

Special rollerbit cutting teeth were developed to best penetrate and break the hard boulders. Matching technique with *in-situ* conditions, however, the contractor decided to excavate the core materials with hydraulic grabs. The cutter was then introduced to excavate the lower part of the trench, except where especially large, mobile boulders (over 42 inches in dimension) required the use of a grab again.

The specified verticality tolerance of 1 percent depth, resulting in a minimum wall thickness of 24 inches, was assured for each panel by preconcreting surveys with the Koden ultrasonic sounding device. A maximum primary panel length of 30 feet was set, "based on anticipated trench stability in the dense embankment and foundation glacial tills" (Gagliardi and Routh 1993, p. 763). The secondaries were built in one bite. There are no reports of massive, sudden slurry loss into the outwash materials, although a very detailed emergency reaction plan was devised with four different response levels. Other significant construction challenges included a narrow working platform (59 feet wide), the remote site location and the short working season (April through September).

4.3.5.5 Twin Buttes Dam, Texas

Twin Buttes Dam is located about 6 miles southwest of San Angelo, Texas and was constructed for the U.S. Bureau of Reclamation from 1960 to 1963 (Dinneen and Sheskier 1997). The dam is 8.2 miles long and extends over three streams. It is an earthfill embankment with a maximum structural height of 134 feet and a crest elevation of 1,991 feet. The dam was built without a positive cutoff in its central 4 miles, where an alluvial Pleistocene gravel deposit, overlain by 10–60 feet of clay, underlies the dam and extends from the reservoir downstream beneath the dam. A cutoff was omitted due to the depth of the sandstone/shale bedrock (average 60 feet, maximum 100 feet) and because of the blanketing influence of the clay over the gravel.

However, outcroppings of the alluvium are exposed throughout the reservoir and, during construction, borrow areas were excavated within 150 feet of the upstream toe of the dam, further exposing the gravel. The absence of a cutoff and the exposure of the gravel to the reservoir unsurprisingly led to significant underseepage, with the potential to fail the dam due to uplift pressures or internal erosion. Given the risks to population, water supply, and economic loss, a reservoir restriction to EL 1,930 feet was imposed in 1991 pending the completion of remedial measures to address the seepage deficiency.

The upper fine-grained material is in fact caliche—an indurated, lean clay rich in calcite. The coarse alluvial comprises mainly clayey gravel of highly variable gradation and cementation, and ranges from zero to 65 feet in thickness. The coarse fraction predominantly consists of limestone, but with chert, and quartzite and is also variably cemented, having UCS of up to 15,000 psi. All sediments were found to be extremely variable and unpredictable laterally and vertically, especially with regard to cementation and permeability. Its measured permeability ranged up to 5×10^{-1} cm/s. The bedrock is practically impermeable, and the upper 1–3 feet was weathered. The site is seismically inactive.

Seepage had been noted one year after completion, with the pool at EL 1,926 feet, and when in 1974 the reservoir reached EL 1,941 feet, a rapid rise in piezometric levels as far as 1 mile downstream was recorded. Drilling and grouting programs from 1976 to 1980 were ineffective due to the limits of the technology employed. A 1984 series of 61 relief wells was also ineffective and by the early 1990s piezometric pressures remained high, with total underseepages estimated at over 25,000 gpm. A remediation alternatives analysis was conducted, which concluded that a cutoff wall should be installed in the 4-mile "gap."

Design requirements for the cutoff included low permeability, resistance to hydraulic gradients, constructability, and cost. A target permeability of 10^{-6} cm/s was set, and the wall had to be of sufficient strength to resist a differential head of up to 120 feet (i.e., 50 psi for a 30-inch-wide wall). Various options were considered for the backfill material, namely plastic concrete, cement-bentonite, soil-bentonite, soil-bentonite with an internal (vertical) membrane, and soil-cement-bentonite. Studies showed that plastic concrete would be too costly. Cement-bentonite was judged not to be technically feasible due to the slow excavation rates predicted, for being incompatible with hydromill technology (not now the case), and for having a specific gravity close to bentonite slurry. Soil-bentonite was also ruled out due to potential for hydrofracture of a wall of typical width (2–5 feet), for settlement-induced horizontal cracking, and for piping potential (via "blowout gradient" tests). The use of a membrane was also ruled out on various fears, including damage during installation. On the other hand, soil-cement-bentonite had been used by the USACE on previous projects at the Sacramento River levees, and at Sam Rayburn Dam, Texas. This was judged adequate to resist hydrofracturing and/or blowout of the backfill into the gravels. The target 28-day UCS was 100 psi (twice the potential 120-foot differential), and the optimum wall thickness was taken as 30 inches.

Trials were to start with a mix comprising 6 percent (+/– 2 percent) cement (or cement plus pozzolan) by dry mass of soil, plus 1 percent (+/– 0.5 percent) bentonite by dry mass of soil (as added in a 4 or 5 percent slurry). The soil "aggregate" was a reasonably well-graded mixture of gravel, sands, and fines with a maximum size of 1½ inches and 10–20 percent fines. This was batched in a continuous mixing plant and transported to site in trucks with agitators. The target slump was 7–10 inches. Sufficient tremies had to be placed in each panel such that the backfill did not have to flow more than 7½ feet from a tremie. The bentonite slurry, prior to panel backfill, had to have a density below 75 pcf, a sand content of less than 5 percent, and a Marsh funnel viscosity of less than 45 seconds.

Based in part on the results of stability analyses, the wall location varied from the upstream toe of the dam to 25 feet upstream of the toe. The reservoir was lowered to EL 1,925 feet, allowing more than half the wall to be excavated "in the dry." The balance required cofferdams and work pads at elevations 5 feet above the lake elevation. The wall was keyed a minimum of 30 inches into rock, as determined by core drilling at 100-foot centers along the alignment at least 10 feet into rock, and water testing. The wall tied in longitudinally into the preexisting cutoff trench, 100 feet upstream of the centerline of the dam.

Cutoff construction began in 1996 with a 1,200-foot-long test section, conducted in 50-foot primaries and 8-foot secondaries using hydraulic and cable clamshells, and a hydromill. The backfill was cored at six locations and subject to permeability and geophysical testing. The remainder of the production work ran until early 1999, with the total work comprising 21,000 lft of wall, as much as 100 feet deep and covering an area of over 1,400,000 sf.

4.3.5.6 West Hill Dam, Massachusetts

This USACE dam is located in Uxbridge, Massachusetts, and was placed in operation in June 1961 (USACE 2004). It is a zoned embankment 2,200 feet long, a maximum of 48 feet high, and was constructed from locally available random fill materials and more limited impervious soils. There were no original foundation seepage-control features, and only limited remedial measures, including shallow toe drains constructed after flood events in 1979 and 1987. The embankment has an upstream inclined impervious zone and a limited upstream blanket. The foundation materials comprise primarily stratified sand and gravel glacial outwash deposits with highly permeable open-work gravel in channels, to depths of over 90 feet under the dam.

The dam experienced serious seepage problems during several moderate to low pools between 1979 and 2001. These induced gradients sufficiently high to cause sand boils and piping of foundation materials. Piezometer data indicated that excessive pressures were present beneath the embankment and the downstream toe area and flows peaked at 650 gpm. These pressures developed with little or no time lag as the reservoir pool rose. Analysis showed that the past remedial measures provided only limited and very localized protection and that much more adverse seepage-related problems could be anticipated when reservoir pool levels would exceed those experienced hitherto. Indeed, the government's studies concluded that the dam and foundation were inadequate to prevent extensive adverse seepage conditions from developing when the pool exceeded the 15-foot stage— a 2.5-year event. The recommended solution was a concrete diaphragm wall that would extend for 2,083 feet, to a maximum depth of 123 feet (including 2 feet into the basal granite gneiss), comprising 143,000 sf of cutoff. The excavated width was 31.5 inches and the wall was constructed using a hydromill with rotating tungsten carbide teeth, in three-bite primaries, 23 feet in length, and 9.2-foot-long secondaries, cutting about 6 inches into each adjacent primary, for a total of 117 panels. This project, although not of great scale, is particularly interesting on three counts: the construction details and problems, the QA/QC and verification program, and the performance of the cutoff immediately after completion.

The target 28-day concrete strength was 4,000 psi, and the target slump was 6–9 inches. Entrained air was 6 percent +/–1.5 percent. The principal mix for the tremie concrete comprised (per cubic yard of mix, and in accordance with the specification):

Type I/II cement	500 lbs
Type F flyash	124 lbs
Fine aggregate	1,300 lbs
Coarse (3/4") aggregate	1,629 lbs
Water	281 lbs
Air-entraining agent	1.0 oz
Water-reducing agent	31.2 oz
Superplasticizer	25 oz

The concrete was batched off-site and delivered in 10-cubic-yard trucks, in journeys taking 20–30 minutes, bearing in mind that concrete had to be placed within 45 minutes after introduction of the cement into the water/aggregate blend. Tremie pipes (10-inch diameter) were raised by a 110-ton crane and had to remain embedded 10–30 feet into concrete except for the initial 11 feet of placement. Pipes were placed at 11-foot centers in the 23-foot-long primaries, while for each secondary panel only one tremie was used. The top surface of the wall was moist-cured with saturated burlap mats. A total of 17,817 cubic yards of concrete was placed during 89 days (equivalent to an overbreak of almost 25 percent), and was subject to testing for compressive strength, slump, air content, and concrete temperature. Actual 28-day strengths ranged from 4,015 to 6,200 psi (σ = 593 psi). Average slump was 8.28 inches (σ = 0.7 inches) and air content averaged 6.1 percent (σ = 0.7 percent).

A test section was first conducted in the fall of 2001 and comprised two primary panels and one secondary panel. Full-length cores were taken from each panel and down each of the two interpanel joints. During production, a cable-suspended clamshell was used for the required pre-excavation, removal of the occasional large boulders, and construction of panels less than 15 feet deep. In general, the production rate through the fills and foundation soils was high, with only one major sudden slurry loss ("a few hundred gallons") into the coarse deposits. The hydromill was significantly slowed when excavating the toe into rock, especially when the rockhead elevation varied abruptly within one hydromill bite. Conventional chiseling was needed to help penetrate up to 10 feet of "slabby" granitic conditions. Following a winter shutdown, the cutoff was completed in late July 2002.

As part of the quality control and assurance program, one NX core was drilled every 200 lft of completed wall, in the middle of the concreted panel, and at least 5 feet into the bedrock. In addition, one 6-inch diameter core of the total panel joint and at least 5 feet into the bedrock was drilled at the same frequency, with the special condition that the joint had to be located at the center of the core for the initial 30 feet. Approximately 1,180 lft of concrete core (Tables 4.4 and 4.5) was consequently recovered from 13 panels, and 14 joints, although only 798 feet satisfied the joint location criterion. Upon completion of each core hole, a water-pressure test was performed at 15 psi for 15 minutes. No "appreciable" water losses were found in any hole.

While it is clear that the quality of the concrete and the joints was high, the real value of these two tables is in the "comments" columns: they encapsulate the typical spectrum of findings that are found when attempting such programs, and illustrate the problems and observations that can always be anticipated, even when the actual surface location of the interpanel joints can be accurately located.

The first significant pool after completion of the cutoff wall occurred in March–April 2003 when the pool peaked at 18.7 feet (EL 246.9 feet). All of the piezometers except two were influenced "primarily" by only tailwater changes and not by pool-level changes. The performance of the other two

Table 4.4 Concrete panel test coring summary, West Hill Dam, Massachusetts

Station	Panel number	Depth boring	Comments
22+70	Panel 10	54.1	Good concrete and contact
24+57	Panel 22	59.6	Good concrete and contact
25+82	Panel 30	74.3	Good concrete and contact
26+76	Panel 36	83.0	10" concrete missing at contact
26+90	Panel 37	78	Good concrete and contact
27+00	Panel 38	75.0	Good concrete and perfect contact
26+96	Panel 38	75.0	Boring near joint 37–38. G.C.C.
28+94	Panel 50	94.5	Good concrete and contact
29+56	Panel 54	105.0	Good concrete and perfect contact
31+12	Panel 64	113.5	Good concrete and contact
32+37	Panel 72	124.6	Good concrete 7" void at contact
34+55	Panel 86	104.6	Good concrete and contact
35+64	Panel 93	74.6	Control concrete panel. Concrete OK
37+98	Panel 108	65.5	Good concrete, 1" void at contact

Source: USACE. "West Hill Dam." Project Completion Report, Permanent Seepage Repairs. USACE New England District, 2004.

Table 4.5 Concrete panel joint coring summary, West Hill Dam, Massachusetts

Station	Panel number	Depth boring	Comments
21+97	5-6 Joint	46.9	Good concrete and contact
23+53	15-16 Joint	55.4	Good concrete and contact
25+00	24-25 Joint	14.1	Abandoned, no joint located
25+00	24-25 Joint 1a	9.5	Abandoned, no joint located
25+00	24-25 Joint 1b	9.5	Abandoned, no joint located
25+00	24-25 Joint 1c	64.0	Good concrete and contact
26+87	36-37 Joint	76.0	Joint visible to 66 feet
29+96	37-38 Joint	52.0	Good concrete and contact
28+21	45-46 Joint	81.2	Joint visible in core to 39 ft
29+77	55-56 Joint	52.0	Joint visible to 29.9 ft; boring abandoned
29+77	55-56 Joint 1a	104.6	Good concrete and core
30+39	59-60 Joint	29.2	Abandoned, joint runs out of core
31+33	65-66 Joint	117.7	Replacement for 69-70; good concrete contact
31+95	69-70 Joint	19.5	Abandoned, no joint located
31+95	69-70 Joint 1a	28.2	Abandoned, no joint located
33+82	81-82 Joint	23.8	Abandoned, no joint located
34+13	83-84 Joint	59.5	Joint visible to 26 ft., boring exit panel at 59.5 ft
34+13	83-84 Joint	110.9	Good concrete and contact
36+23	96-97 Joint	14.3	Abandoned, no joint located
36+23	96-97 Joint	18.9	Abandoned, no joint located
36+31	97-98 Joint	19.7	Replacement for 96-97, abandoned, no joint loc.
36+31	97-98 Joint 1a	89.5	Good concrete, encountered both panel corners, about seven feet difference in panel depths Bentonite and gravel in corners

Source: USACE. "West Hill Dam." Project Completion Report, Permanent Seepage Repairs. USACE New England District, 2004.

piezometers was rationalized as not being indicative of the performance of the wall and seepage was negligible. Very close analysis of the characteristics of certain piezometers adjacent to the wall revealed that the increasing pore pressures at their tips led to a flushing out of the bentonite lost during construction of the wall. Overall, it was concluded that "no deficiencies were detectable in the cutoff wall based on the piezometer responses" (USACE 2004, p. 18).

4.3.5.7 Mississinewa Dam, Indiana

This USACE dam is located in northern Indiana, about 65 miles northeast of Indianapolis (Hornbeck and Henn 2000, Henn and Brosi 2005).[*]

---

[*] This project I also referred to in Section 2.2.7.1.2 wherein the pregrouting of the rock mass to facilitate the safe construction of the diaphragm wall is detailed ("composite wall concept"). In this section only the diaphragm walling activities are discussed.

It comprises an 8,100-foot-long compacted earthfill embankment with a maximum height of 140 feet (EL 797 feet). It was completed in 1967 and placed in full operation in 1968.

During construction of the outlet works and left abutment, deeply karstified rock was found to be especially prevalent, and two very large clay-filled solution channels were uncovered. The construction records indicated a highly jointed, open-bedded, fractured foundation with substantial clay infilling. The two major channels were oriented at about 90 degrees to the centerline of the conduit running through the dam. A cutoff trench, grouting, and dental treatments were required in these areas. However, the right abutment had been almost completed by the time the extent of the Silurian limestone karstification at the site had been fully appreciated: "It appears the option to de-construct the right abutment was waived, based on the amount of funding it would take for the effort. It was also assumed the sands and gravels would act as a conduit for seepage waters flowing under the dam and filter any embankment materials eroded from beneath" (Hornbeck and Henn 2001, p. 4). The right part of the embankment was thus founded on 5–20 feet of coarse glacial outwash materials overlying unprotected and untreated karstic limestone, and had no cutoff to rock. The unweathered limestone strength reached 25,000 psi.

In 1988, project personnel noticed a depression ("significant and abnormal") in the guard rail on the right abutment. Re-evaluation of the data from crest displacement monuments revealed that a stretch of the embankment 300 to 400 feet long was continuing to settle at an average rate of about 0.035 feet per year (Figure 4.12), with no indications of stabilization. By 1999 the total crest elevation decrease in the settlement zone was almost 10 inches compared to 3 inches of post-construction settlement along the remainder of the dam. Furthermore, two aluminum casings for slope inclinometers had been found to be crushed and destroyed at depth in the area of distress, as a result of the settlement.

Incidentally, turbid seepage (up to 700 feet downstream) and boils had been observed as early as December 1966 (when the dam had just been completed), along the embankment toe and up the right abutment. An intensive grouting operation was immediately conducted with very large grout takes common. Not unusually for the times, "grouting was halted after several times the initial cost estimate was spent" (Henn and Brosi 2005, p. 6) with no discernible benefit. Other remedial efforts included an upstream seepage blanket and additional relief wells on the right abutment, and these appeared to be providing adequate security until the 1988 settlement observation. Logically, and based on a very detailed evaluation of the construction records, subsurface investigation, instrumentation analyses, and other observations, the government concluded that Mississinewa Dam was experiencing a progressive failure of the foundation, which, by subduction, would lead to an embankment failure, with the potential to occur both rapidly and early. Pool

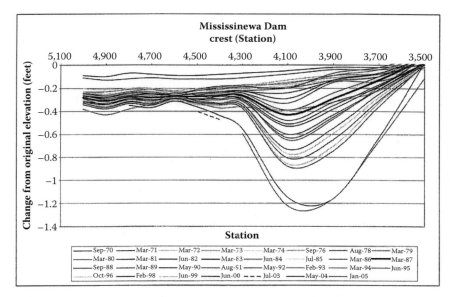

Figure 4.12 Change in vertical crest movement, 1970–2005, Mississinewa Dam, Tennessee. (From Henn, K., and B. E. Brosi, "Mississinewa Dam—Settlement Investigation and Remediation," *Association of State Dam Safety Officials 22nd Annual Conference*, Orlando, FL, 2005. With permission.)

restrictions, emergency action plans, additional instrumentation, and further explorations were all quickly implemented. The exploration holes "revealed increasingly negative and alarming information about the bedrock" (Henn and Brosi 2005, p. 7) and in particular one core hole encountered a 24-foot-deep solution feature incised into the bedrock. Karstic features were either infilled with very soft clay or had been washed open.

A concrete cutoff wall was selected to safely maintain future flood storage pools and stop the progressive deterioration of the foundation, and hence the threat to the overlying embankment. It was designed to extend approximately 2,600 feet along the length of the center valley and right abutment, reaching to depths of 147 to 185 feet, that is, at least 5 feet into a competent, unweathered limestone foundation. The wall tied into the original conduit dental concrete and was located about 10 feet downstream of the dam centerline, for logistical reasons and with considerations for preserving the existing instrumentation as much as possible.

Both hydraulic clamshells and hydromills were anticipated for building the wall, while "heavy duty cable-clamshell buckets [Photo 4.11] and 27-foot-long, 12-ton chisels" (Henn and Brosi 2005, p. 10) were also to be used to excavate difficult overburden conditions containing boulders. Primary panels were 25 to 26 feet long, separated by 9-foot-long secondaries spaced to provide a nominal 7.5-inch overlap. A minimum continuous

Photo 4.11 Heavy-duty cable-suspended and operated clamshell, Mississinewa Dam, Indiana. (Courtesy of Bencor-Petrifond JV and USACE.)

width of 18 inches was specified, which, anticipating reasonable panel deviation, involved the construction of 30-inch-wide panels.

The work was to begin with a 100-foot-long test section in the extreme right section of the right embankment. To the surprise of all parties, five sudden, massive slurry losses (up to 30,000 gallons each) were recorded at various depths in the bedrock during several attempts to excavate the first panels; the contractor successfully implemented the appropriate responses and was able to extract the hydromill on each occasion while assuring that the dam's security was not compromised. No evidence of the slurry was ever found downstream. It was at this point that the systematic pregrouting of the embankment/rock contact, and the rock itself, was conducted (Section 2.2.7.1.2). Not only did this operation seal the ground to the target residual permeability goals, so eliminating any subsequent slurry loss, but it discovered that there were *two* deep karstic features under the embankment in the critical zone, each of extremely complex geometry. The wall was therefore deepened locally to about 230 feet, and in this area individual panel lengths were limited to 10 feet, single bite.

The actual installed cutoff area was over 427,000 square feet, composed as follows:

| | Area of wall | |
	Embankment	Rock
Test section	6,825 sf	6,528 sf
Production wall	323,302	90,653

The overbreak in the production section was almost 29 percent. Concrete mix designs are provided in Table 4.2; the average strengths at 7, 28, and 90 days were 2,290, 4,100, and 4,875 psi on the test section, respectively, and 828, 2,887, and 3,230 psi in the production wall, respectively.

To measure panel verticality and hence assure that the minimum wall thickness was obtained, three independent verticality and continuity measurements were made on each trench:

1. An inclinometer and gyroscope were installed in the hydromill frame to provide a real-time evaluation of verticality and torsion.
2. A Koden ultrasonic monitor was used at the open-panel end points and panel midpoints to measure panel verticality and shape.
3. A 500-pound plumb-bob almost the same diameter as the panel width (Photo 4.12) was suspended from a crane to measure transverse deviations every 5 feet along each panel.

Photo 4.12 Traditional plumb-bob method as used in concert with modern methods to check panel deviation. (Courtesy of Bencor-Petrifond JV and USACE.)

Verticality was judged by comparing and studying data from all the sources prior to concreting. Also, no concrete verification hole (drilled to an accuracy of 0.13 degrees) exited any panel, the maximum depth being 230 feet. The concrete cores were "almost flawless" with only minor and infrequent honeycombing.

An immediate impact was noted on the piezometers, with those on the upstream rising by about 20 feet, and those on the downstream dropping about 15 feet. The relief wells in the valley center also confirmed the immediate effectiveness of the cutoff. "All instrumentation appears to be acting properly and the project as a whole appears to be working within the designed parameters" (Henn and Brosi 2005, p. 15).

4.4 CUTOFFS CONSTRUCTED BY SECANT PILES

4.4.1 Design and construction

Inter-element joints are always potentially a source of concern in cutoff walls in that they create a structural interface that may be contaminated with bentonite slurry and/or may be of less than desirable overlap or thickness, due to the natural tendency of the individual elements to deviate. One huge advantage of being able to construct cutoff walls with the panel method is that the number of inter-element joints is reduced to one every 10 to 25 feet or so. Whereas the significance of this threat is eliminated in backhoe walls or Category 2 walls constructed with the TRD method (Section 3.3), it is intensified when building cutoffs with secant piles, since an interpanel joint is to be found every 70 percent or so of the diameter of the pile. Such diameters are typically in the range of 24–40 inches. So why even consider building a secant pile cutoff wall at all, especially when the construction costs are typically higher than for panel walls?

The answer is, quite simply, that sometimes you just have to, in the face of the geological circumstances. The record shows that these "sometimes" are indeed infrequent; as shown in Section 4.1, secant piling has only been used as a *successful* remedial cutoff choice on U.S. projects since 1975, although rotary piling methods have been used as part of a composite (pile-panel) approach on two others during that period. It is also important to note that the issue of interpile continuity was addressed in a different way when building the cutoff for Arapuni Dam, New Zealand (Section 4.4.2.4). Instead of rotary drilling being used to create discrete primary and secondary columns, overlapped to form the cutoff, the rotary method was used to create slots of length equivalent to several individual pile diameters; each adjacent hole in each slot was drilled using the previous hole as a guide. When each slot had reached the maximum length compatible with dam safety calculations, it was concreted, to be later connected with adjacent slots.

This section deviates from the structure followed in the prior sections dealing with backhoe and panel walls. Since each application of secant piling, or rotary drilling, has had a unique set of driving factors, the technique is described through detailed evaluations of the few memorable case histories, including one project in Thailand and one in New Zealand; relevant because currently the principles of both are being employed in ongoing U.S. dam remediations.

4.4.2 Case histories

4.4.2.1 Khao Laem Dam, Thailand

4.4.2.1.1 Background

The Khao Laem multipurpose project is located on the Quae Noi River in Kanchanaburi Province, west-central Thailand, 270 km northwest of Bangkok (Alfonso 1984; Siepi, personal communication, Khao Laem Dam, Thailand, 2011). It involved the construction of a concrete-faced rockfill dam, 114 m high, with a crest length of 1,020 m and a total embankment volume of approximately 9 million cubic meters (cu.m). On the left abutment at the base of the dam, there is a surface power station with three 100 MW units.

The project was planned and designed by the Snowy Mountains Engineering Corporation (SMEC) of Australia for the Electricity Generating Authority of Thailand (EGAT). The contract for the construction was divided into three subcontracts: the drilling and grouting works of contract C2, the main concrete cutoff wall, C3, and the conventional grout curtain works. Construction began in 1980 and was completed in 1984.

4.4.2.1.2 Geology

The Thung Song Group formation outcrops on the left abutment and part of the right abutment, and the Ratburi Group outcrops only on the right abutment. The Three Pagodas Fault marks the contact zone between these two geological groups and traverses the karstic limestone of the Ratburi Group. The hydrogeological investigations therefore indicated adequate rock conditions to be present only on the left abutment. The Thung Song Group in the central area of the foundation for the main embankment and in part of the right abutment contained interbedded 0.01–0.5 m-thick layers of shale, siltstone, limestone, and sandstone, dipping 30–50 degrees to the south-southwest. The subvertical foliations of the beds had high permeability, due to the presence of cavities and fissures as deep as 60 m below the original river bed.

4.4.2.1.3 Choice of construction method

The karstic nature of the bedrock required the installation of a cutoff to a maximum depth of 180 m, using mostly conventional drilling and grouting techniques. However where large cavities were present, a concrete cutoff wall up to 55 m deep was necessary. Reflecting an almost total lack of previous experience in similar projects, the bid documents specified only the required nominal thickness of the wall and the concrete placement method. Based on an assumed rock compressive strength of 21.7 to 57.1 MPa, the contractor proposed the use of a hydromill to construct a panel wall, but further tests carried out on core samples provided compressive strengths much higher than the 100 MPa limit that typically renders the use of the hydromill unpractical. A secant pile wall alternative was therefore proposed by the contractor.

The drilling equipment consisted of a 30-m-high tubular-steel mast, mounted on a 70-tonne crawler crane, acting as both guide and support for the rotary head mounting a 686-mm-diameter down-the-hole (DTH) hammer, with a 762-mm bit. This equipment was designed with regard to the straightness and verticality requirements of the holes. Four test piles carried out in December 1980 demonstrated the suitability of the drilling method; these achieved an industrial drilling rate of 3.35 meters per hour (m/h), deviation from the vertical of about 0.25 percent, and no tendency to cause hole collapses.

4.4.2.1.4 Construction of the cutoff

A cutoff wall using the secant pile method was necessary in contract C3 (central section) and in part of contract C2 (from inside the grouting tunnel). The total cutoff wall length constructed in these working areas was 431 linear meters, with depths ranging from 15 m to 55 m.

For the C3 contract work, a 700-mm-thick concrete platform was first established, including the preset position of the cutoff wall axis, to facilitate correct rig setting up. The 762-mm-diameter secant piles were drilled at 508–615-mm centers (Figure 4.13) to form a continuous concrete cutoff wall having a thickness between 450 and 568 mm. The piles were installed in a primary-secondary sequence.

Drilling was performed using an Ingersoll-Rand DHD 130 DTH (Photo 4.13), having an air consumption of 90–100 cubic meters per minute (m^3/min) at 10.5 bar pressure. The drill string was mounted on a SoilMec drill tubular EC-80 mast fitted on an industrial excavator (Photo 4.14). Compressed air, cooled to improve efficiency, was supplied from a fixed battery of compressors delivering 180 m^3/min at the target pressure.

After at least a 36-hour setting time between two adjacent concreted primary piles, the overlapping secondary pile was drilled and concreted. Drilling operations were stopped if a cavity were encountered. After removing the drill string, the clay filling the cavity was removed using a bucket mounted on a Kelly bar. When the rock was found, the bucket was removed and replaced by a jetting torpedo with four radial coaxial nozzles for air (10.5 bars) and

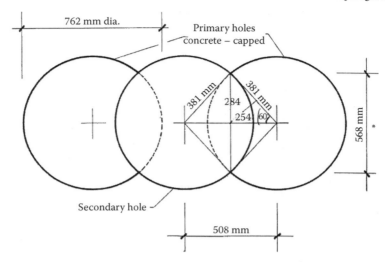

* Chord = min diaphragm thickness

Figure 4.13 Spacing and diameter of secant piles, Khao Laem Dam, Thailand. (From Rodio, Khao Laem Multipurpose Project—C3—Concrete Diaphragm Wall—Construction Method Analysis, Internal Report, 1980. With permission.)

water (2,000 liters per minute [l/min] at 20 bars), slowly rotated in the hole to complete the cleaning of the cavity. The cavity was then filled with tremied concrete. Once the concrete had set, drilling resumed until the required depth was reached in the classic "downstage" fashion. In many piles, multiple cavity treatments were necessary before the final depth could be reached. Some

Photo 4.13 Ingersoll-Rand DHD 130 DTH hammer, Khao Laem Dam, Thailand. (From Rodio, Khao Laem Multipurpose Project—C3—Concrete Diaphragm Wall—Construction Method Analysis, Internal Report, 1980. With permission.)

Photo 4.14 The secant pile rig showing the rod-loading system, Khao Laem Dam, Thailand. (From Rodio, Khao Laem Multipurpose Project—C3—Concrete Diaphragm Wall—Construction Method Analysis, Internal Report, 1980. With permission.)

cavities were up to 50 m³ in volume and, in several piles, the volume of concrete averaged over five times the theoretical neat-hole volume.

The average instantaneous drilling rate was 3.6 m/h, with peaks of up to 6 m/h. The overall mean drilling rate was 46 m/day. All drilling was conducted underwater as the water table was about 1 m below the working platform level. The total volume of placed concrete was about 18,700 m³, giving an average consumption of 0.73 m³/m length of pile.

For the C2 contract, the six horizontal, superimposed galleries for the grouting on the right abutment (named from top A to bottom F) were excavated with a vertical spacing of approximately 10 m between each other. The galleries, 3 m in diameter, extended as far as 3,900 m into the abutment, for a combined length of 22 km. The galleries were specifically targeted to find and treat the cavities and fractures in the karstic limestone. The original pattern of treatment was to drill 3-m-long grout holes radially around the galleries, but up to 50 m long in the deepest gallery (tunnel F). Unfortunately, during tunnel excavation, the quality of the rock was found to be much worse than expected, and so given the success of the secant pile method adopted in contract C3, the designer selected to adopt three additional cutoff methods designed to suit the specific rock conditions:

1. A continuous mined concrete cutoff wall was built across the micro-fractured limestone that could not be grouted due to the presence of blocky material surrounded by clay infilling.
2. A 300-mm-diameter secant pile cutoff wall was selected in zones of major karst below the lowest gallery where shaft construction and trench excavation would have been impracticable due to the high water table.
3. In zones of minor karst between the galleries, which also could not be grouted, 165-mm-diameter holes were drilled at 330-mm centers and backfilled with tremie concrete.

The six drill rigs, mounted on bogeys and customized to work in gallery conditions, were equipped either with a top hammer or a DTH hammer. In order to grout a total of 300,000 linear meters of holes, drilled to a maximum depth of 100 meters, 20 piston pumps were mobilized to site. Where the holes intercepted clay-filled cavities, the area was more intensely treated, by drilling 165-mm-diameter holes spaced at 330-mm centers. The cavities were cleaned with water jets, and then backfilled with mortar. Three additional rigs were used to drill a total of 45,000 linear meters of holes.

Where the rock conditions were worse, the designer decided to install a secant pile cutoff, replicating what was done in contract C3, but with smaller holes, of diameter 300 mm, drilled to a maximum depth of 15 m (Photo 4.15). The holes were cleaned with water jets before backfilling with concrete. A total of 6,000 linear meters were drilled using DTH hammers.

4.4.2.1.5 QA/QC

To ensure adequate overlap for their full depth, each pile had to be drilled to a tolerance of about 0.2 percent. Verticality was checked after drilling, using a neatly fitting metal cage suspended on a wire from a large steel tripod set over the hole. Hole deflection was determined by measuring wire movement for varying cage depth on a scale at the base of the tripod. The original method to check the actual overlap of the secondary pile with the two adjacent primary piles foresaw the use of a television camera lowered into the secondary hole, but this turned out to be impractical. Checks revealed that deviations in the piles were of the order of 150 mm per 50-m depth (0.3 percent). The quality and continuity of the cutoff wall was further confirmed when drilling for grouting through the area, and by the diamond core drilling of the cutoff. A pump-out test was performed about 100 m to the left of the riverbed, to verify the effectiveness of the cutoff.

4.4.2.1.6 Observations

The choice of the secant pile method assured the efficient cleaning and removal of the clay infilling the karstic limestone in the cutoff. In fact,

Photo 4.15 Drilling with 300-mm DTH hammer in the galleries, Khao Laem Dam, Thailand. (From Rodio, Khao Laem Multipurpose Project—C3—Concrete Diaphragm Wall—Construction Method Analysis, Internal Report, 1980. With permission.)

during the following grouting works in an area where four outer rows of grout holes were installed up to depths of 100 m, severe hole collapses confirmed the extremely poor geological conditions. The final closure was achieved by a 100–130-m-deep row of holes drilled through the concrete of the cutoff wall itself. Cement takes, using heavily sanded thick mortar mixes, of up to 225 kg/m of grout hole were recorded, with the cutoff wall already in place. These high grout takes suggest that the flushing effect of the hammer promoted the subsequent replacement of the compressible infill materials with grout. Moreover, the subterranean water flows discovered by geological investigation would have required the use of chemical grouting; the secant pile concept was much more efficiently implemented.

The definitive account of the construction of the dam concludes:

> With the experience gained at Khao Laem, the use of the overlapping piles cutoff produced by large diameter DTH methods has no practical limits regarding hardness of rock, whereas the depth is limited only by the available air pressure and machine torque. Apart from the need

for quite large flat working platforms imposed by the overall dimensions and weight of the drill rig, it is believed that this method of cutoff construction could be used, without resorting to impractical power requirements, up to depths of about 100 m. (Alfonso 1984)

Given the experience gained in similar projects that followed in the United States and New Zealand, this was a remarkably prescient statement.

4.4.2.2 Beaver Dam, Arkansas

4.4.2.2.1 Background

Beaver Dam is located on the White River in northwest Arkansas (Bruce and Stefani 1996). It was constructed for the U.S. Army Corps of Engineers between November 1960 and June 1966. It consists of a concrete gravity section 1,332 feet long, rising to a maximum height of 228 feet above the stream bed, flanked successively to the north by a main zoned embankment 1,242 feet long, and three smaller saddle dikes. The top elevation of the flood control pool was originally 1,130 feet, and the maximum pool elevation is 1,137 feet.

Dike 1 is adjacent to the north end of the main embankment. During design, a graben beneath Dike 1 had been identified as a potential problem source, due to the resultant presence of very permeable, highly weathered karstic Mississippian limestone with clay infilling (Boone Formation). A grout curtain was therefore originally installed along the centerline using contemporary practices. However, soon after initial filling of the reservoir, seepage was observed at several exit points on the downstream face of Dike 1, totaling 800 gpm. Remedial grouting in 1968–71 succeeded in reducing the flow to about 500 gpm. Clearly, the presence in the Boone limestone of many open and clay-filled cavities and channels, porous beds, and deep, intensely weathered permeable zones, allied to the difficulty of grouting in dynamic water-flow conditions, had limited the effectiveness of the grouting operation.

The seepage had traditionally remained clear, but a new muddy spring appeared in December 1984 after a long period of unseasonably heavy rains. Fearing material loss from the dike, the USACE decided to lower the flood-control pool level to 1,128 feet. This markedly reduced the rate of clear seepage, but hardly influenced the new dirty flow. In addition, the reduced pool elevation directly affected flood-management capacity and restricted generating capacity in the powerhouse in the concrete gravity section.

By February 1988 the USACE had designed a concrete cutoff wall to be installed in the bedrock upstream of the dike's crest, with a depth varying from 80 to 185 feet. The first attempt to construct a panel-type cutoff using a hydromill failed. Apparently, those beds of fresh rock had *in-situ* compressive strengths of over 25,000 psi and rendered the use of a hydromill "practically impossible," given the time and cost overrun projections.

In August 1990, the USACE's Resolicitation of Request for Proposals led to the award of a contract based on the concept of constructing the wall by large-diameter secant concrete piles. One of the contractor's Joint Venture partners (Rodio) had the experience of constructing a similar cutoff at Khao Laem Dam, Thailand (Section 4.4.2.1). Construction of the wall itself began in October 1992 and lasted for 22 months.

4.4.2.2.2 Geology

The graben underlaid Dike 1 and the contiguous 200 feet of the northern main embankment (Figure 4.14), and was downfaulted about 200 feet between NE/SW trending faults characterized by zones of disturbed material. Some planes were infilled by competent breccias or solution deposits, while others were open and clean.

Under variable thickness of relatively impermeable overburden (typically 15–40 feet) the deeply weathered siliceous and cherty Boone overlaid sound limestone. The Boone was mainly spongy and chalklike, containing highly irregular tubular and sheet-like cavities, mostly infilled with soft clay containing rock fragments and chert concentrations. The sound rock contained a network of interconnecting cavities that locally extended down to EL 974 feet, about 170 feet below the dike's crest elevation.

Prior to drilling for the cutoff, the upper layers of the work platform, embankment, and overburden materials were excavated by clamshell under slurry, to the top of weathered rock, and replaced by concrete. This was intended to act as a 4-foot-thick "casing" for the piles when subsequently

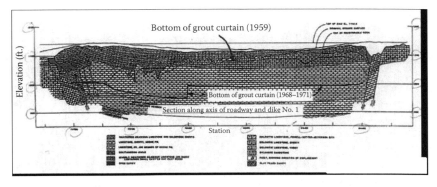

Figure 4.14 The inferred geology of the graben area underlying Dike 1, Beaver Dam, Arkansas. (From Llopis, J. L., D. K. Butler, C. M. Deaver, and S. C. Hartung, "Comprehensive Seepage Assessment: Beaver Dam, Arkansas." *Proceedings of the 2nd International Conference on Case Histories in Geotechnical Engineering*, St. Louis, MO, 1988. With permission.)

passing through these upper layers. This overburden replacement covered 4,713 square yards and consumed 7,011 cubic yards of concrete, mainly of 3,000 psi strength. Figure 4.15 shows the recorded profile of overburden depth, and the subdivision of the wall, into four "areas," based on the different geological and construction conditions subsequently encountered:

Area	Pile number	Dike station
A	0–496	62+00–72+43
B	497–638	72+45–75+25
C	639–687	75+27–76+22
D	688–738	76+24–77+22

4.4.2.2.3 Construction of the cutoff

The cutoff wall extended for a total length of 1,475 feet from Sta 62+00. It was offset 65 feet upstream of the embankment centerline and was built from a 65- to 80-foot-wide work platform, benched into the upstream face of Dike 1 at EL 1,130 feet.

The wall depth varied in response to the geological conditions from 80 to 185 feet although Pile 572 was extended to 215 feet for exploratory purposes. The individual 34-inch-diameter piles were located at 24-inch

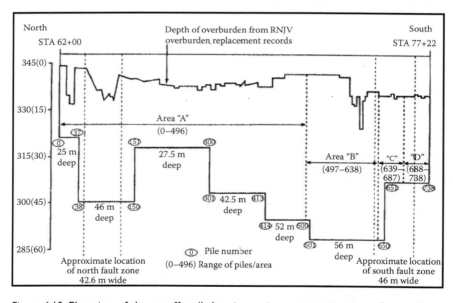

Figure 4.15 Elevation of the cutoff wall showing main construction areas, Beaver Dam, Arkansas. (From Bruce, D. A., and S. Stefani, "Rehabilitation of Beaver Dam: A Major Seepage Cutoff Wall," *Ground Engineering*, 29, 5, 1996. With permission.)

centers, yielding a nominal chordal joint width of 24 inches. They were installed in classic primary-secondary sequence. The total wall area was 207,700 square feet. A total of 24 additional ("conforming") piles were installed, mainly in areas A and D to assure the required pile overlap at full depth. Coring of piles, and their contacts, was executed for QA/QC purposes at a total of 40 locations.

The following general rules were observed to avoid disturbing nearby piles being drilled or that had been recently concreted:

1. Drilling was permitted only beyond a distance of 30 feet from an adjacent open pile not entirely in rock.
2. A minimum elapse of 48 hours was specified after completion of concreting in a primary pile before drilling the next successive primary pile.
3. Drilling of a secondary pile could begin only when the concrete of the two adjacent primaries had reached at least 2,000 psi unconfined compressive strength.

Two rigs (Photo 4.16) employed 32-inch-diameter drill rods in 30-foot lengths. While most of the drilling was conducted with a conventional DTH, successful experiments were made with a Fisher-Soppe "cluster" DTH comprising five 8-inch hammers on a 24-inch diameter casing. Drill penetration rates for primary holes ranged from 8 to 21 feet per hour (average 14.5), and from 13.5 to 23 feet per hour (average 18) in secondaries. These rates varied considerably between areas and between rigs.

At each pile location, the drilling rig was set up using conventional survey and laser systems. The verticality profile of each pile was measured

Photo 4.16 Secant pile operations at Beaver Dam, Arkansas. (Courtesy of Rodio-Nicholson JV.)

using a "submersible reverse plumb-bob," a very simple device, but yet one of exceptional accuracy.

The various concrete mixes used during construction were produced by an automatic batching plant in the immediate project area, rated at 200 cyds/hr. Transport to the cutoff was by means of 9-cyd truck mixers. Mixes were varied during the work in response to experiences gained, and strong QA/QC measures were enforced both at the batching plant and at the point of placement for both fluid and set properties. The most commonly used mixes had the following composition (per cubic yard):

Coarse aggregate	1,600–1,660 lbs
Fine aggregate	1,280–1,363 lbs
Cement	485–400 lbs
Flyash	100–130 lbs
Water	33.0–27.5 gal
Water reducer "A"	12–15 oz
Reducer "B"	9–0 oz
Air-entraining agent	3.8–2.25 oz
Calcium chloride	0–28 oz (use very limited)

Water was heated or chilled, depending on other material and ambient temperatures. Actual concrete volumes are illustrated in Figure 4.16. In total, the cutoff comprised 739 piles, almost 104,000 lft of drilling, 29,000 lft of redrilling, and almost 30,000 cubic yards of concrete. These numbers exclude work conducted on prestabilization, downstaging, additional piles (the 24 conforming elements), and the initial overburden replacement.

Several changes to the foreseen method statement had to be made during the work in response to geotechnical challenges and in the interests of progressive improvement and efficiency. The more remarkable changes were as follows:

Downstaging: Some problems of ground instability were foreseen while drilling through the weathered rock, that is, below the overburden replacement and above the bedrock. When these instabilities prevented continuous drilling to full depth, the rods were extracted and the pile depth sounded. Following removal, whenever necessary, of appreciable amounts (more than 2 feet) of loose material, the hole was backfilled from the surface by concrete. No earlier than 24 hours later, the hole was redrilled through the unstable zone. A total of 71 holes in Area "A" were completed in this way, some requiring three successive treatments. These piles involved 6,087 feet of redrilling and 2,111 cubic yards of concrete.

Ground pretreatment by pressure grouting: For a 300-foot-long section in Area A, a layer of coarse gravel was encountered, and a test-grouting operation undertaken over a 120-foot-long section. Two rows of holes were installed,

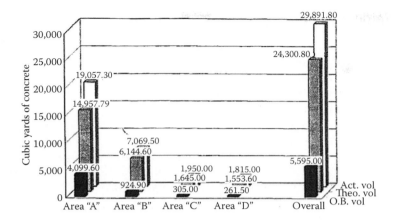

Overbreak percentage weighted
Total overbreak = 21.01%

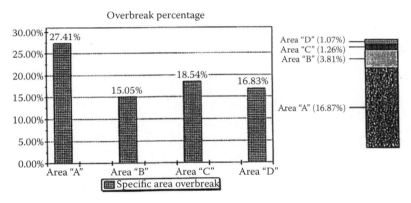

Figure 4.16 Quantities of final concrete used in the cutoff wall, Beaver Dam, Arkansas. (FromBruce,D.A.,andS.Stefani,"RehabilitationofBeaverDam:AMajorSeepage Cutoff Wall," *Ground Engineering*, 29, 5, 1996. With permission.)

4 feet apart, respectively, 1 foot upstream and downstream of the cutoff. Seven-inch-diameter steel casing was drilled to the level of "recoverable rock" and cementitious mixes injected during their withdrawal. Totals of 3,483 ft of drilling and 282 cubic yards of grout were involved. Thereafter, 14 piles were installed in this grouted zone, without the need for downstaging. This trial showed that the principles of grouting could be well employed to fill voids and permeate coarser, loose, cohesionless materials in the weathered zone.

Pile hole stabilization by grouting: During wall construction in areas B and D, severe problems were posed by the instability of the weathered rock. In Area D, one consequence was settlement of the work platform and the appearance of sinkholes, notably near piles 690 and 714. The sinkholes were excavated, examined, and then backfilled with lean mix concrete, while activities on several other piles under construction in this area were suspended. After much discussion, a modified grouting-based method was selected. The work platform, embankment, overburden, weathered rock, and the top 3–5 feet of sound rock were to be treated. Basically, the percussive drilling was interrupted at various depths and the resultant (partial) hole visually inspected. Once the maximum achievable stable depth was identified, the rods were reintroduced, but with a 32-inch-diameter rock roller bit at their tip. Grout was then pumped through the rods and bit and simultaneously mixed with the unstable material, while also filling voids. In this way, efficient stabilization was achieved in this section of hole. The method was repeated where necessary until stable bedrock was reached, and it proved highly successful in permitting the standard construction methods to then be used to hole completion. A total of 51 piles in Area D were treated in this way, some requiring as many as six (Pile 714) successive treatments. In total, over 1,470 cubic yards of grout were injected plus 132 cubic yards of a concrete used in the more conventional downstaging process in certain piles in less problematical sections.

In Area B, the early piles showed the existence of particularly unstable, weathered rock, containing cavities, voids, and very soft clay pockets in places. This zone ranged from 15 to 90 feet deep. Basically the same method proved in Area D was used there also, with similar success. The process was needed in all the holes in this area, at least once, and as often as six times (Piles 534 and 550) and, on one occasion (Pile 584), ten times. A total of 4,971 cubic yards of grout and 2,553 cubic yards of downstage concrete were used, both volumes considerably in excess of neat drilled hole volume, emphasizing the very cavitated nature of the ground in this area.

4.4.2.2.4 Cutoff effectiveness

Data were recorded from the existing seepage monitoring instrumentation. Of five seepage areas (SA-1 to SA-5), the area of most concern was SA-1, located in a natural gully 310 feet downstream of the centerline of Dike 1 at Sta 71+00 between EL 1,052 and 1,032 feet. Besides the piezometric network in the three embankment, specific measuring devices were installed to monitor the seepage exiting downstream of Dike 1. The most significant of these were:

Seepage Area SA-1

Parshall Flume No. 1: measures flow from SA-1
Parshall Flume No. 2: measures flow from all the seepages from Dike No. 1
V-Notch Weir: measures surface water seepage from SA-1

Seepage Area SA-2

French Drain Weir: connects to Parshall Flume No. 2
Artesian Well

Major observations on these instruments were:

- Parshall Flumes no. 1 and no. 2: The rise in flows in mid-June 1993 was associated with pumping excess surface water from the work platform to the other side (downstream) of the dike. By mid-March 1994, Area D had been completed and the grout stabilization of Area B commenced; flows decreased. By mid-September 1994, after surface pumping had ceased, the remnant seepage was barely 4 gpm, compared to a maximum of over 1,270 gpm in September 1993.
- V-Notch Weir: A sharp decrease (in both seepage and pumped water) also occurred from mid-March 1994 but by late August 1994, when the cutoff was completed, it had totally dried up.
- French Drain Weir: Until September 1993 the flow was related to lake level, stabilizing at 3.25 gpm when the level fell below 1,118 feet. At the beginning of grout stabilization in Area D in late January 1994, a sharp increase in flow occurred—greater than attributable to lake-level fluctuation alone. This suggested a redistribution and concentration of flow paths by the treatment. This flow peaked at 11.64 gpm on March 14, 1994, three days after the completion of Area D. However, by the end of March 1994, the flow had dropped to 2.06 gpm and dried up totally two weeks later.
- Artesian Well: Reacted, with delay to lake level, but showed a 41.35-foot drop in March 1994, despite the rise in lake level. In mid-June 1994, it reached its "drying up" elevation of 1,066.2 feet.

In general, these measurements showed that, prior to construction, the seepage was several hundred gallons per minute, varying with lake level. In mid-March 1994, when Area D was completed and grout stabilization in Area B was commenced, all devices showed a sharp decrease. This trend continued until the completion of the whole wall in August 1994, when all five seepage areas had dried up and the total underseepage was barely 4 gpm.

4.4.2.3 Walter F. George Dam, Alabama

4.4.2.3.1 Background

The Walter F. George Lock and Dam comprises a 460-m-long concrete structure housing the spillway, a non-overflow portion, the four units for 3.2 megawatt (MW) turbines, and two embankment wing dams, for a total length of 3,660 m (Photo 4.17) (Ressi di Cervia 2003, 2005; Siepi, personal

Photo 4.17 W. F. George Lock and Dam, Alabama. (M. Siepi, personal communication, 2011.)

communication, Walter F. George Dam, AL, 2011). The navigation lock is 25 m × 137 m with a lift of 26.8 m. The dam was constructed for the USACE on the Chattahoochee River, between Georgia and Alabama, from 1955 to 1963.

Immediately after the reservoir was impounded, it became apparent that excessive underseepage was occurring. Efforts to stop the leakage were only partially successful; in 1981 and 1985 shallow cutoff walls were built through the east and west embankments to control the seepage. While this greatly reduced the seepage under these embankments, the seepage under the concrete structures increased. In 1982 a flow of over 8,000 liters per minute developed under the powerhouse. This was mitigated by locating the connection on the lake bottom and filling it with concrete. Grouting was also carried out to seal seepage through discrete channels in the bedrock, particularly under the powerhouse but they did not resolve the problem of the seepage below the concrete structure, with peak flows estimated at almost 2,000 liters per second.

The decision to construct a 60-cm-thick positive cutoff upstream of the concrete structure was taken when it became clear that this situation compromised the generating potential of the dam as well as potentially jeopardizing its stability. This is the first project in which the USACE had designed a cutoff wall in front of an active concrete dam, and it was the first installation of a cutoff wall through approximately 30 meters of water.

4.4.2.3.2 Geology

The cutoff was installed through three units of the Clayton formation. The uppermost is an argillaceous limestone followed by a shelly limestone and a sandy limestone, and the lowermost stratum, 1.2–2.1-m thick, has a compressive strength in excess of 130 MPa. The cutoff wall was to extend into the Providence sand formation at EL −5 feet.

4.4.2.3.3 Contract procurement

The contractors invited to bid were requested to propose construction methods consistent with the delicate environment of the lake. The specification requirements were minimal and required the bidders to submit separate technical and, later, cost proposals. This process allowed the USACE to award the contract to the bidder whose proposal was in the best interest of the government and not necessarily the lowest bid. This is also called "Best Value" contracting.

Only proposals from three contractors were found acceptable, and then matched with the prices. On August 2001 the USACE awarded the contract to the Joint Venture of TreviIcos South, a subsidiary of Trevi SpA, and Rodio, Inc., a subsidiary of Rodio SpA, which in turn was acquired by Trevi in 2005. Both companies had participated in the secant pile works at Khao Laem Dam, Thailand (Section 4.4.2.1) and Rodio had been the senior Joint Venture partner at Beaver Dam, Arkansas (Section 4.4.2.2).

The JV's proposal—not unsurprisingly, given their provenance—for installing the cutoff wall consisted of secant piles using reverse circulation drilling and panels excavated by a hydromill.

4.4.2.3.4 Exploration and pregrouting

Before cutoff wall construction began, an exploration and grouting campaign was completed along the entire centerline of the wall (Photo 4.18), in order to fill any large cavities or solution channels that may have had a negative impact on the installation of the cutoff wall, and to minimize the fluid losses. Although the grout takes on the 100 holes drilled from a barge were generally modest, totaling 350 cubic meters (Figure 4.17), the grouting program increased the level of confidence in, and knowledge of, the rock in which the cutoff had to be installed.

Photo 4.18 The exploratory and grouting campaign being conducted from a barge, W. F. George Dam, Alabama. (M. Siepi, personal communication, 2011.)

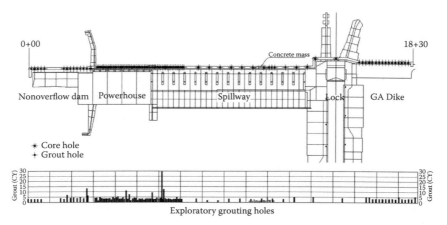

Figure 4.17 Preliminary grouting results, W. F. George Dam, Alabama. (Ressi di Cervia, A., "Construction of the Deep Cutoff at the Walter F. George Dam," *United States Society on Dams, 25th Annual Conference*, Salt Lake City, UT, 2005. With permission.)

4.4.2.3.5 Cutoff wall construction

In preparation for the secant pile portion of the work, the lake bottom in front of the powerhouse, spillway, and lock of the dam was dredged to remove obstructions and debris and to form a trench in which a flowable fill was placed. A known obstruction in the line of the cutoff wall was a remnant of a sheet pile coffer cell, used in the original construction of the dam. The operation to remove this structure involved a team of 5 divers completing over 100 dives, in water 33 m deep, so it required an onsite decompression chamber.

The secant piles were to be installed in the main portion of the wall in the classic primary/secondary sequence. Once the working apron was in place, the contractor installed a system of modular mobile templates secured to the concrete dam to place and drive the temporary casings into the apron with the necessary accuracy. These templates were positioned so that the cross frames cantilevered out far enough to avoid the toe of the dam (Photo 4.19 and Figure 4.18). Modification and customization of the main support had to be completed on site, to follow the profile of the dam, and to avoid the obstructions found on the anticipated alignment. This leveraged the flexibility of the secant pile method, which naturally allows virtually any alignment to be followed. Fixed positions were provided on the frames at the same spacing as the piles, acting as guides to install the 35-m-long temporary casings. Each casing, fitted with a positioning sleeve, was then positioned using a survey instrument on each axis to assure plumb installation, and then driven into the apron using a vibratory hammer, or an impact hammer.

Photo 4.19 Template supports out from the buttresses of the dam, W. F. George Dam, Alabama. (M. Siepi, personal communication, 2011.)

The main equipment for the excavation of the piles was two Wirth reverse-circulation drill rigs, assisted by two cranes mounted on large barges (Photo 4.20). The piles were drilled to a final depth of about 70 m from top of casing, using water from the lake as the drilling fluid.

The airlifted cuttings were delivered to a hopper barge fitted with silt curtains around the perimeter, and deposited to the bottom of the lake. Water quality was constantly monitored using a remote underwater automatic system, which posted the information on an FTP site.

Upon completion of drilling, the pile was surveyed for verticality by using a biaxial inclinometer. The results of these measurements were plotted to assure that the requirement of the specifications were met, and the positive cutoff assured. Figure 4.19 shows an example of the as-built conditions of a section, where the position of each pile is shown, successively, at top of apron, at EL 40 feet, 25 feet, 10 feet, and −5 feet (final elevation).

A plastic concrete was tremied to an elevation corresponding to the bottom of the lake. This mix had a 28-day unconfined compressive strength of 7 MPa.

4.4.2.3.6 Cutoff wall construction

All work was conducted within 20 feet of the dam and all of the operations required close cooperation and coordination with the USACE. The power

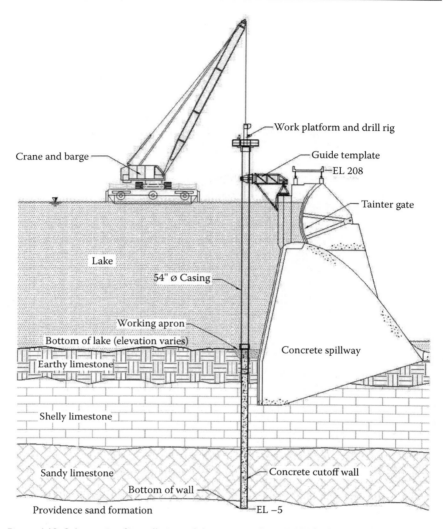

Figure 4.18 Schematic of installation of the secant pile wall, W. F. George Dam, Alabama. (Ressi di Cervia, A., "Construction of the Deep Cutoff at the Walter F. George Dam," *United States Society on Dams, 25th Annual Conference*, Salt Lake City, UT, 2005. With permission.)

generation schedule was reviewed and adjusted on a daily basis to accommodate activities in front of the generators. Opening of the 14 spillway gates during high water delayed the contractor for a few days but typically the USACE was able to keep closed those gates that were adjacent to the work.

A diaphragm wall was installed from land, in primary and secondary panels, 800 mm thick and up to 64 m deep, and tied into the shallower existing

Photo 4.20 Top pile rig excavating with reverse-circulation method, W. F. George Dam, Alabama. (M. Siepi, personal communication, 2011.)

walls on the east and west embankments. The two walls of the navigation lock needed to be perforated in order for the secant pile cutoff to be continuous. A hydromill was used to cut a slot 1.8 m thick in the two concrete lock walls (33.5 m deep and up to 26.8 m thick at their base), and a submerged retaining wall (Figure 4.20). Several alternatives were studied for cutting of the lock walls but all utilized divers extensively: by utilizing the hydromill, divers were not required. The crane carrying the hydromill was mounted on the concrete

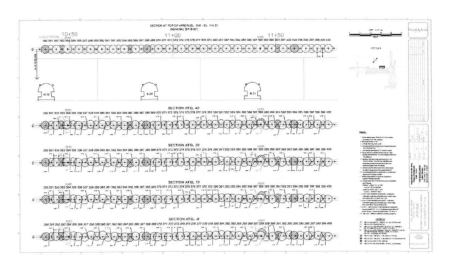

Figure 4.19 As-built drawing of the secant pile cutoff wall at different elevations, W. F. George Dam, Alabama. (Ressi di Cervia, A., "Construction of the Deep Cutoff at the Walter F. George Dam," *United States Society on Dams, 25th Annual Conference*, Salt Lake City, UT, 2005. With permission.)

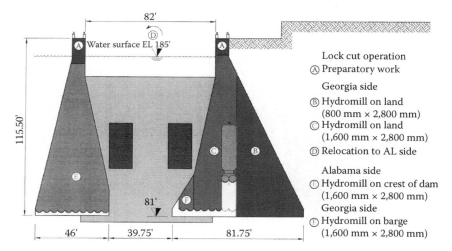

Figure 4.20 Navigation lock cut, W. F. George Dam, Alabama. (M. Siepi, personal communication, 2011.)

platform (Photo 4.21) or on a barge. In the west area, an underwater retaining wall had to be penetrated before installation of the cutoff wall and soil improvement by jet grouting was necessary at the back of the wall (Figure 4.21) since the mass excavation needed in this area would have created a risk of collapse.

After 500,000 working hours with no accident, in October 2003 a total of about 470 each 1.25-m-diameter secant piles and about 3,800 square meters of hydromill wall had been installed to complete the 600 linear meter of cutoff wall.

4.4.2.3.7 Cutoff wall effectiveness

The final quality control of the wall was performed by using coring. Thereafter, the tops of the piles were trimmed and a capping beam placed to

Photo 4.21 Hydromill cutting the lock walls standing on the concrete dam, W. F. George Dam, Alabama. (M. Siepi, personal communication, 2011.)

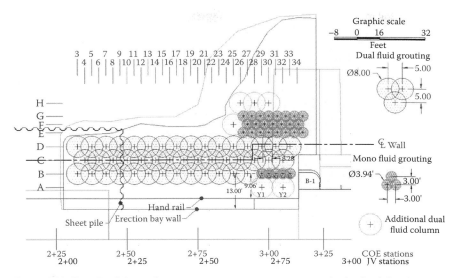

Figure 4.21 Sketch of the underwater jet grouting treatment at the back of the retaining wall, W. F. George Dam, Alabama. (M. Siepi, personal communication, 2011.)

seal the cutoff against the concrete of the dam. A series of piezometers was installed by the USACE just downstream of the wall to monitor the effectiveness of the cutoff wall. These immediately showed a considerable drop in levels where the wall was complete, indicating a very successful cutoff.

4.4.2.4 Arapuni Dam, New Zealand

4.4.2.4.1 Background

Arapuni Dam is a 64-m-high curved concrete gravity structure, with a crest length of 94 m. It spans the Waikato River south of Hamilton, New Zealand (Photo 4.22), and is owned and operated by Mighty River Power Ltd. (Amos et al. 2008; Gillon and Bruce 2003; Siepi, personal communication, Arapuni Dam, NZ, 2011). It was completed in 1927. A concrete-lined diversion tunnel runs through the right abutment around the dam, with separate gate and bulkhead shafts. A series of 600-mm × 600-mm underdrains was imbedded as the main uplift control at the dam/foundation interface. The original cut off walls extended to a depth of 65 m below the dam crest and also into the left and right abutments of the dam. No grout curtain was constructed initially.

4.4.2.4.2 Geology

The bedrock on which the dam is founded is composed of multiple ignimbrite flows from volcanic eruptions that have occurred over the

Photo 4.22 Arapuni Dam, New Zealand. (From Amos, P. D., D. A. Bruce, M. Lucchi, N.
Watkins, and N. Wharmby, "Design and Construction of Deep Secant Pile
Seepage Cutoff Walls Under the Arapuni Dam in New Zealand," *USSD 2008
Conference*, Portland, OR, 2008. With permission.)

last 2 million years (Figure 4.22). The upper part of the Ongatiti unit
has a UCS of 2 to 6 MPa, while the "hard zone" of the Ongatiti has a
strength up to 28 MPa. Subvertical cooling cracks trending north-south
are present for the full depth of Ongatiti, with apertures of up to 80 mm.
The fissure infill is nontronite, a very weak smectite clay that is highly
susceptible to erosion.

4.4.2.4.3 Historical performance

Since first impounding in June 1930, seepage pressures correlating to
reservoir level were present in some areas of open joints under the dam,
and a series of seepage events occurred—all related to the erosion and
piping of the weak clay infilling. Thereafter, the reservoir was completely
emptied after a large crack was discovered in the headrace channel near
the powerhouse. During the repair works, a single-row vertical grout
curtain was constructed along the upstream toe of the dam and along the
front of both abutment cutoff walls (Figure 4.23). Grouting appeared to
be initially successful, since leakage at the dam was considerably reduced
when the reservoir was refilled in 1932. However, it was realistic that
grouting would not have intersected all vertical rock joints or displaced
clay in the joints, and that therefore future leakage incidents were probable. Also it was clear that the curtain was not in physical contact with
the heel of the dam.

The most recent seepage incident developed in 1995 and required grouting (completed in December 2001) to fill an open void within a foundation
joint. During drilling of two drainage holes into a fissure under the dam,

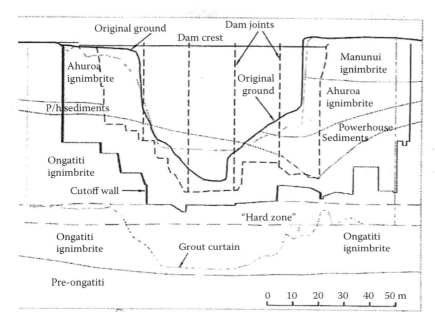

Figure 4.22 Elevation of Arapuni Dam, New Zealand, looking downstream. (Amos, P. D., D.A.Bruce,M.Lucchi,N.Watkins,andN.Wharmby,"DesignandConstructionofDeep Secant Pile Seepage Cutoff Walls Under the Arapuni Dam in New Zealand," *USSD 2008 Conference*, Portland, OR, 2008. With permission.)

the holes intersected a discharge of approximately 600 liters per minute at near-lake pressure.

4.4.2.4.4 Investigation and design

An extensive investigation program, involving the coring of 122 holes from downstream was initiated between 2000 and 2005 to determine the extent of foundation features requiring treatment. The cores allowed the preparation of a detailed tridimensional map of the foundation, based on which four fractured and potentially erodible zones beneath the dam were identified as requiring a permanent cutoff (Figure 4.24).

With the safety of the dam temporarily secured by the 2001 grouting program, the owner decided to upgrade the dam by installing a high-quality and verifiable cutoff solution to definitively arrest the seepage. On the Waikato River, Mighty River Power has eight dams, sited to allow water to flow directly from one lake into the next, and it was essential that the new cutoff be constructed while the reservoir remained in service; dam safety during construction was therefore a vital consideration in the selection of the remediation technique.

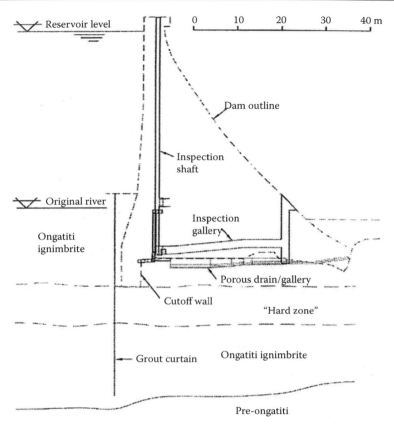

Figure 4.23 Cross-section of Arapuni Dam, New Zealand, looking west. Note the spatial separation of the grout curtain from the dam. (From Amos, P. D., D. A. Bruce, M. Lucchi, N. Watkins, and N. Wharmby, "Design and Construction of Deep Secant Pile Seepage Cutoff Walls Under the Arapuni Dam in New Zealand," *USSD 2008 Conference*, Portland, OR, 2008. With permission.)

A preliminary selection of possible positive cutoff solutions was proposed to the bidders; a conventional grouting solution was not deemed sufficient due to the presence of the nontronite. The options included the installation of secant piles, a diamond wire saw to create thin panels, and water knifing, in which high pressure water is injected to displace and remove the infill of the fractures. Each of these options proved to be beyond the limits of technology, given the depth and the site conditions. The decision process required thorough consideration of risks, both technological and to dam safety and, given the difficulties to be tackled during the project, the owner decided to create an alliance between himself and the selected specialty contractor in order to identify cutoff options, develop them, and

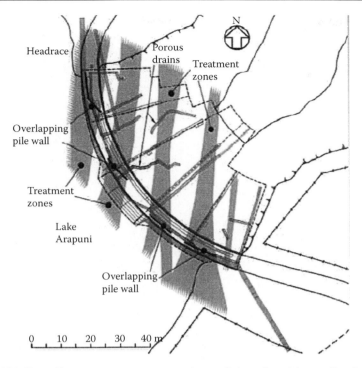

Figure 4.24 Plan of long-term seepage control remedial works, with cutoff walls, treatment zones, and underdrain, Arapuni Dam, New Zealand. (From Amos, P. D., D. A. Bruce, M. Lucchi, N. Watkins, and N. Wharmby, "Design and Construction of Deep Secant Pile Seepage Cutoff Walls Under the Arapuni Dam in New Zealand," *USSD 2008 Conference*, Portland, OR, 2008. With permission.)

implement the selected methodology (Carter and Bruce 2005). The preferred specialized contractor was therefore involved in the long process wherein every proposed solution was analyzed in depth to determine the best for the project.

4.4.2.4.5 Construction concepts

An overlapping pile option was chosen considering the technical objectives, the cost, and the safety of the dam during construction. The following considerations were paramount:

- The "positive" cutoff concept offered by the overlapping bored piles was fundamentally the closest to a concept that would be used if the dam were to be built today.

- The chosen method was the simplest to construct, which provided confidence that the treatment objective would be met.
- The method met all the technical requirements for construction with a full reservoir since the drilling equipment physically linked the hole being drilled to the previous hole, the resulting panels provided assurance of a continuous cutoff.
- The methodology selected scored the lowest construction risk when compared to the other options considered, while not preventing construction alternatives if the methodology failed.
- The selected methodology had the lowest risks of construction cost overruns.

The four separate cutoffs required 134 piles of 400 mm diameter, spaced 350 mm on centers to form a conceptual overlap of 200 mm, and a total drilling depth of 11,600 m (Figure 4.25). Given the depth of the cutoff walls, the thickness of the cutoff to be installed, and the intensity of the existing features in the dam (such as drains, tunnels, and joints), drilling accuracy was of paramount importance. The most challenging aspect of the project was the ability to install the piles without having one hole deviating into the previously installed, adjacent one in the very weak Upper Ongatiti. An innovative guide system was developed, using a frame formed by two pods to hold the drill bit in position to drill a continuous open slot of up to 8 piles (Figure 4.26), assuring the continuity of the cutoff.

Drilling was conducted from the dam crest (i.e., above reservoir level) to minimize construction and personnel safety risks. The drilling sequence began by drilling a small-diameter (150-mm) pilot hole, using a mud-motor fitted with a bent sub and using measuring while drilling (MWD) technology. The hole was surveyed using a biaxial inclinometer to confirm its position. The hole was then enlarged to its final diameter of 400 mm, using reverse circulation and a "hole seeker" finger attached to a tricone bit. On completion of the hole, a biaxial inclinometer was used inside the drill rods to survey the position of the hole. Even though this was not the fastest drilling method, it provided a suitably rough concrete finish and was "gentle" enough to limit the risk of foundation damage compared to a DTH hammer. The overlap between holes (and hence continuity within a slot) was then created by the use of the 400-mm-diameter guide piece attached to the drill string but running in the adjacent completed hole (Photo 4.23).

A "reverse circulation" drilling method was adopted, as it was necessary to remove cuttings from the drill hole while having the adjacent hole open. Reverse circulation removes cuttings by using the airlift to vacuum the bottom of the hole, using the differential head between the drill bit and the interior of the drill-string. The drill flush was then transported to settling ponds, and clean water was recirculated to the drilling location. The reverse-circulation

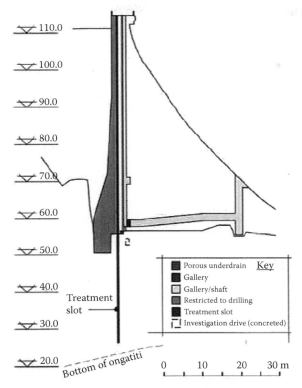

Figure 4.25 Typical cross-section of Arapuni Dam, New Zealand, at a contraction joint show-ing cutoff location and relationship with shafts, gallery, and underdrain. (From Amos, P. D., D. A. Bruce, M. Lucchi, N. Watkins, and N. Wharmby, "Design and Construction of Deep Secant Pile Seepage Cutoff Walls Under the Arapuni Dam in New Zealand," *USSD 2008 Conference*, Portland, OR, 2008. With permission.)

method was deemed also to be the safest approach and the most environmen-tally friendly, since only water was used as the flushing medium.

4.4.2.4.6 Full-scale field test

Given the difficulty of the project, and considering that the innovative method of the continuous "slot wall" had not been tested before, it was decided to perform a preliminary trial in Italy, the native country of the drill-ing contractor, before shipping the equipment to the other side of the world. In Nepi, close to Rome, a tuff quarry was identified that had geological similarities with the geology at the dam. The test consisted of the execution of a slot of five holes (Photo 4.24) to approximately 60 m depth. This gave a sufficient level of confidence in the functionality of the system and allowed some modifications to the originally conceived method to be made.

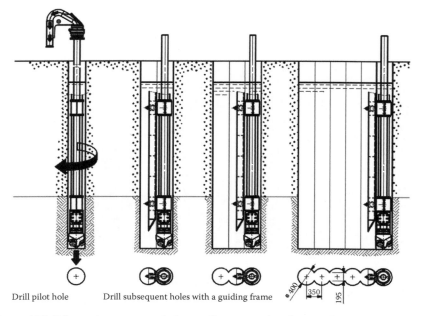

Drill pilot hole   Drill subsequent holes with a guiding frame

Figure 4.26 Schematic sequence of pile installation as a slot, Arapuni Dam, New Zealand. (M. Siepi, personal communication, 2011.)

4.4.2.4.7 Construction details

Drilling rates were lower than anticipated, especially in the concrete of the dam (UCS in the range 40–45 MPa), most probably because of the aggregate, a greywacke quarried locally and ranging in size from 10 to 100 mm. The guide frequently jammed in the hole, causing high friction

Photo 4.23 The guide connected to the drill-string, Arapuni Dam, New Zealand. (M. Siepi, personal communication, 2011.)

Photo 4.24 Looking down into the drilled slot, test panel, Nevi, Italy. (M. Siepi, personal communication, 2011.)

and increased torque. Jamming of the guide was mainly influenced by out of verticality and/or straightness of the original directionally drilled pilot hole, or of the "guide hole" in which the guiding pod of the device ran. In most cases drilling the holes in a continuous series was successful, and therefore the continuity between holes was mechanically guaranteed. In other cases, a new pilot hole had to be drilled, and the sequence repeated.

One of the concerns during drilling was that the water pressure of the lake could break the thin concrete cover between the open slot and the lake (a minimum of 1.2 m). Therefore the number of consecutive holes that could be open at a time before backfilling the slot with concrete was typically limited to seven or eight holes. The method of drilling the slot in contiguous piles allowed the level of the water in the slot to be easily verified, near the pile being drilled, as being at lake level. This was vital to avoid pressure differentials in the foundation when intercepting fractures at high pressures. On completion of each slot, the base was cleaned by airlifting to remove the residual cuttings.

The concrete backfilling rate was limited to avoid imposing excessive stresses to the original concrete of the dam. The placing rate was based on the length of the slot, the stage of the pour, and the temperature of the drilling water and the dam concrete. The compressive strength of concrete for backfilling was selected as 40–45 MPa at 28 days to match the concrete within the dam, and to assure a durable cutoff.

4.4.2.4.8 Monitoring, quality control, and quality assurance

Dam safety during construction was of primary importance. The instrumentation (a total of 62 electronic piezometers, 18 weir flow transducers installed in the dam foundation, turbidity, and pH-measuring transducers located in each weir box to identify fracture infill erosion or cement ingress into drains), was closely monitored in real time by a dedicated team. This team was permanently in contact with the designer and with the contractor throughout the construction period following a detailed monitoring program developed for the project. The data, recorded as time-dependent variables, were stored in a database and analyzed for trends outside preset alarm limits.

Piezometric levels were dynamic during drilling of the piles. The dam safety team was working in the same offices as the construction team on site so that coordinating the activities and exchanging information about any change to the state of the foundation were facilitated. Contingency plans were in place to respond to any rapidly deteriorating condition in the dam foundation.

It was vital for the project to obtain a high degree of confidence on the quality of the installed cutoff wall. It was therefore important to have tools and methods available to verify the as-built location and quality of the drilled holes for each slot and subsequent treatment panel. To survey each pile, a precision inclinometer probe was used. The inclinometer was centralized inside the drill pipes with spring-loaded wheels, measuring cumulative displacement of the drill-string. Survey data were then plotted in an AutoCAD model to produce an as-built drawing.

Figure 4.27 summarizes the deviation of all the piles (pilot holes and following piles), while recalling that the special slot-drilling method adopted allowed a continuous cutoff to be built even with holes not perfectly vertical. Figure 4.28 shows the as-built results from Panel B, where the position of each pile is plotted at different depths, to scale.

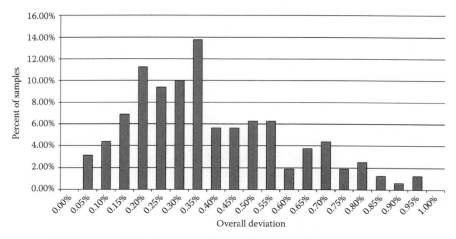

Figure 4.27 Summary of pile deviations, Arapuni Dam, New Zealand. (M. Siepi, personal communication, 2011.)

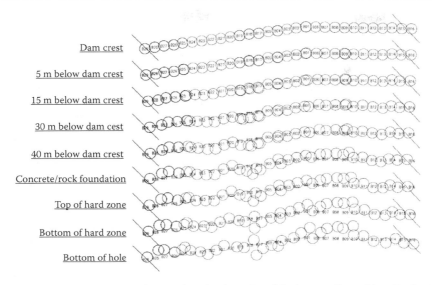

Dam crest

5 m below dam crest

15 m below dam crest

30 m below dam cest

40 m below dam crest

Concrete/rock foundation

Top of hard zone

Bottom of hard zone

Bottom of hole

Figure 4.28 As-built result of verticality checks in Panel B, Arapuni Dam, New Zealand. (M. Siepi, personal communication, 2011.)

In order to check the continuity of each panel, a simple steel frame, 75 mm wide (the required minimum width of the cutoff) 1.2 m long, and 0.6 deep, was lowered down the slot with a tape measure attached to each end. In case discontinuities existed, one possibility was to redrill either of the holes concerned, while an alternative was to use a heavy chisel to remove the remaining rock.

Water velocity flow tests were systematically performed in the slots when drilling into areas of fractures, to prewarn if the concrete would need to contain antiwashout admixtures. A horizontal-flow impeller and wire centralizer were lowered in the hole, and the signal and direction of the flow transmitted to the surface via cable. Another advantage of drilling with recirculated water was the ability to use a DTH camera for visual inspection of the holes and of the slot. In particular, the camera was used to check the continuity between adjacent slots and to check any fractures encountered during drilling.

Final verification was carried out by core drilling the completed cutoff wall, using a wire-line system. The cores showed a perfect bonding between the new concrete and the original concrete of the dam. In one case the coring was successfully completed to the foundation level. It was possible to distinguish the ignimbrite, the concrete of the dam, and the concrete of the pile, identifiable because of the smaller size of the aggregates. Even in this case the bonding between the three materials was excellent.

4.4.2.4.9 Conclusions

The construction of the cutoff panels was completed in September 2007. Foundation drains responded to drilling activities in some areas under the dam, but with preventative actions such as closure of dam drains during drilling and concreting, there were no instances of concentrated seepage through open slots that would have affected backfill concrete quality. The quality of the completed concrete cutoff has met or exceeded the high standards set in the specifications. Postconstruction instrumentation demonstrated that flow through the former high-pressure fissure zones into the downstream area beneath the dam has decreased by approximately 90 percent, while pressures downstream of panels in the high-pressure fissure zones have decreased by approximately 14 m. Furthermore, at the end of the project, the underdrain remains serviceable, even though cutoff construction intersected the underdrain several times.

The outcome of this project is the formation of four robust and verifiable cutoff walls beneath Arapuni Dam that will greatly reduce the risk of future foundation leaks. The collaborative design process and use of the alliance procurement model delivered a mechanism for problem solving and equitable risk sharing and rewarding. All parties (owner, contractor, and designer) concluded that the collaborative design and alliance delivery model contributed to the outstanding execution of the solution. The project design and remedial works were reviewed throughout by independent specialists to ensure that the dam met internationally recognized dam safety standards.

4.5 OVERVIEW OF CATEGORY I CUTOFFS

Most remedial cutoffs in dams have been installed using Category 1 walls due to the nature of the material comprising the foundation (i.e., rock, most of which will be hard and competent), and its depth beneath the working platform (typically hundreds of feet). As illustrated in Chapter 3, structures built with Category 2 walls are typically more cost-effective for installing cutoffs in the fills and foundation soils of levees, and for providing seismic mitigation in similar materials in and under the toes of embankment dams.

There is currently an unprecedented level of activity in the construction of Category 1 walls, with an annual volume several times that of anything seen in the previous three decades. This largely reflects the government's desire to remediate with alacrity its DSAC-1 dams—large and vital structures under dire threat from the results of seepage through their karstified foundations. When the current phase of work has been completed by 2014, then it is most likely that some at least of the DSAC-2 structures will be similarly remediated, although the scale, complexity, and delicacy of these repairs will most probably not quite match those of the earlier phase.

It is accurate to state, and vital to note, that dam remediation has generated a unique set of "lessons learned" (Bruce et al. 2006), and it is highly likely that more such lessons remain to be learned in due course. In the meantime, however, the following summary has been prepared to provide a contemporary perspective.

4.5.1 Particular and notable advantages of Category 1 cutoffs

- The backfill material can be engineered to provide specific properties in order to optimize construction techniques and satisfy specific service performance requirements.
- A method can be found and/or developed to create cutoffs through all types of soil and fill and rock (to depths of over 400 feet).
- In conditions favoring the backhoe method, unit costs can be very low (less than $10/sf). However, deeper walls and more challenging geotechnical conditions (requiring, say, a hydromill or secant piles), will drive unit costs many times higher.
- All the methods of excavation, and all the types of backfill, have extensive history of use and are supported by a long and comprehensive literature base of successful case histories.
- In very favorable conditions, industrial productivities can be very high (over 3,000 sf per shift for backhoe and over 1,500 sf per shift for clamshell and hydromill). When excavating in very hard rock, productivities will be much lower—by as much as one order or more.
- There is an excellent but relatively shallow pool of experienced specialty contractors in North America, who have principally learned their craft in France, Germany, and Italy.
- A vital key to success is to ensure that slurry used to support excavations prior to concreting must not be allowed to escape rapidly. This may well lead to loss of the excavation equipment, and can have the potential to seriously and fundamentally compromise dam safety. Prevention of this scenario now routinely features the use of pregrouting, as in the "composite wall" concept described in Chapter 5.

4.5.2 Particular potential drawbacks of Category 1 cutoffs

- More spoil is created than for Category 2 cutoff construction, and the displaced bentonite or polymer slurry must be handled and stored appropriately. On certain projects in particularly sensitive areas, this is indeed a major challenge and concern.

- Backhoe walls are commercially very attractive and are somewhat of a construction "commodity." However, QA/QC is always a concern, and a backhoe will not be viable in obstructed, very dense, or hard ground conditions, or to depths in excess of about 100 feet.
- For the other methods, a main concern is the lateral continuity of the wall in deeper cutoffs, that is, as reflected in the verticality control of each panel or pile. Furthermore, slurry contaminated joints may exist if proper care has not been exercised during concrete placement. (This is not an issue, of course, for SHS walls.)
- Poor backfilling procedures may result in pockets of trapped slurry and/or segregation of the backfill. Such flaws can be predicted by appropriate real time and QA/QC measures and then proved (or disproved) by subsequent verification drilling and associated testing.
- Sudden loss of supporting slurry into the formation during excavation or drilling may occur and can potentially create an embankment safety situation.
- Clamshell, hydromill, and secant piling operations need substantial working platform preparations and unrestricted overhead access conditions.

4.5.3 Unit costs

It is extremely difficult to provide definitive guidance, given the huge range of methods, materials, and project requirements (such as depth and geological conditions). For example, the backhoe is only used in favorable conditions to moderate depths, whereas the hydromill is typically called upon for cutoffs of relatively great depth and/or to penetrate into resistant bedrock conditions, and secant piles are used only in response to especially challenging geological conditions or dam safety circumstances. The data in the following table are therefore to be used with caution and understanding.

	Clamshell	Hydromill	Backhoe	Secant pile
Mob/ Demob	$100,000–$250,000	$250,000–$500,000	$25,000–$50,000	$100,000–$500,000
Unit Cost	$30–$100/sf	$50–$250/sf	$6–$12/sf	$150–$300/sf

4.5.4 Overall verdict

Category 1 walls have a long and successful history of usage throughout the United States. They cover a wide variety of excavation methods and

Table 4.6 Summary of characteristics of various cutoff wall methodologies

Method	Principle	Wall dimensions			Costs		Relative benefits/problems	
		Depth	Width	Properties of backfill	Mobl/demob	Unit prod.	Pros	Cons
Category I—Excavate and replace (i.e., under bentonite slurry)								
Clamshell	Various types of clams can be used to remove soil in panels	Max 250' (typically <150')	16–66" (typically 24–36')	Wide range of materials can be used for backfill depending on project requirements	Low–Moderate	Moderate–High	Experience Adaptability Wide range of backfills Considerable depth capability	Spoils handling Boulders Rock Verticality control Loss of slurry Access/headroom
Hydromill	Large frame with cutting wheels and reverse circulation mud pump to remove soil in panels	Max over 400'	24–60" (typically 33–39')	Wide range of materials can be used for backfill depending on project requirements	Moderate–Very High	High–Very High	Greater depth capability Can penetrate all conditions Can be very fast Verticality control	Cost base Loss of slurry during excavation Spoils handling Access/headroom
Backhoe	Modified long-reach excavator to create continuous trench	Max 100' (typically <80')	30–36"	Mainly used with SB backfill, but feasible with SCB and CB	Very Low	Very Low	Very high productivity Low cost Experience Many practitioners Practical in tight access/low headroom conditions	Depth limitation QA/QC Obstructions/dense or stiff soil/rock

continued

Table 4.6 Summary of characteristics of various cutoff wall methodologies (Continued)

Method	Principle	Wall dimensions		Properties of backfill	Costs		Relative benefits/problems	
		Depth	Width		Mob/demob	Unit prod.	Pros	Cons
Overlapping Piles	Large-diameter drilled shafts are installed to overlap in strict sequence to form "scalloped" wall of certain minimum chord thickness	Max 300'	24–36"	Mostly conventional concrete	High–Very High	Very High	May be only solution in certain conditions Very sophisticated construction, QA, and QC	High cost Overlap verification Low–moderate productivity

Method	Principle	Wall dimensions		Properties of backfill		Costs		Relative benefits/problems	
		Depth	Width	UCS	K	Mob/demob	Unit prod.	Pros	Cons
Category 2—Mix in place									
Conventional DMM	Vertically mounted shafts are rotated into the soil creating panels of soilcrete	Max practical about 110'	20–40"	100–1,500 psi	5×10^{-6} to 1×10^{-8} cm/s	Moderate–High	Low to Moderate	Speed Experience Several practitioners High productivity	Large equipment needs good access and virtually unlimited headroom Depth limits Very sensitive to obstructions Variable homogeneity with depth Cost base

| TRD | Vertical chainsaw providing simultaneous cutting and mixing of soil to produce continuous soilcrete wall | Max 170' | 18–34" | 100–3,000 psi | 10^{-6} to 10^{-8} cm/s | Moderate–Very High | Moderate–High | Continuity of cutoff is automatically assured Homogeneity Speed Quality Adaptability to wide range of ground conditions Low noise and vibrations Low headroom potential Inclined diaphragms possible Wide range in cutoff properties can be engineered | Difficult wall geometries Medium-hard rock, and boulder nests Currently only one U.S. contractor Requires very specialized equipment Cost base |

continued

Table 4.6 Summary of characteristics of various cutoff wall methodologies (Continued)

Method	Principle	Wall dimensions		Properties of backfill		Costs		Relative benefits/problems	
		Depth	Width	UCS	K	Mob/demob	Unit prod.	Pros	Cons
CSM	Cutting and mixing wheels mounted on horizontal axes create vertical soilcrete panels (Deeper and wider with CT jet variant)	Max 180'	20–47"	100–3,000 psi	10^{-6} to 10^{-8} cm/s	Low–Moderate	Moderate	Panel continuity/verticality Homogeneity Speed Adaptable to conventional carriers Wide range in cutoff properties can be engineered Can accommodate sharp geometry changes	Rock, boulders, and other obstructions Cost base

Key to costs

Mob/demob		Unit costs (cost per square foot of cutoff)	
<$50,000	Very low	<$10	Very low
$50,000–$150,000	Low	$10–$20	Low
$150,000–$300,000	Moderate	$20–$50	Moderate
$300,000–$500,000	High	$50–$100	High
>$500,000	Very high	>$100	Very high

backfill materials and so provide a huge range of options relating to constructability and performance. They include the cheapest (backhoe) and the most expensive (secant pile) cutoffs that can be built for levee or dam remedial purposes. A summary of characteristics is provided in Table 4.6.

REFERENCES

Alfonso A. D. B. 1984. "Overlapping Piles Form Cutoff Wall at Thai Dam." *Water Power and Dam Construction*, June.

Amos, P. B. 2007. "Design and Construction of Seepage Cutoff Walls under a Concrete Dam in New Zealand with a Full Reservoir." *Proceedings of Dam Safety Conference*, Austin, TX, September 9–13. Lexington, KY: Association of Dam Safety Officials (ADSO).

Amos, P. B. 2008. "Design and Construction of Seepage Cutoff Walls under a Concrete Dam with a Full Reservoir." *Journal of Dam Safety (ADSO)*, 32–43.

Amos, P. B., D. A. Bruce, M. Lucchi, N. Watkins, and N. Wharmby. 2008. "Design and Construction of Deep Secant Pile Seepage Cutoff Walls Under the Arapuni Dam in New Zealand." Paper presented at the USSD 2008 Conference, Portland, OR, April 28–May 2.

Anastasopoulos, K., Ch. Economidis, D. A. Bruce, and G. Rizopoulos. 2011. "Construction of Papadia Dam Secant Pile Plastic Concrete Cutoff Wall." Paper presented at the 15th European Conference on Soil Mechanics and Geotechnical Engineering, Athens, Greece, September.

API Standards. 1991. "Recommended Standard Procedure for Field Testing Water-Based Drilling fluids," RP 13B-1. 90 pp.

ASTM. 1992. *Slurry Walls: Design, Construction and Quality Control*. Edited by D. B. Paul, R. R. Davidson, and N. J. Cavalli. West Conshohoken, PA: ASTM.

ASTM D-1140. 2010. *Standard Test Materials for Amount of Material in Soils Finer Than No. 200*. West Conshohoken, PA: ASTM.

ASTM D-2216. 2010. *Standard Test Method for Laboratory Determination of Water (Moisture) Content of Soil and Rock by Mass*. West Conshohoken, PA: ASTM.

ASTM D-4832. 2010. *Standard Test Method for Preparation and Testing of Controlled Low Strength Material (CLSM) Test Cylinders*. West Conshohoken, PA: ASTM.

ASTM D-5084. 2010 *Standard Test Methods for Measurement of Hydraulic Conductivity of Saturated Porous Materials Using Flexible Wall Permeater*. West Conshohoken, PA: ASTM.

Barret, B. C., and M. Bliss. 2008. "Emergency Remedial Actions at A.V. Watkins Dam." Paper presented at the USSD 2008 Conference, Portland, OR, April 28–May 2.

Bruce, D. A., B. DePaoli, C. Mascardi, and E. Mongilardi. 1989. "Monitoring and Quality Control of a 100 Meter Deep Diaphragm Wall." Paper presented at the DFI International Conference on Piling and Deep Foundations, London, May 15–18, 10 pp.

Bruce, D. A., T. L. Dreese, and D. M. Heenan. 2008. "Concrete Walls and Grout Curtains in the Twenty-First Century: The Concept of Composite Cutoffs for Seepage Control." Paper presented at the USSD 2008 Conference, Portland, OR, April 28–May 2, 35 pp.

Bruce, D. A., and G. Dugnani. 1996. "Pile Wall Cuts Off Seepage." *Civil Engineering* 66 (7): 8A–11A.

Bruce, D. A., A. Ressi di Cervia, and J. Amos-Venti. 2006. "Seepage Remediation by Positive Cut-Off Walls: A Compendium and Analysis of North American Case Histories." Paper presented at the ASDSO Dam Safety, Boston, MA, September 10–14.

Bruce, D. A., and S. Stefani. 1996. "Rehabilitation of Beaver Dam: A Major Seepage Cutoff Wall." *Ground Engineering* 29 (5): 40–45.

Carter, J., and D. A. Bruce. 2005. "Enhancing the Quality of the Specialty Contractor Procurement Process: Creating an Alliance." pp. 76–87 in *Geo3 GEO Construction Quality Assurance/Quality Control Conference Proceedings*, edited by D. A. Bruce and A. W. Cadden. Dallas, TX: ADSC.

Choi, H., and D. E. Daniels. 2006. "Slug Test Analysis in Vertical Cutoff Walls I: Analysis Methods." *Journal of Geotechnical and Geoenvironmental Engineering*, 132 (4): 429–38.

Christman, C. W., L. G. Bromwell, and M. A. Schwartz. 2009. "Soil-Bentonite Slurry Walls for Seepage Control at L-8 Reservoir." Paper presented at Association of State Dam Safety Officials Annual Dam Safety Conference, Hollywood, FL, September 27–October 1.

Cyganiewicz, J. M. 1988. "Construction of a Cutoff Wall at Fontenelle Dam." *Proceedings of 16th ICOLD Conference* 5:227–30.

Davidson, L. 1990. "Performance of the Concrete Diaphragm Wall at Navajo Dam." Paper presented at the 10th Annual USCOLD Conference, New Orleans, LA, March 6–7, 21 pp.

Davidson, R. R., G. Dennis, B. Findlay, and R. B. Robertson. 1992. "Design and Construction of a Plastic Concrete Cutoff Wall for the Island Copper Mine." Pp. 271–88 in *Slurry Walls: Design, Construction, and Quality Control*, edited by D. B. Paul, R. R. Davidson, and N. J. Cavalli. ASTM Publication Code Number 04-011290-38. STP 1129.

Davidson, R. R., J. Levallois, and K. Graybeal. 1992. "Seepage Cutoff Wall for Mud Mountain Dam." Pp. 309–23 in *Slurry Walls: Design, Construction, and Quality Control*, edited by D. B. Paul, R. R. Davidson, and N. J. Cavalli. ASTM Publication Code Number 04-011290-38. STP 1129.

Demars, B., C. Pledger, and B. Barrett. 2009. "A.V. Watkins Dam Modification: Cement-Benonite Slurry Cutoff Wall." Paper presented at USSD 29th Annual Conference, Nashville, TN, April 20–24. 12 pp.

Dewey, R. 1988. "Installation of a Deep Cutoff Wall at Navajo Dam." *Proceedings of 16th ICOLD Conference*, 5:230–37.

Dinneen, E. A., and M. Sheskier. 1997. "Design of Soil-Cement-Bentonite Cutoff Wall for Twin Buttes Dam." *Proceedings of USCOLD Annual Meeting*, pp. 197–211.

Dunbar, S. W., and T. C. Sheahan. 1999. "Seepage Control Remediation at Hodges Village Dam, Oxford, Massachusetts." *Journal of Geotechnical and Geoenvironmental Engineering* 125 (3): 198–206.

Eckerlin, R. D. 1993. "Mud Mountain Dam Concrete Cutoff Wall." Paper presented at ASCE Specialty Conference on Geotechnical Practice in Dam Rehabilitation, North Carolina State University, Raleigh, NC, April 25–28, pp. 284–98.

Engineering News Record. 1990. "Seepage Cutoff Wall Is Deepest Yet." April 12, pp. 31–32.

Erwin, E. D. 1994. "Major Rehabilitation of Wister Dam for Seepage Control." Paper presented at 14th Annual USCOLD Meeting, Phoenix, AZ, June, 16 pp.

Erwin, E. D., and J. M. Glenn. 1992. "Plastic Concrete Slurry Wall for Wister Dam." Pp. 251–67 in *Slurry Walls: Design, Construction, and Quality Control*, edited by D. B. Paul, R. R. Davidson, and N. J. Cavalli. ASTM Publication Code Number 04-011290-38. STP 1129.

Fetzer, C. A. 1988. "Performance of the Concrete Wall at Wolf Creek Dam." *Trans. of 16th ICOLD* 5:277–82.

Fisher, M. J., K. B. Andromalos, and D. N. Johnson. 2005. "Construction of a Self-Hardening Slurry Cutoff Wall at Taylorsville Dam, OH." Paper presented at the U.S. Society on Dams, 25th Annual Conference, Salt Lake City, UT, June 6–10, pp. 105–12.

Forster, D. N. 2006. "Quality Remediation Working in the Dark at Arapuni Dam" [CD]. *12th Hydro Power Engineering Exchange*. Tasmania.

Gagliardi, J., and R. Routh. 1993. "Geotechnical Modifications at Meeks Cabin Dam." Paper presented at the ASCE Specialty Conference on Geotechnical Practice in Dam Rehabilitation, North Carolina State University, Raleigh, NC, April 25–28, 15 pp.

Gillon, M., and D. A. Bruce. 2003. "High Pressure Seepage at Arapuni Dam, New Zealand: A Case History of Monitoring, Exploration and Remediation." Paper presented at the U.S. Society on Dams, San Diego, CA, June 24–28, 10 pp.

Graybeal, K. D., and J. Levallois. 1991. "Construction of a Cut Off Wall with the Hydrofraise Through the Core of Mud Mountain Dam (USA)." Paper presented at the 17th International Congress on Large Dams (ICOLD 17), Vienna, 22 pp.

Henn, K., and B. E. Brosi. 2005. "Mississinewa Dam—Settlement Investigation and Remediation." Paper presented at the Association of State Dam Safety Officials 22nd Annual Conference, Orlando, FL, September, 15 pp.

Hillis, R. M., and H. W. Van Aller. 1992. "Cement-Bentonite Cutoff Wall Remediation of Small Earthen Dam." Paper presented at the Association of State Dam Safety Officials, 9th Annual Conference, Baltimore, MD, September 13–26, 7 pp.

Hornbeck, S., and K. Henn. 2001. "Mississinewa Dam Settlement Investigation and Remediation." Paper presented at the Association of State Dam Safety Officials, Snowbird, UT, September 9–12.

Hvorslev, M. J. 1951. "Time Lag and Soil Permeability in Ground-Water Observation." Bulletin No. 36, Waterways Experiment Station, USACE, Vicksburg, MS.

Khoury, M. A., A. L. Harris Jr., and P. Dutko. 1989. "Construction of a Cement-Bentonite Cutoff Wall to Control Seepage at the Prospertown Lake Dam." Paper presented at the Association of State Dam Safety Officials, 6th Annual Conference, Albuquerque, NM, October 1–5, 6 pp.

Kulesza, R., K. Pflueger, C. Gardner, and P. Yen. 1994. "Investigations and Design of Diaphragm Wall for Seepage Control at Wells Dam." Paper presented at the 14th Annual USCOLD Meeting, Phoenix, AZ, June, 20 pp.

Llopis, J. L., D. K. Butler, C. M. Deaver, and S. C. Hartung. 1988. "Comprehensive Seepage Assessment: Beaver Dam, Arkansas." *Proceedings of the 2nd International Conference on Case Histories in Geotechnical Engineering*, pp. 519–26.

Millet, R. A., J.-Y. Perez, and R. R. Davidson. 1992. "USA Practice Slurry Wall Specifications 10 Years Later." Pp. 42–66 in *Slurry Walls: Design, Construction, and Quality Control*, edited by D. B. Paul, R. R. Davidson, and N. J. Cavalli. ASTM Publication Code Number 04-011290-38. STP 1129.

Murray, B. 1994. "Serious Seepage Problems at McAlpine Locks and Dam Lead to Concrete Cutoff Wall Remediation." Paper presented at USCOLD, 14th Annual Conference, Phoenix, AZ, June, 13 pp.

Pagano, M. A., and B. Pache. 1995. "Construction of the Meeks Cabin Cut Off Wall." Paper presented at the Association of State Dam Safety Officials, 11th Annual Conference, Atlanta, GA, September 17–20, 19 pp.

Parkinson, J. L. 1986. "Use of the Hydrofraise to Construct Concrete Cutoff Walls." Procedure. REMR Workshop, USACOE, Vicksburg, MS, October 21–22, pp. 61–70.

Ressi di Cervia, A. 2003. "A Better Barrier." *Civil Engineering* 73 (7): 44–49.

Ressi di Cervia, A. 2005. "Construction of the Deep Cutoff at the Walter F. George Dam." Paper presented at the U.S. Society on Dams, 25th Annual Conference, Salt Lake City, UT, June 6–10, 16 pp.

Rice, J. D. 2009. "Long-Term Performance of Seepage Barriers, Findings from Case Histories and Analyses." Paper presented at the 29th Annual USSD Conference, Nashville, TN, April 20–24, 16 pp.

Rice, J. D., and M. J. Duncan. 2010a. "Deformation and Cracking of Seepage Barriers in Dams Due to Changes in the Pore Pressure Regime." *ASCE Journal of Geotechnical and Geoenvironmental Engineering* 133 (1): 2–15.

Rice, J. D., and M. J. Duncan. 2010b. "Findings of Case Histories on the Long-Term Performance of Seepage Barriers in Dams." *ASCE Journal of Geotechnical and Geoenvironmental Engineering* 133 (1): 16–25.

Roberts, D. L., and D. Ho. 1992. "Construction of Diaphragm Wall at Wells Dam East Embankment." Pp. 235–50 in *Slurry Walls: Design, Construction, and Quality Control*, edited by D. B. Paul, R. R. Davidson, and N. J. Cavalli. ASTM STP 1129.

Rodio. 1980. Khao Laem Multipurpose Project—C3—Concrete Diaphragm Wall—Construction Method Analysis. Internal Report.

Ryan, C. R., and S. R. Day. 2003. "Soil-Bentonite Slurry Wall Specifications." Paper presented at the Pan American Conference on Soils Mechanics and Geotechnical Engineering, Geo-Institute, and MIT, Cambridge, MA, June, 8 pp.

Singh, N. K., R. Douglas, S. R. Ahlfield, and M. Grant. 2005. "Construction of a Roller Compacted Concrete Upstream Blanket and Plastic Concrete Cutoff Wall at Cleveland Dam." Paper presented at the Association of State Dam Safety Officials, New Orleans, LA, September 25–29.

Soletanche Bachy. 1999. Promotional material.

USACE. 1984. Internal Letter Report, March 22.

USACE. 2004. "West Hill Dam." Project Completion Report, Permanent Seepage Repairs. USACE New England District. 29 pp.

USACE. 2005. "Hodges Village." Project Completion Report, Permanent Seepage Repairs. USACE New England. 24 pp.

USACE. 2006. "Waterbury Dam" Detailed Documentation Report, Final Draft. USACE Baltimore District. 109 pp.

Washington, D., D. Rodriguez, and V. Ogunro. 2005. "An Effective Approach to Rehabilitate Damaged Barrier System Against Piping and Containment Flow." Paper presented at the International Conference on Energy Environment and Disasters, INCEED 2005, Charlotte, NC, July 24–30, 11 pp.

Xanthakos, P. P. 1979. *Slurry Walls*. New York: McGraw-Hill.

Chapter 5

Composite cutoff walls

Donald A. Bruce

5.1 BACKGROUND

As illustrated in Chapter 2, grout curtains have been used in the United States to control seepage in rock masses under and around dams of all types since the 1890s. For a variety of understandable, if not always laudable, reasons, the long-term performance of many of these curtains has not been satisfactory, especially in lithologies containing soluble and/or erodible materials. Foundation remediation in such instances traditionally involved regrouting, often, of course, using the same means, methods, and materials whose defects contributed to the inadequacy in the first place.

Disillusionment on the part of owners and engineers with these traditional grouting practices to provide a product of acceptable efficiency and durability led to the chorus of "grouting doesn't work" voices in the industry from the mid-1970s onward. The fact that effective and durable grout curtains were being installed successfully elsewhere in the world, using different perspectives on design, construction, and contractor procurement processes, largely escaped the attention of the doubters who, for all their other and obvious qualities, exhibited technological xenophobia.

Partly as a result of the antigrouting lobby, partly in response to indisputable geological realities and challenges, and building on technical advances in slurry wall techniques, the concept and reality of "positive cutoffs" became the mantra for major embankment dam foundation rehabilitation in North America from 1975 onward. Such Category 1 walls (Chapter 4, this volume), built through and under existing dams by either the panel wall technique, or secant large-diameter piles, comprise some type of concrete, ranging from high strength to plastic. In contrast to grout curtains, where well over 90 percent of the cutoff is, in fact, the virgin *in-situ* rock, these positive cutoffs are, conceptually, built of 100 percent pre-engineered material of well-defined properties. The necessity for such positive cutoff walls remains today in certain geological conditions, and the list of successful projects is extremely impressive as is illustrated in Chapter 4.

From the mid-1980s—albeit in Europe (Lombardi 2003)—a new wave of dam grouting concepts began to emerge. Given that most of the leading North American practitioners had close corporate and/or professional and personal links with this insurgency, it is not surprising that their heretofore moribund industry began to change. By the time of the seminal 2003 ASCE grouting conference in New Orleans, the revolution in North American practice for dam foundation grouting had been clearly demonstrated (e.g., Wilson and Dreese 2003; Walz et al. 2003) and this status was confirmed in the 2012 New Orleans Conference. The concept of a qualitatively engineered grout curtain was established. Differences in opinion and philosophies with the great European practitioners such as Lombardi, the architect of the Grouting Intensity Number (GIN) method, were not necessarily resolved; they were debated between equals and the respective opinions fairly acknowledged.

Chapters 2 and 4 clearly illustrate the unprecedented levels of expertise and experience in both grout curtains and concrete cutoff walls in North America. This is particularly serendipitous given that the dollar requirement for the application of both technologies—in federal dams alone in the next five years—is of an order equivalent to the aggregate of the preceding forty years (Halpin 2007). It is therefore surprising that it is only in recent years that the concept of "composite" walls has been formalized (Bruce, Dreese, and Heenan 2008). In essence, the cutoff features both techniques, with the grouting facilitating the diaphragm wall construction and providing a cost-effective barrier in those parts of rock masses without clay infilling, while the concrete wall assures durability in such potentially erodible horizons and features. As illustrated in Section 5.5.2, this formalization was in fact precipitated by events at Mississinewa Dam, Indiana, in the early 2000s, but has now been employed systematically for the construction of huge remedial cutoffs in carbonate terrains in Missouri (Clearwater Dam), Tennessee (Center Hill Dam), and Kentucky (Wolf Creek Dam), in addition to the completed project at Bear Creek Dam, Alabama (Section 5.5.2). Prior to this time, grouting conducted on large concrete cutoff wall projects was typically of a very minor scale and conducted primarily as a scouting or investigatory tool, as was the case at W. F. George Dam, Alabama (Ressi 2003).

In this chapter, critical aspects of grouting and concrete cutoff walls are reviewed—as related to the *composite cutoff philosophy*. The review offers a different perspective from the systematic treatment devoted to these subjects in Chapters 2 and 4. Additional background may be found in Bruce, Dreese, and Heenan (2008, 2010).

5.2 GROUT CURTAINS

5.2.1 Design

Design of grout curtains based on rules of thumb without consideration of the site geology is not an acceptable practice or standard of care.

Contemporary approaches are based on the concept of a quantitatively engineered grout curtain (QEGC), which provides criteria for the maximum acceptable residual permeability and minimum acceptable dimensions of the cutoff (Wilson and Dreese, 1998, 2003). Prerequisite geological investigations and other work required to perform this quantitative design include:

- thorough geologic investigations identifying structure, stratigraphy, weathering, solutioning, and permeability of the foundation rock
- establishment of project performance requirements in terms of seepage quantities and seepage pressures (design requirements should consider dam safety, cost, and political acceptability or public perception as they relate to residual seepage)
- seepage analyses to determine the need for grouting, the horizontal and vertical limits of the cutoff, the width of the curtain, and the location of the curtain
- specifications written to assure best practice for field execution of every element of the work
- where relevant, the value of the lost water should be compared to the cost of more-intensive grouting in a cost-benefit analysis

Quantitative design of grouting requires that the curtain be treated in seepage analyses as an engineered element. The specific geometry of the curtain in terms of depth and width must be included in the model, and the achievable hydraulic conductivity of the curtain must also be assumed. Guidance on assigning grout curtain design parameters and performing seepage analyses for grout curtains is covered in detail by Wilson and Dreese (2003). More substantial and complete guidance on flow modeling of grouted cutoffs is included in the update to USACE EM 1110-2-3506 issued in 2008.

5.2.2 Construction

Many aspects of the construction of QEGCs have also changed greatly in the last ten years or so, driven by the goals of achieving improved operational speed and efficiency; satisfying lower residual permeability targets; enhancing QA/QC, verification, and real-time control; and assuring long-term durability and effectiveness. Particularly important advances are as follows:

- The traditional concepts of stage grouting (up—or down—depending on the stability and permeability of the rock mass) and closure (primary-secondary-tertiary phases) still apply. However, construction in two initial rows, with the holes in each inclined in opposite directions, has become standard practice.

- Multicomponent, balanced, cement-based grouts are used to provide high-performance mixes, which provide superior stability and rheological and durability properties. The use of "neat" cement grouts with high water:cement ratios and perhaps nominal amounts of super-plasticizer or bentonite is simply not acceptable (Chuaqui and Bruce 2003).
- The current state of the art in grouting monitoring and evaluation is a fully integrated system where all field instruments are monitored in real time through a computer interface, all necessary calculations are performed automatically, grouting quantity information is tabulated and summarized electronically, program analyses are conducted automatically by the system using numerous variables, and multiple custom as-built grouting profiles are automatically generated and maintained. This level of technology provides the most reliable and highest-quality project records with minimal operator effort. In fact, the advent of such technology has been found to substantially decrease grouting program costs while providing unprecedented levels of assurance that the design goal is being met (Dreese et al. 2003).
- Modern drilling recording instruments and borehole imaging technology allow for better investigation and understanding of subsurface conditions than was previously possible. Measurement while drilling (MWD) instrumentation provides additional geological information during the drilling of every hole on a grouting project (Bruce and Davis 2005) and not only from the limited number of cored investigatory holes. Specific energy and other recorded data can be evaluated and compared to the subsequent grouting data to extract as much information as possible from every hole drilled. Each hole on a grouting project is thereby treated as an exploration hole, and the data gathered are utilized to increase the understanding of subsurface conditions. After a hole has been drilled, borehole imaging can be performed to obtain a "virtual core." This equipment is especially useful for destructively drilled production holes where recovered core is not available for viewing and logging, and it provides invaluable data such as *in-situ* measurements of fracture apertures and bedrock discontinuity geometry. These are then utilized in designing or modifying the grout methods and materials. Borehole images are mapped by qualified personnel, and the data may be further analyzed using stereonet analyses.

5.2.3 Verification and performance

Successfully achieving a cutoff closure is a three-step process: achieving closure on individual stages and holes, achieving closure on individual lines, and achieving closure on the entire curtain. Proper closure on individual stages and holes is primarily a function of the following six items: (1) drilling a properly flushed hole, effectively washing the hole, and understanding the geology of the stages being grouted; (2) applying that knowledge, along with the results of water-pressure testing, to determine technically effective and cost-effective stage selection; (3) selecting appropriate starting mixes; (4) real-time monitoring of the grouting and assessment of the characteristics of each grouting operation; (5) making good and informed decisions regarding when to change grout mixes during injection within a stage; and (6) managing the hole to completion (i.e., refusal to further grout injection) within a reasonable amount of time. The key during grouting is to gradually reduce the apparent Lugeon value of the stage to practically zero. The apparent Lugeon value is calculated using a stable grout as the test fluid, taking into account the apparent viscosity of the grout relative to water.

Pumping large quantities of grout for an extended period of time without any indication of achieving refusal (i.e., a reduction in the apparent Lugeon value) is generally a waste of precious time and good grout. Unless a large cavity has been encountered, the grout being used in this case has a cohesion that is too low and is simply traveling a great distance through a single fracture. Grout mixes need to be designed properly for economy and value, especially in karstified conditions.

Each row of a grout curtain, and the completed curtain, should be analyzed in detail. Each section of the grout curtain should be evaluated, and closure plots of pregrouting permeability for each series of holes in the section should be plotted. As grouting progresses, the plots should show a continual decrease in pregrouting permeability for each successive series of holes. For example, the results for the exploratory holes and primary holes from the first row within a section represent the "natural permeability" of the formation. Secondary holes on each row should show a reduced permeability compared to the primary holes due to the permeability reduction associated with grouting of the primaries. Similarly, the pregrouting permeability of tertiary holes should show a marked decline relative to the secondary holes, and so on.

In addition to performing the analyses described previously, it is also necessary to review profiles indicating the geology, water testing, and grouting results. Review of the profiles with the water Lugeon values displayed on each zone or stage gives confirmation that the formation behavior is consistent with the grouting data, and permits rapid evaluation of any trends or problem areas requiring additional attention. In addition, this review permits identification of specific holes, or stages within a hole, that behaved abnormally and that could be skewing the results of the closure analysis. For example, the average pregrouting permeability of tertiary holes that appear on a closure analysis plot may be 10 Lugeons, but that average may be caused by one tertiary hole that had an extraordinarily high reading; averages are interesting, but spatial distributions are critical.

Review of the grout row profiles with the grout takes displayed is also necessary along with comparison of the average grout takes compared to the average Lugeon values reported by the closure analysis. Areas of abnormally high or low grout takes in comparison to the Lugeon values should be identified for further analysis. The grouting records for these abnormal zones should be reviewed carefully, along with the pressure testing and grouting records from adjacent holes.

5.3 CONCRETE CUTOFF WALLS

5.3.1 Investigations, design, specifications, and contractor procurement

Intensive, focused site investigations are essential as the basis for cutoff design and contractor bidding purposes. In particular, these investigations must not only identify rock mass lithology, structure, abrasivity, and strength ("rippability"), but also the potential for loss of slurry during panel excavation. This has not always been done, and cost and schedule have suffered accordingly on certain major projects. Special considerations have had to be made when designing cutoffs that must abut existing concrete structures, or that must be installed in very steep-sided valley sections, or that must toe in to especially strong rock.

"Test sections" have proven to be extremely valuable, especially for permitting contractors to refine their means, methods, and quality-control systems. Such programs have also given the dam safety officials and owners the opportunity to gain confidence and understanding in the response of their dams to the invasive surgery that constitutes cutoff wall construction. Furthermore, such programs have occasionally shown that the foreseen construction method was practically impossible (e.g., a hydromill at Beaver Dam, Arkansas) or that significant facilitation works were required (e.g., pregrouting of the wall alignment) as discussed in Section 5.5.

Every project has involved a high degree of risk and complexity and has demanded superior levels of collaboration between designer and contractor. This situation has been best satisfied by procuring a contractor on the basis of "best value," not "low bid." This involves the use of RFPs (Requests for Proposals) with a heavy emphasis on the technical submittal and, in particular, on corporate experience, expertise, and resources, and the project-specific method statement. These projects are essentially based on performance, as opposed to prescriptive, specifications. Partnering arrangements (which are postcontract) have proven to be very useful to both parties when entered into with confidence, enthusiasm, and trust.

5.3.2 Construction and QA/QC

The specialty contractors have developed an impressive and responsive variety of equipment and techniques to ensure cost-effective penetration and appropriate wall continuity in a wide range of ground conditions. More than one technique (e.g., clamshell followed by hydromill) has frequently been used on the same project and especially where bouldery or obstructed conditions have been encountered (Bruce, Ressi di Cervia, and Amos-Venti 2006).

Cutoffs can be safely constructed with high lake levels, provided that the slurry level in the trench can be maintained a minimum of three feet above these levels. In particularly challenging geological conditions, this may demand pretreatment of the embankment (e.g., Mud Mountain Dam, Washington) or the rock mass (Mississinewa Dam, Indiana) to guard against massive, sudden slurry loss. For less severe geological conditions, contractors have developed a variety of defenses against slurry losses of smaller volume and rate by assuring large slurry reserves, using flocculating agents and fillers in the slurry, or by limiting the open-panel width.

Very tight verticality tolerances are necessary to ensure continuity and especially in deeper cutoffs. Such tolerances have been not only difficult to satisfy, but also difficult to measure accurately (to within 0.5 percent of wall depth) and verify.

The deepest panel walls have been installed at Wells Dam, Washington (223 feet, clamshell) and at Mud Mountain Dam, Washington (402 feet, hydromill). The hydromill has proved to be the method of choice for large cutoffs in fill, alluvial soils, and in rock masses of unconfined compressive strengths less than 10,000 psi (massive) to 20,000 psi (fissile and therefore rippable).

Secant pile cutoffs are, by comparison, expensive and intricate to build. However, they are the only option in certain conditions (e.g., heavily karstified, but otherwise hard limestone rock masses) that would otherwise defeat the hydromill. The deepest such wall (albeit a composite pile/panel wall) was the first—at Wolf Creek, Kentucky, in 1975. It reached a maximum of 280 feet. The most recent pure secant pile wall in carbonate terrain was constructed at Beaver Dam, Arkansas (1992–1994), while a secant/panel combination is currently being installed at Wolf Creek Dam, Kentucky.

A wide range of backfill materials has been used, ranging from low-strength plastic concrete to conventional high-strength concrete. This is a critical design decision.

The preparation and maintenance of a stable and durable working platform has proven always to be a beneficial investment, and its value should not be underestimated. The highest standards of real-time quality assurance/quality control (QA/QC) and verification are essential to specify and implement. This applies to every phase of the excavation process, and to each of the materials employed.

Enhancements have progressively been made in cutoff excavation technology, especially to raise productivity (particularly in difficult geological conditions), to increase the mechanical reliability of the equipment, and to improve the practicality and accuracy of deviation control and measurement.

5.3.3 Potential construction issues with cutoffs

Satisfactory construction of positive cutoff walls requires experience, skill, and dedication to quality in every aspect of the construction processes, including site preparation, element excavation, trench or hole cleaning, concrete mixing, and concrete placement. A positive cutoff requires the elements of the wall to be continuous and interconnected.

The following issues are possible concerns that must be taken into account in wall construction to prevent defects:

Element deviation: Misalignment of the equipment or inability to control the excavation equipment can cause significant deviation of elements and can therefore result in a gap in the completed wall.

Uncontrolled slurry loss: Although bentonite slurries are proven in creating a filter cake in soils, their ability to form a filter cake in rock fractures is limited. As a general rule of thumb, if water is lost during exploration drilling, one should assume that slurry losses in rock will occur during element excavation. If the rock mass is sufficiently permeable, uncontrollable and complete slurry loss can occur. Slurry losses in embankments have also occurred on past projects due to hydrofracturing of susceptible zones. This is a particularly sensitive issue when excavating through epikarstic horizons and major karstic features lower in the formation. In this regard, epikarst is defined as the transition/interface zone between soil and the underlying, more competent, if still karstified, rock. Epikarst typically contains very fractured and solutioned conditions, and much residual material and voided areas. Epikarst usually plays an extremely important role in the hydrogeological regime of karst aquifers.

Trench stability: The factors of safety of slurry-supported excavations in soil are not high. Movement of wedges into the trench or "squeeze in" of soft zones can occur.

Concrete segregation: Mix design and construction practices during backfill must be optimized so as to prevent segregation or honeycombing within the completed wall.

Soil or slurry inclusions: The occurrence of soil- or slurry-filled defects or inclusions in completed walls has been recognized. These defects are not critical if small or discontinuous, but they become significant if they fully penetrate fully across the width of the wall.

Panel joint cleanliness: Imperfections or pervious zones along the joints between elements are sources of leakage through completed walls. Cleaning of adjacent completed elements by circulating fresh slurry is necessary to minimize the contamination of joints. In extreme cases, mechanical cleaning with "brushes" has to be conducted.

5.3.4 Performance

Surprisingly little has in fact been published to date describing the actual efficiency of cutoff walls after their installation; most of the publications describe design and construction and have usually been written soon after construction by the contractors themselves. The research into this matter conducted by the Virginia Tech team of Rice and Duncan (Rice and Duncan 2010a, 2010b) is, therefore, of particular significance. Although there is some published evidence (e.g., Davidson 1990) that the walls have not always functioned as well as anticipated, it can be reasonably assumed that the majority of the remediations have been successful, provided that (1) the wall has been extended laterally and vertically into competent, impermeable and nonerodible bedrock; (2) there is full lateral continuity between panels with no clay contamination; and (3) the panels themselves contain no concrete segregations or slurry/soil inclusions. It may also be stated that the capabilities of the technology of the day have not always been able to satisfy the depth criterion. EM 1110-2-1901 published in 1986 by the USACE states that the experienced efficiency of cutoff walls calculated based on head reduction across the wall was 90 percent or better for properly constructed walls. The term "properly" is not defined, and no update to this information has since been published.

There is also the case of the original diaphragm wall at Wolf Creek Dam, the length and depth of which were restricted by the technology and/or funds available at the time (1975–1979). As a result, the new wall, deeper and longer, is being built to finally cut off the flow, which has resumed through the deep, heavily karstified limestones under and beyond the existing wall.

5.4 "COMPOSITE" CUTOFFS

5.4.1 The basic premise

In recent years, there has been a number of projects, both completed and in planning, that have featured the construction of a concrete cutoff wall installed through the dam and into karstified carbonate bedrock. The basic premise of such a positive cutoff is clear and logical; the presence of large clay-filled solution features in the bedrock will defeat the ability of a grout curtain—even when designed and built using best contemporary practices—to provide a cutoff of acceptable efficiency and durability. This is particularly important when permanent "walk-away" solutions are required that must be robust, reliable, and durable. There is no question that rock-fissure grouting techniques are incompatible with satisfying that long-term goal in the presence of substantial clayey infill materials. However, the benefits of a concrete cutoff come at a substantial financial premium over those provided by a grout curtain. A typical industry average cost for a grouted cutoff is of the order of $30–$50 per square foot. The cost of a concrete cutoff is anywhere up to 5 times this figure, depending on the technique (panel or secant), the ground conditions, the depth of the cutoff, and the challenges of the site logistics. Furthermore, the construction of a concrete cutoff wall through the typical karstified limestone or dolomite rock mass will involve the excavation of the rock (which in the main part will be in fact very hard, impermeable, and competent with unconfined compressive strength values in excess of 20,000 psi) and backfilling that relatively thin diaphragm with a material of strength 5,000 psi or less. In effect, great effort and expense are expended to provide a membrane through the greater part of the project which is of lower strength than the rock mass excavated to construct it.

Another practical factor that has often been overlooked historically is that construction of a concrete cutoff wall may simply not be feasible in ground conditions that permit the panel trench-stabilizing medium (bentonite or polymer slurry) or the drill flush medium (air or water) to be lost into the formation; in extremis, either of these phenomena could create a dam safety threat, let alone the loss of very expensive excavation or drilling equipment at depth. The solution, not surprisingly, in such situations has been to suspend the wall construction and to systematically and intensively pretreat the formation by grouting.

In doing so, however, it has not been always the case that the designer of the wall has appreciated that, in addition to this campaign of drilling, water-pressure testing, and grouting (constituting a facilitating improvement to the rock mass), such work also constitutes a most detailed site investigation—at very close centers—of the whole extent of the originally foreseen concrete cutoff area. It is reasonable, therefore, to deduce that the data from these pretreatment programs can be used to review the true required extent of the subsequent concrete wall and thereby reduce overall project costs with sound justification.

The concept may then be taken a stage further. Instead of drilling and grouting being conducted only as a remedial/facilitating operation under emergency conditions, it can be specified as a rigorous design concept to

- precisely identify the location and extent of the major karstic features that are actually required to be cutoff with a concrete wall
- pretreat the ground, and especially the epikarst, to an intensity that bentonite slurry or drill flush will not be suddenly lost during the concrete wall construction (a typical acceptance criterion is 10 Lugeons)

- grout—to a verified engineered standard—the rock mass that does not contain erodible material in its fissures around and under the karstic features (a typical acceptance criterion is in the range of 1–3 Lugeons)

By embracing these precepts, it is therefore logical to define the concept of a "composite cutoff": an expensive concrete wall, where actually required for long-term performance certitude, plus a contiguous and enveloping grout curtain to provide acceptable levels of impermeability and durability in those portions of the rock mass with minimal erodible-fissure infill material.

5.4.2 Conceptual illustrations

With one eye on the immediate future requirements of seepage remediation involving cutoffs under existing dams, it may be stated that karst is usually either stratigraphically driven, or structurally related. Figure 5.1 shows a case where the major horizon of concern for long-term seepage and erosion is limited to the 30 feet or so of epikarst, Figure 5.2 is the case where the seepage and erosion concern is in a certain deep stratigraphic member, and Figure 5.3 shows the condition where the karstification has developed along discrete, vertical structural discontinuities. For the sake of illustration, it may be assumed that the final cutoff has to be 1,000 feet long, the cost of drilling and grouting is $30 per square foot, the concrete wall costs $120 per square foot, and the maximum vertical extent of the cutoff is 110

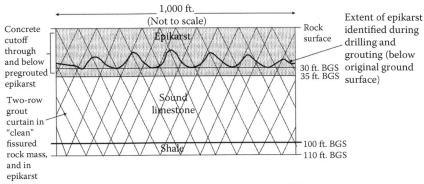

- Area of grout curtain (including pretreatment of epikarst) = 1,000 ft. × 110 ft. = 110,000 sq. ft.
- Area of subsequent concrete wall = 1,000 ft. × 35 ft. = 35,000 sq. ft.

Figure 5.1 Epikarst is found during pregrouting to an average of 30 ft. BGS. Therefore, the concrete cutoff is installed only to 35 ft. BGS, and the grouting provides the cutoff in the "clean" rock below.

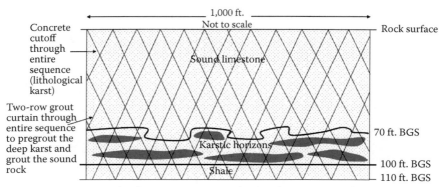

- Area of grout curtain = 1,000 ft. × 110 ft. = 110,000 sq. ft.
- Area of subsequent concrete wall = 1,000 ft. × 110 ft. = 110,000 sq. ft.

Figure 5.2 Heavily karstified horizons are found at depth during predrilling and grouting. Therefore the concrete cutoff is required for the full extent. The grouting has pretreated the karstic horizons to permit safe concrete cutoff construction.

feet since a shale aquiclude exists at 100 feet below ground surface (BGS). The dam itself is "invisible" in this exercise.

In the configuration of Figure 5.1, the original design featured a concrete cutoff wall extending 10 feet into the aquiclude. The cost would therefore be 1,000 ft × 110 ft × $120 = $13.2 million. This would, of course, assume (or worse, ignore) that construction of the wall through the epikarst would

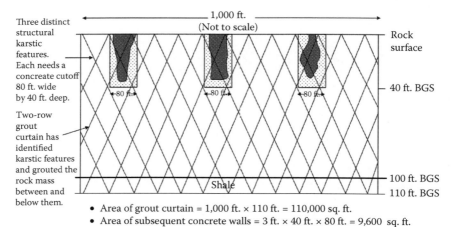

- Area of grout curtain = 1,000 ft. × 110 ft. = 110,000 sq. ft.
- Area of subsequent concrete walls = 3 ft. × 40 ft. × 80 ft. = 9,600 sq. ft.

Figure 5.3 Discrete karstic features have been found during the drilling and grouting, driven by major structural lineations. Thus, individual concrete cutoff panels can be installed, after drilling and grouting have confirmed the extent of these features and have pretreated them to permit safe concrete cutoff construction.

be feasible without its pretreatment by grouting. Alternatively, if the entire alignment were to be predrilled and pregrouted, it would be revealed that there was no need to construct the wall deeper than, say, 35 feet. The total cost of this composite cutoff would therefore be:

Drill and grout: 1,000 ft × 110 ft × $30/square foot = $3.3 million

Concrete wall: 1,000 ft × 35 ft × $120/square foot = $4.2 million

TOTAL $7.5 million

This represents a cost savings of $5.7 million on the original estimate.

For the configuration of Figure 5.2, the cost of the predrilling and grouting would be the same, $3.3 million. However, in this case, the concrete wall would still have to be $13.2 million, since the critical zone is at depth. The overall cost of the composite cutoff would therefore be $16.5 million. However, the pretreatment in advance of the concrete wall would assure that the wall could in fact be built in a cost-effective, safe, and timely fashion, i.e., without interruptions caused by massive slurry loss. The overall (high) project cost would simply be a reflection of a uniquely challenging geological situation, i.e., a continuous horizon of erodible material at depth.

For the configuration of Figure 5.3, the pretreatment cost would be the same ($3.3 million). It would result in the identification of three discrete zones of structurally defined karst of combined area 3 × 80 ft × 40 ft = 9,600 square feet. Therefore, the cost of the concrete wall actually needed to cut off these features would be 9,600 square feet × $120/square feet = $1,152,000. The total cost of the composite wall is $3,300,000 + $1,152,000 = $4.5 million, which would represent a savings of $8.7 million on the original "full cutoff" cost estimate.

Thus, the investment in the predrilling and grouting program in this exercise generates significant savings in the cases of Figures 5.1 and 5.3, whereas for the case of Figure 5.2 it assures that the wall, which must be built to full depth, can be installed without massive delays, difficulties, or—at worst—creating dam safety issues.

5.4.3 Recommendations for grouting as a component of a composite cutoff wall

5.4.3.1 Site investigation assessment and design

The most important elements of this phase are as follows:

- Research and utilize all the historical data (including original construction photographs) that may have bearing on the development of a tentative geostructural model for the site. An excellent example is provided by Spencer (2006) for Wolf Creek Dam, Kentucky.

- Conduct a new, thoughtful, and focused site investigation to test the tentative geostructural model and so provide prospective bidders with the kinds of information they truly need to estimate construction productivity and to quantify other construction risks.
- Develop an initial estimate of the extent of the composite cutoff and its contributory components, i.e., the concrete wall and the grout curtain.
- Assess the adequacy of the existing dam and foundation instrumentation, and design and install additional monitoring arrays as appropriate. Revise the reading frequency protocols as appropriate, especially in the vicinity of construction activities.

5.4.3.2 Preparation of contract documents and contractor procurement methods

Major recommendations are as follows:

- Draft a performance specification (as opposed to a prescriptive specification), and clearly define the methods and techniques that are not acceptable. Performance goals must be explicitly defined, together with their means of verification.
- Procure the specialty contractor on the "best value" basis, not "low bid."
- Mandate "partnering" as a minimum; favor "alliancing" as the goal (Carter and Bruce 2005).
- Perhaps separate general construction activities (e.g., office modifications, relocation of existing utilities and services) into a different contract, but always leave the design and construction of the working platform to the specialist contractor.

5.4.3.3 Technical aspects

The following items are particularly important:

- If flush water has been lost during investigatory drilling, slurry will certainly be lost during wall excavation, without pretreatment of those same horizons.
- The minimum pretreatment intensity will feature two rows of inclined holes, one on either side of the subsequent wall location. The rows may be 5–10 feet apart, and the holes in each row will typically close at 5- to 10-foot centers (i.e., after all successive orders of holes are installed). The inclination (typically 15 degrees off vertical) will be oppositely orientated in each row.
- The curtain should be installed to at least 50 feet below and beyond the originally foreseen extent of the cutoff to ensure adequate

coverage and to identify unanticipated problems. The treatment is to be regarded as an investigatory tool, equally as much as a ground pretreatment operation and as a sealing of clean rock fissures.

- MWD (measurement while drilling) principles are to be used, the philosophy being that every hole drilled in the formation (not just cored investigations) is a source of valuable geotechnical information.
- Special attention must be paid to the epikarst, which will typically require special grouting methods such as MPSP (multiple packer sleeve pipe) (Bruce and Gallavresi 1988), descending stages, and different grout mixes.
- A test section at least 500 feet long should be conducted and verified to allow finalization of the method statement for the balance of the grouting work. A residual permeability of 10 Lugeons or less should be sought in the area that is later to accept the cutoff, and 1–3 Lugeons in the "clean" rock below the future cutoff toe. Conversely, a falling head test in vertical verification holes, using bentonite slurry as the test fluid, is an appropriate test. Verification holes should be cored and the holes observed with a televiewer to demonstrate the thoroughness of the grouting.
- In terms of the details of execution, the principles previously detailed to create quantitatively engineered grout curtains should be adopted. Thus, one can anticipate the use of stage water tests; balanced, modified, stable grouts; and computer collection, analysis, and display of injection data. When drilling the verification holes (at 25- to 100-foot centers between the two grout rows), particular care must be taken to ensure that no drill rods are abandoned within the alignment of the wall, since this steel will adversely impact subsequent wall excavation techniques.
- Grouting pressures at refusal should be at least twice the foreseen maximum slurry pressure to be exerted during panel construction.

5.4.3.4 Construction

Every project is different, and the following basic recommendations must be supplemented on a case-by-case basis:

- The work must be conducted in accordance with the contractor's detailed method statement. This document, in turn, must be in compliance with the minimum requirements of the performance specification unless otherwise modified during the bidding and negotiation process. At the same time, modifications to the foreseen means and methods can be anticipated on every project in response to unanticipated phenomena. Prompt attention to and resolution of these challenges are essential.

- Special attention is merited to the details of the design and construction of the working platform. The contractor's site support facilities (e.g., workshop, offices, slurry storage and cleaning, concrete operations) can be completed and the utilities extended along the alignment (water, air, electricity, light, slurry) during the building of the work platform.
- The test section should be established in a structurally and geologically noncritical area that does not contain the deepest extent of the foreseen concrete wall. The test section should, however, be integrated into the final works if it is proven to have acceptable quality.
- The concrete wall excavation equipment must have adequate redundancy and must be supported by appropriate repair/maintenance facilities. A variety of equipment is usually necessary (clamshell, hydromill, chisels, backhoe) to best respond to variable site conditions and construction sequences. Standard preinstalled mechanical features, such as the autofeed facility on hydromills, must not be disabled in an attempt to enhance productivity.
- Special protocols should be established to ensure that the flow of real-time construction data (e.g., inclinometer readings from a hydromill) is regular, uncontaminated, and of verifiable provenance.
- The site laboratory must be capable of accurately and quickly conducting the whole range of material tests required. In addition, the contractor's technical/quality manager, who is a vital component in any such project, must be fully conversant with all the principles and details involved in the monitoring of the construction, and of the instrumentation of the dam itself. In particular, expertise with panel or pile verticality and continuity measurement is essential, as is an awareness of the significance of piezometric fluctuations or changes.
- Emergency response plans must be established to satisfy any event that may compromise dam safety.

5.4.3.5 Assessment of cutoff effectiveness

The protocols established for observations and instrument readings during remedial construction must be extended after remediation, although usually at a somewhat reduced frequency. The data must be studied and rationalized in real time so that the remediation can be verified as meeting the design intent. Alternatively, it may become apparent that further work is necessary, a requirement that becomes clear only when the impact of the remediation of the dam/foundation system is fully understood. Finally, owners and designers should publish the results of these longer-term observations so that their peers elsewhere can be well informed prior to engaging in their own programs of similar scope and complexity.

5.5 CASE HISTORIES

5.5.1 Mississinewa Dam, Indiana

This 8,100-foot-long, earthfill, flood-control structure is located near Peru, Indiana, and went into operation in 1968 (Henn and Brosi 2005). Its maximum height is 140 feet at the tailwater, with normal seasonal pools varying between EL 712 and 737 feet. In 1988 significant and abnormal crest settlement was first noted over a 400-foot-long stretch of the right abutment, of a magnitude 0.5 feet above the predicted consolidation amount. Minor seepage at the toe in this area had been noted in this vicinity in 1966 with the pool at EL 740, prompting the installation of a traditional, one-row grout curtain. This effort was abandoned when closure could not be achieved. Even more worrisome, it was most likely that certain holes were drilled but neither pressure-grouted nor backfilled. This leads to the unfortunate conclusion that a breach had probably been created in the upstream impervious blanket allowing for direct hydraulic communication from the lake to the foundation. A multiyear series of intensive geotechnical investigations, instrumentation, and monitoring was undertaken to determine the cause and extent of the subsidence. It was determined that the karstic foundation was, in fact, in an active failure mode. By 2001 a contract had been let to build a concrete cutoff wall through the right abutment. This was 2,600 feet long, with actual depths ranging from 147 to 230 feet, and a final area of almost 430,000 square feet, of which over 97,000 square feet was to be in rock. The centerline of the 30-inch-wide wall was 8 feet downstream of the dam centerline, for access reasons and to preserve upstream embankment instrumentation (Photo 5.1).

Work commenced with a 100-foot-long test section at the shallowest end of the cutoff, toward the extreme right abutment. However, during excavation of the first panel, there was sudden and complete loss of slurry with the hydromill 9 feet into rock. Prompt action involving rapid panel backfill prevented significant embankment damage. A second test panel 90 feet away experienced a similar phenomenon with the hydromill 18 feet into rock. The government then directed a pregrouting program, initially in the test section, but later extended over the whole extent of the cutoff wall.

The grouting contractor used the same advanced methodologies previously developed at Penn Forest Dam, Pennsylvania, and Patoka Lake Dam, Indiana, among others. These featured the use of sonic drilling, balanced, stable grouts, and real-time computer monitoring and analysis ("Intelligrout" system). The design philosophy of the grouting program was to prevent total and complete sudden slurry losses, and not the partial loss of slurry due to natural filtration, or seepage into minor rock features. To rationally establish the grouting performance goal, the required residual permeability and the dimensions of the grouted zone had to be balanced.

Photo 5.1 View of the remediation works at Mississinewa Dam, Indiana. (Courtesy of ACT, Gannett Fleming, and the USACE.)

Bearing in mind also the contractor's slurry reserve capacity, the boundary conditions set for analysis were a sudden loss of 1,970 gpm to a 2-foot loss over 10 minutes, equating to 80 gpm. From the loss analyses, a target maximum residual permeability of 10 Lugeons was set, within a grouted zone encompassing the cutoff wall and extending 5 feet upstream and downstream of it.

For each of the two rows of 15° inclined holes, 12-foot-long stages were used, with a final closure, involving tertiaries, at 6-foot centers. Most of the work could be conducted with upstages. Holes were extended at least 5 feet below the proposed cutoff wall depths, and at least 5 feet below specific deep solution features. In one feature area, additional lines of holes were installed to better define its geometry and extent, and to treat it more intensely.

Very careful monitoring of the drilling processes led to a clear delineation of the soil-rock interface, the degree and depth of bedrock weathering, and a confirmation of what constituted "competent" rock. In combination with the water-pressure test and grout-injection data, the designed depth of the cutoff wall was validated, and modified as appropriate. Particular attention was paid to treatment of the contact, using a modified MPSP system.

A suite of 5 grouts of progressively increasing Marsh cone value were developed, ranging between water:cement ratios of 2.0 and 0.66 (by weight). Prehydrated bentonite, type F flyash, welan gum, and superplasticizer were also used in all mixes. The properties of the respective mixes are shown in Table 5.1. No Mix A was used since the program goals did not require a mix of very low Marsh cone value (typically around 35 seconds).

The maximum effective grouting pressure was 170 psi and refusal was determined to be 0.5 liters of grout per minute, held over a 10-minute period. After each 600-foot section of the curtain had been grouted to apparent refusal, one or two verification holes were drilled, water-tested and grouted to confirm the target residual permeability had been met. The majority of tests gave values of less than 2 Lugeons.

The program was conducted on a two-shift-per-day basis, and lasted about 12 months, during which time 64,050 feet of overburden drilling and 22,396 feet of rock drilling were conducted, together with the injection of 52,313 cubic feet of grout into 2,306 stages. The efficiency and quality of the works were greatly improved by the construction of a concrete work platform all along the alignment.

Within days of completion of the grouting in the test section, diaphragm walling resumed; no significant slurry loss occurred, and panels installed in previously "voided areas" that were grouted successfully retained slurry with only minimal seepage. The pregrouting was also found not to have, in any way, adversely affected either the bentonite slurry properties during excavation, or the rate of excavation progress in rock. Likewise, during the

Table 5.1 Grout mix properties, Mississinewa Dam, Indiana

Grout Mix Properties						
Mix		B	C	D	E	F
Specific gravity		1.38	1.44	1.53	1.57	1.57
Marsh funnel (sec)[a]		41	48	55	–	–
Flow cone (sec)[a]		–	–	–	14	24
Cohesion (g/mm^2)		0.180	0.348	0.490	0.630	0.630
Bleed	1 hr	0	0	0	0	0
	2 hrs	0	0	0	0	0
	3 hrs	0.5	0	0	0	0
Pressure filtration coefficient (min$^{-1/2}$)		0.045	0.042	0.040	0.036	0.031
Initial gel time	(hrs:mins)	3:00	2:00	1:30	1:00	1:00
Final gel time		7:00	5:00	4:00	3:00	3:00
Initial set time		17:00	15:30	13:30	12:30	12:30
Final set time		22:30	19:00	15:00	13:30	13:30
Compressive Strength[b] (psi)		260	360	770	920	870

[a] Marsh funnel versus flow cone depended on viscosity of mix because thicker mixes had infinite Marsh funnel time, and thus flow cone results were used.

[b] The compressive strength for B and C mixes are nine-day results, and D, E, and F mixes are seven-day results.

subsequent production works, no significant slurry losses were encountered and in part due to the improved ground condition, the contractor was permitted to increase the number of panels open at any time to four, thereby affording a significant scheduling advantage. Wall depths were modified, based on an examination of the grouting data, to help confirm that the cutoff was taken at least 3 feet into rock of sufficient quality.

5.5.2 Bear Creek Dam, Alabama

Bear Creek Dam is a 1,385-foot-long, homogeneous fill embankment dam constructed in the late 1960s by the Tennessee Valley Authority (TVA) and first filled in 1969 (Charlton, Ginther, and Bruce 2010; Ferguson and Bruce 2010). The dam's crest elevation is 618 feet and it has a maximum height of 85 feet. The dam is equipped with a reinforced-concrete ogee crest overflow chute spillway (crest elevation, 602 feet) and a gated intake tower to a 9-foot-diameter sluiceway tunnel and stilling basin that are used to control lake levels under normal conditions.

The dam was constructed with a traditional single-line grout curtain and key trench for approximately two-thirds of the embankment foundation. During the initial construction, numerous solution features in the Mississippian limestone foundation were encountered and backfilled, large volumes of extremely weathered rock were removed, and large grout takes were common. The aforementioned treatment procedures were not performed on a section of the foundation from the left abutment at the spillway that extended 300 feet across the foundation. Upon the first filling in 1969, seepage was discovered along the toe of the embankment and was the subject of various studies and treatment programs thereafter. Sustained foundation seepage flows captured and measured near the surface—on the order of 800 gpm at normal summer pool levels—indicate the existence of higher flows through the untreated cavernous subsurface near the left abutment. Subsequent grouting programs have been successful at temporarily reducing flows to approximately half of the historical maximum. However, the grouting efforts were never brought to closure, and over time, flows returned to previous seepage rates.

In December 2004 a high-headwater event resulted in the appearance of numerous boils, small sinkholes, and new seepage flows from the toe. A study comprising piezometer installation, coring of the foundation rock, and cone penetration testing in the embankment confirmed the left abutment foundation to be the pathway of the majority of the seepage.

The dam provides flood control, water supply, and recreational benefits to the area. In order to preserve these benefits, the TVA elected to embark on an extensive rehabilitation effort, and a permanent solution for the dam's deficiencies was designed. Following an exploratory drilling program and preliminary design phase, and based on input from the TVA and its independent review board (IRB), the following rehabilitation scheme was selected as the best solution for the remediation of Bear Creek Dam:

1. To prevent loss of the dam as a result of overtopping of the embankment during the potential maximum flood (PMF), a downstream roller-compacted concrete (RCC) reinforcement structure or berm would be constructed.
2. Initial seepage cutoff concepts included the construction of a full-length, secant pile wall through the limestone and toed into underlying shale. However, the benefit of further site investigation information and insight into successful case histories of composite wall applications (such as at Mississinewa Dam), a much more economic approach was developed. This featured a two-line curtain, designed and constructed in accordance with best contemporary standards, and the subsequent installation of concrete panels into those discrete karstic features, defined by the drilling and grouting program and observations made during surface preparation (Photo 5.2).

In addition, during the foundation treatment phase, a large solution feature that intersected the sluiceway tunnel was encountered that required an additional grouting program to be conducted approximately perpendicular to the two-line grout curtain, to ensure the continuity of the seepage barrier.

The initial subsurface exploration program for the new RCC structure included core drilling, borehole pressure-testing, limited soil-sampling, geophysical borehole-logging, surface geophysics, field and laboratory testing, and groundwater-flow analysis. Site investigation began in June of 2007 and was completed in September 2007.

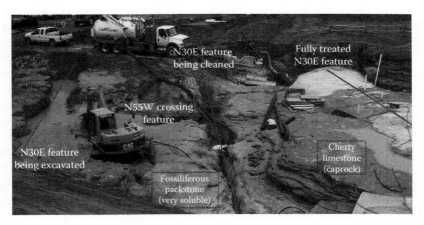

Photo 5.2 Cleaning of N30E features in the foundation of the new Bear Creek Dam, Alabama. (Courtesy of Paul C. Rizzo Associates, Inc.)

The objectives of the site investigation included:

- Determination of the hydraulic properties of the rock mass, including extent and nature of karst development.
- Determination of the "groutability" of the karst features: features containing significant amounts of detrital and residual material that provide potential erosive zones that could compromise the completion of the grout curtain.

An extremely high standard of core was exercised in exposing, mapping, and preparing the foundation surface of the new RCC structure. Preparation of the foundation required excavation of approximately 40,000 cubic yards of residual soil, 25,000 cubic yards of alluvium, 6,000 cubic yards of fill, and 10,000 cubic yards of moderately to intensely weathered rock. Approximately 100 cubic yards of existing detritus was removed from solution cavities. Five thousand five hundred cubic yards of minimum 3,000-psi dental concrete were placed in irregularities in the foundation, and an additional 1,200 cubic yards were placed in order to prepare a more level working surface for drill rigs and to provide a surface conducive to RCC placement.

In order to further develop understanding of the foundation conditions, a comprehensive system of logging both exploratory (HQ size core) and production (rotary percussive drilling) borings, including down-hole geophysical methods, was included for the drilling and grouting program.

A total of 34 exploratory HQ-size core holes were placed on 80-foot centers on both lines of the grout curtain. These borings were logged conventionally by a geologist in the field as the core was recovered and were then subjected to geophysical logging after being washed thoroughly when the coring was completed. Geophysical logging included photographic logging of the walls of the core hole with a down-hole optical televiewer camera capable of identifying bedding features and fractures and producing a 360-degree view of boring sidewalls, gamma logging to assist in delineating bedding features (primarily shale lenses), and caliper logging to measure spatial deviations in the sidewalls of the core hole. The addition of the geophysical logs to conventional logging practice enhanced understanding of the subsurface fracture patterns and solution mechanisms and proved very valuable to the generation of an accurate portrait of the site stratigraphy. Figure 5.4 is a portion of a log produced by the optical televiewer showing the camera shot of a vertical fracture encountered in the core hole within a cherty zone in the limestone and the corresponding mapping data recorded by the televiewer.

Upon completion of coring and logging of the exploratory borings, the borings were water-pressure tested using five-step tests and grouted as production grout holes.

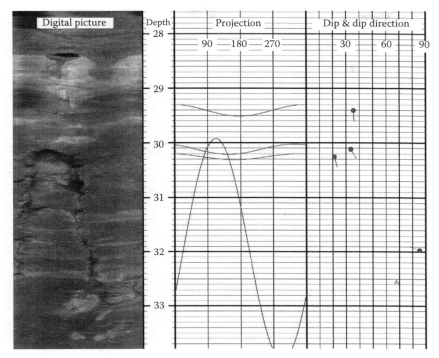

Figure 5.4 Section of optical televiewer log showing mapped fracture. Depth in feet. (Charlton, J. E., C. H. Ginther, and D. A. Bruce, "Comprehensive Foundation Rehabilitation at Bear Creek Dam," *Environmental and Engineering Geoscience*, 16, 3, 2010.)

Upon completion of the exploratory borings in a given area, production drilling (starting with primary holes) was performed using a rotary percussive drill rig, with water used as the flushing medium. In order to gain maximum information from these destructive drilling techniques, a drilling parameter recorder (DPR) was installed on the drill rig; this DPR recorded drilling rate, thrust pressure, drilling torque, and water flow through the drill-string for every boring performed. Through the course of the project, the DPR logs proved to be a valuable resource for identifying areas of fractured rock, clay infill, and changes in stratigraphy particularly as the driller's understanding of site-specific ground conditions improved.

Other features of the program may be summarized as follows:

- Computer-controlled, real-time data monitoring of water-pressure test and grouting parameters was mandated for all stages.
- A suite of three HMG grouts were developed for routine treatment (Table 5.2), while the contractor also developed a medium-mobility grout (MMG), comprising Mix C plus flyash. This mix was a very

Table 5.2 High-mobility grout properties, Bear Creek Dam, Alabama

Parameter (unit)	Mix A	Mix B	Mix C	Purpose of requirement
Bleed (percent)	≤3.0	≤3.0	≤3.0	Low bleed prevents voids caused by grout settlement (stability)
Pressure filtration K_{pf} (min$^{-1/2}$)	≤.040	≤.040	≤.040	Low-pressure filtration corresponds to less mix water being pressed out of the grout, and so promotes long-distance penetration into fractures
Marsh viscosity (sec)	35	50–55	80+	Provide range of viscosities to adjust, as appropriate to subsurface conditions
Initial stiffening time (hrs)	≥3	≥3	≥3	Provide enough time for mix, injection, and travel prior to initial set

effective "closer" for large take stages. In addition, LMG was used in particularly open karstic features.

- The two-grout lines, 10 feet apart, had opposed inclined holes (15°) with primary-secondary spacings in each row being 20 feet. Tertiary holes were directed depending on data analyses, in certain locations and to certain depths. Stages were typically 10 feet in length, but increased in tighter conditions. Upstaging was conducted wherever geologically practical.
- A target residual permeability of 5 Lugeons was set, verified by HQ-cored verification holes between the two grout lines, at 100-foot centers. Further confirmation of grouting effectiveness included visual observations of reduced downstream flows into the spillway tailrace, and the fact that post-treatment excavations (below lake elevation) were performed in the dry.

Four cutoff wall panels were prescribed in light of the results of the drilling and grouting program and a supplemental exploratory drilling program consisting of rotary percussive borings on 2- or 3-foot centers (Table 5.3). DPR drilling logs from the supplemental drilling were used to clearly iden-tify the vertical and lateral extents of clay infill for panels 1 and 2.

Cutoff panels were centered between the two lines of the drilling and grouting program in order to make best use of the pretreatment afforded by the previously performed grouting. Several construction methods were evaluated for the construction of these cutoff panels, including drilling and blasting prior to excavation, use of a secant pile wall system, and the use of an excavator-mounted hoe-ram and long-reach excavator to remove the material from the cutoff wall sections. Drilling and blasting were even-tually discounted as viable construction methods given the potential for

Table 5.3 Cutoff panel details, Bear Creek Dam, Alabama

Cutoff panel no.	Station extents	Expected maximum depth	Geologic rationale for panel delineation
1	8+00 to 8+67	35 ft.	Clay infill/void activity at depths of 25–30 ft.
2	7+00 to 7+40	35	Clay infill at depths of up to 30 ft.
3	3+10 to 4+77	35	Cut off very weathered zones in the Bangor shale at the maximum section of the new structure
4	2+40 to 2+50	23	Cut off the continuation of N32E sluiceway solution feature, act as test panel for construction method

damage to the thinly bedded overlying limestone and because of concerns about the effectiveness of blasting in the shale unit. Secant pile installation was the technically preferred method of installation of the cutoff panels, as it was the method least likely to cause damage to the foundation and previous grout treatments, and because it offered relative ease of installation at depth and the relatively low volume of excavation to be backfilled. However, the high cost of mobilization of a secant pile contractor, in relation to the small area treatment, in addition to schedule availability issues, prevented the use of the secant pile method. In the end, the construction method utilizing excavation with a hoe-ram and long-reach excavator was selected because of the availability of the necessary equipment within TVA's Heavy Equipment Division.

Construction of the cutoff panels began with Panel 4, which served as a proof-of-method test. Panel 4 was chosen to test the hoe-ram and excavation construction method because of its smaller size and shallower depth. Based on successful performance in the installation of Panel 4, the method was approved for the remaining sections of cutoff wall. Prior to construction of Panels 1 and 2, an exploratory program consisting of rotary percussive borings on 2- to 3-foot centers around the upstream and downstream faces of the planned panel locations was performed to clearly delineate the extent of the clay infill to be treated by these panels. The DPR logs recorded during this additional exploration provided a basis for depth reduction along the panels at several locations. Cutoff panel construction generally consisted of an excavation phase, followed by thorough washing of the sidewalls and floor of the panel (similar to the specifications of the previously performed dental concrete treatments), survey of the surface extent of the panel, mapping of the excavation sidewalls by a geologist, and then backfill of the cleaned, mapped excavation with concrete. While a minimum 2-foot panel width was specified, the construction method resulted in widths at the bottom of each panel ranging from 6 to 8 feet and widths at the top of each panel ranging from 8 to 10 feet. Dewatering issues during panel excavation were minimal, as the two-line curtain was complete at the time of panel excavation. The only notable seepage occurred in Panel 3B at the interface of the lower limestone and shale. At this interface, an estimated 5–7 gpm seeped into Panel 3B between RCC centerline stations 3+20 to 3+40. This area coincided with a weathered zone in shale. Panels 1, 2, and 4 exhibited no stability issues, while Panel 3 had two significant areas in which shale periodically sloughed off into the excavation. As a result, a geologist had to map the upstream and downstream surface panel walls from the foundation surface at the edge of the panel while tied off to a loop anchored in the rock.

Construction of these cutoff panels was completed in December 2008. Table 5.4 summarizes the final extents, depths, and total concrete volumes placed to backfill the panel excavations.

After several days of set time, verification core borings were drilled along the centerline of each panel at 20–30-foot spacing, and five-step water pressure was tested to verify the integrity of the panel. Boring locations were chosen to intercept the abutments of the panels, the bottom contact of the panel with the foundation, or in some cases areas of interest or concern based on foundation conditions noted during the mapping process. The acceptance criterion for the cutoff panels was 5 Lugeons, the same as for the two-line grout curtain. In fact, all verification tests performed through the cutoff panels yielded "no take" (0 Lugeon) results. After completion of the water-pressure testing and acceptance of the panel, verification boreholes were backfilled with a high-strength cement grout.

5.6 PERSPECTIVE

The previous sections of this chapter, together with the contents of Chapters 2 and 4, clearly illustrate that U.S. engineering practice in rock grouting and concrete cutoff wall construction has reached high levels of competence. However, even the best grouting practices cannot assure a robust, durable seepage barrier in terrains containing significant amounts of potentially

Table 5.4 Cutoff panel construction details, Bear Creek Dam, Alabama

Cutoff panel no.	Station extents	As-built maximum depth	Cutoff panel area	Concrete volume placed
1	8+00 to 8+67	32 ft.	2,013 sq. ft.	594 cu. yards
2	7+00 to 7+40	22	754	276
3	3+10 to 4+77	32	5,490	1,416
4	2+40 to 2+50	23	250	100

erodible materials, particularly when these are concentrated in discrete features of considerable dimension and extent. Similarly, diaphragm walling operations will be vulnerable to voided conditions that have the potential to cause sudden and complete loss of the supporting slurry during excavation. This can have serious dam safety implications, quite apart from the prospect of losing extremely valuable equipment trapped hundreds of feet down in collapsed trenches. Furthermore, diaphragm walls, especially in rock, are costly, which is particularly galling when it is noted that oftentimes large volumes of excellent rock of appreciable strength and low permeability are being replaced with an engineered material (concrete) perhaps half its strength and of equivalent permeability.

It is time to squash the false debate as to which method—grouting or diaphragm wall—is best. The obvious way forward is to take the best from each camp: drill, water test, and grout (relatively cheaply) to prepare the ground for a concrete wall (relatively expensive), the extent of which is now properly defined. Then build, in improved ground conditions, the definitive concrete wall only in those areas where the grouting cannot be expected to be effective in the long term.

Our dams must be repaired in a way that must be conceived to be "permanent." However, the goal remains that we should ensure that our designs and implementations are cost effective. Furthermore, there is simply insufficient industrial capacity in the United States to build the foreseen volume of cutoffs solely by concrete wall construction techniques in the time frame available. The concept of the "composite cutoff" is therefore logical, timely, and the obvious choice. This argument was expressed in somewhat different form by the irrepressible instrumentation specialist John Dunnicliff (1991):

> "Equal rights for grouters,"
> cries Donald Bruce with glee.
> He challenges the doubters,
> with pungent repartee.
> Slurry wall or grouting?
> Which method works the worst?
> The brotherhood is touting
> that grouting should be first.
> Casagrande's basis
> for sealing every crack
> was "use both belt and braces"
> to hold the water back.
> So let's stop all the shouting
> and use them, one and all:
> the wall to seal the grouting;
> the grout to seal the wall.
> The brothers will be wealthy.

The grapevine will be sweet.
The dams will all be healthy,
and flow nets obsolete.

ACKNOWLEDGMENTS

The grouting contractor at Mississinewa Dam was Advanced Construction Techniques, working with Gannett Fleming, Inc. The cutoff wall contractor was Bencor-Petrifond JV. At Bear Creek Dam, the Engineer-of-Record was Paul C. Rizzo & Associates, and the grouting contractor Geo-Con, Inc. It is TVA's policy to state that, while granting permission to publish information about its projects, it does not endorse any entity or firm associated with the work.

Permission was granted by the Association of Environmental and Engineering Geologists to make generous reference to the contents of their special edition on dams, published in August 2010.

REFERENCES

Bruce, D. A., and J. P. Davis. 2005. "Drilling through Embankments: The State of Practice." Paper presented at the USSD 2005 Conference, Salt Lake City, UT, June 6–10, 12 pp.

Bruce, D. A., T. L. Dreese, and D. M. Heenan. 2008. "Concrete Walls and Grout Curtains in the Twenty-First Century: The Concept of Composite Cut-Offs for Seepage Control." Paper presented at the USSD 2008 Conference, Portland, OR, April 28–May 2, 35 pp.

Bruce, D. A., T. L. Dreese, and D. M. Heenan. 2010. "Design, Construction, and Performance of Seepage Barriers for Dams on Carbonate Foundations." *Environmental and Engineering Geoscience* 16 (3): 183–93.

Bruce, D. A., and F. Gallavresi. 1988. "The MPSP System: A New Method of Grouting Difficult Rock Formations." Pp. 97–114 in *Geotechnical Aspects of Karst™ Terrains*. ASCE Geotechnical Special Publication 14. Reston, VA: ASCE.

Bruce, D. A., A. Ressi di Cervia, and J. Amos-Venti. 2006. "Seepage Remediation by Positive Cut-Off Walls: A Compendium and Analysis of North American Case Histories." Paper presented at the ASDSO Dam Safety Conference, Boston, MA, September 10–14.

Carter, J., and D. A. Bruce. 2005. "Enhancing the Quality of the Specialty Contractor Procurement Process: Creating an Alliance." Pp. 76–87 in *Geo3 GEO Construction Quality Assurance/Quality Control Conference Proceedings*, edited by D. A. Bruce and A. W. Cadden. Dallas, TX: ADSC.

Charlton, J. E., C. H. Ginther, and D. A. Bruce. 2010. "Comprehensive Foundation Rehabilitation at Bear Creek Dam." *Environmental and Engineering Geoscience* 16 (3): 211–27.

Chuaqui, M., and D. A. Bruce. 2003. "Mix Design and Quality Control Procedures for High Mobility Cement Based Grouts." Pp. 1153–68 in *Grouting and*

Ground Treatment, Proceedings of the Third International Conference, edited by L. F. Johnsen, D. A. Bruce, and M. J. Byle. Geotechnical Special Publication 120. Reston, VA: ASCE.

Davidson, L. 1990. "Performance of the Concrete Diaphragm Wall at Navajo Dam." Paper presented at the 10th Annual USCOLD Conference, New Orleans, LA, March 6–7, 21 pp.

Dreese, T. L., D. B. Wilson, D. M. Heenan, and J. Cockburn. 2003. "State of the Art in Computer Monitoring and Analysis of Grouting." Pp. 1440–53 in *Grouting and Ground Treatment, Proceedings of the Third International Conference*, edited by L. F. Johnsen, D. A. Bruce, and M. J. Byle. Geotechnical Special Publication 120. Reston, VA: ASCE.

Dunnicliff, J. 1991. "More on the Grouters' Rallying Call." *Geotechnical News*.

Ferguson, K. A., and D. A. Bruce. 2010. "The Bear Creek Dam, Alabama." Paper presented at the 30th Annual USSD Conference, Sacramento, CA, April 12–16, 16 pp.

Halpin, E. 2007. "Trends and Lessons in Assessing Risks Posed by Flood Damage Reduction Infrastructure." Paper presented at the ORVSS XXXVIII, Ohio River Valley Soils Seminar, Louisville, KY, November 14.

Henn, K., and B. E. Brosi. 2005. "Mississinewa Dam—Settlement Investigation and Remediation." Paper presented at the Association of State Dam Safety Officials 22nd Annual Conference, Orlando, FL, September, 15 pp.

Lombardi, G. 2003. "Grouting of Rock Masses." Pp. 164–97 in *Grouting and Ground Treatment, Proceedings of the Third International Conference*, edited by L. F. Johnsen, D. A. Bruce, and M. J. Byle. Geotechnical Special Publication 120. Reston, VA: ASCE.

Procurement Process: Creating an Alliance," Pp. 76–87 in *Geo3 GEO Construction Quality Assurance/Quality Control Conference Proceedings*, Editors D. A. Bruce and A. W. Cadden, Dallas/Ft. Worth, TX, November 6–9.

Ressi di Cervia, A. 2003. "A Better Barrier." *Civil Engineering* 73 (7): 44–49.

Rice, J. D., and M. J. Duncan. 2010a. "Deformation and Cracking of Seepage Barriers in Dams Due to Changes in the Pore Pressure Regime." ASCE *Journal of Geotechnical and Geoenvironmental Engineering* 133 (1): 2–15.

Rice, J. D., and M. J. Duncan. 2010b. "Findings of Case Histories on the Long-Term Performance of Seepage Barriers in Dams." ASCE *Journal of Geotechnical and Geoenvironmental Engineering*, 133 (1): 16–25.

Spencer, W. D. 2006. "Wolf Creek Dam Seepage Analysis and 3-D Modeling." Paper presented at the ASDSO Dam Safety, Boston, MA, September 10–14, 36 pp.

Walz, A. H., D. B. Wilson, D. A. Bruce, and J. A. Hamby. 2003. "Grouted Seepage Cutoffs in Karstic Limestone." Pp. 967–78 in *Grouting and Ground Treatment, Proceedings of the Third International Conference*, edited by L. F. Johnsen, D. A. Bruce, and M. J. Byle. Geotechnical Special Publication 120. Reston, VA: ASCE.

Wilson, D. B., and T. L. Dreese. 1998. "Grouting Technologies for Dam Foundations." Proceedings of the 1998 Annual Conference Association of State Dam Safety Officials, October 11–14, Las Vegas, NV. Paper No. 68.

Wilson, D. B., and T. L. Dreese. 2003. "Quantitatively Engineered Grout Curtains." Pp. 881–92 in *Grouting and Ground Treatment, Proceedings of the Third International Conference*, edited by L. F. Johnsen, D. A. Bruce, and M. J. Byle. Geotechnical Special Publication 120. Reston, VA: ASCE.

Chapter 6

Prestressed rock anchors

Donald A. Bruce and John S. Wolfhope

6.1 BACKGROUND ON U.S. PRACTICE

Although the history of prestressed rock anchors for dams dates from 1934 and the raising of the Cheurfas Dam in Algeria, current research indicates that the first U.S. dam to be stabilized by high-capacity prestressed rock anchors was the John Hollis Bankhead Lock and Dam, Alabama. On that project, the first 6 test anchors and 16 production anchors were installed in 1962. This project was completed for the U.S. Army Corps of Engineers who therefore had sufficient confidence in the technology that they were the sponsor for most of the half dozen or so similar applications in the six years that immediately followed. The U.S. Bureau of Reclamation first used anchors to stabilize appurtenant structures at dams in 1967, while the Montana Power Company was also an early proponent. In those days, the technology was largely driven by the post-tensioning equipment suppliers, employing the same principles and materials such as used in prestressed/post-tensioned structural elements for new buildings and bridges. The "geotechnical" inputs, that is, the drilling and grouting activities, were typically subcontracted to drilling contractors specializing in site investigation and dam grouting in the west, and to "tieback" contractors in the east.

Since those early projects, North American practice has evolved substantially through emphasis on technology and refinements in construction techniques. The engineering community has become familiar with the materials and equipment and has gained an understanding of the important aspects of successfully designing and specifying prestressed rock anchor systems for dams. The post-tensioning equipment manufacturers have invested substantial resources into refining the manufacturing processes for the fabrication of prestressed rock anchor tendons and the equipment for their stressing and testing. An industry of sophisticated geotechnical construction contractors has emerged with significant experience and expertise in the drilling, installation, and testing of prestressed rock anchors in dams. These specialists now act as the general contractor on most projects.

The early engineering and construction practices were documented by Littlejohn and Bruce (1977). This publication provides a comprehensive overview of prestressed rock anchor design, construction, stressing and testing, including establishing the state of the art in anchor capacity and sizing, spacing of anchor tendons, selection and control of drilling equipment, storage and handling of anchor tendons, grout-mix designs and grouting methods, and stressing procedures. Although very little change has occurred over the past thirty-five years in certain aspects, such as our assumptions regarding the grout to steel and grout to rock bond interaction, particular progress has been made in the areas of corrosion protection, quality control, and stressing/testing procedures. Since the first project in 1962, prestressed rock anchors have been used successfully in North America on over 400 dams.

6.2 THE NATIONAL RESEARCH PROGRAM

During the period 2005–6, Phase 1 of a national research program into the use of rock anchors for North American dams was completed. This work had three goals:

1. Develop a bibliography of all technical papers published on the subject of dam anchoring in North America;
2. Create a database containing as much information as possible on each dam anchored in North America; and
3. Conduct a comparative review of each of the five successive versions of the national "recommendations" documents which have been published in the United States since 1974.

This project was funded by a consortium of American and Japanese interested parties. The co-principal investigators relied heavily on the cooperation of specialty contractors and specialist post-tensioning suppliers who provided access to historical records.

6.2.1 Literature survey

As the first task of the research, a comprehensive literature survey was completed to identify published dam anchoring case studies and various publications documenting the evolution of North American dam anchoring practices and construction methods. A total of 230 technical papers were compiled relating to North American post-tensioned rock anchor projects. Hard copy and electronic versions of each paper were collected for use in

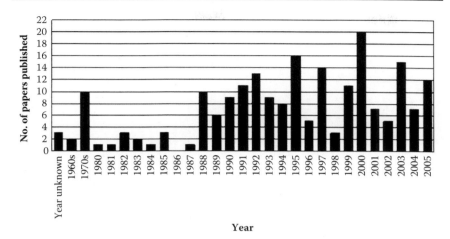

Figure 6.1 Numbers of technical papers on dam anchoring published per year. (From Bruce, D. A., and J. Wolfhope, "Rock Anchors for North American Dams: The National Research Program Bibliography and Database," Paper presented at Institution of Civil Engineers, Ground Anchorages and Anchored Structures in Service, London, England, 2007. With permission.)

further analysis of the anchoring industry. Figure 6.1 shows the number of publications by year, which indicates that over the first five years of the twenty-first century industry has been publishing at a rate of about 13 papers per year. These papers relate to more than 200 different dams.

6.2.2 Case history database

Figure 6.2 presents a histogram of North American dam anchor projects identified during the data-collection process. Since the completion of the initial data research, another dozen case studies have been identified to take the total to over 400 projects completed between 1962 and 2004, ranging from as few as one dam per year during the initial decade for North American practice to a maximum of 23 dam anchoring projects in 1989 when many small- to medium-sized projects were conducted under federal mandate on older hydropower structures. As recently as 2002, 18 anchoring projects were constructed in a single year. Over this 40-year period, more than 20,000 anchor tendons were installed in North American dams, averaging over 500 anchors installed and tested per year. Figure 6.3 shows the number of dams anchored between 1962 and 2004 compared to the number of candidate masonry and concrete dams located in each U.S. state. Figure 6.4 shows the number of dams anchored between 1962 and 2004 compared to the number of candidate "large" dams located in each Canadian province. For comparison and perspective, Figure 6.5 provides a histogram of the ages of concrete and masonry dams in the United States.

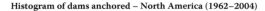

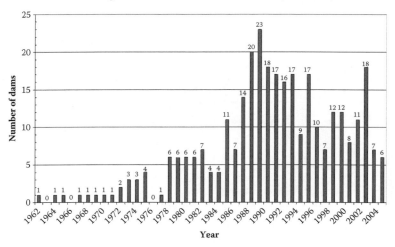

Figure 6.2 Histogram of dams anchored by year (1962–2004). Note: Total number of dams shown = 323; does not include 70 anchor case studies where year anchored not reported or as yet ascertained. (From Bruce, D. A., and J. Wolfhope, "Rock Anchors for North American Dams: The National Research Program Bibliography and Database," Paper presented at Institution of Civil Engineers, Ground Anchorages and Anchored Structures in Service, London, England, 2007. With permission.)

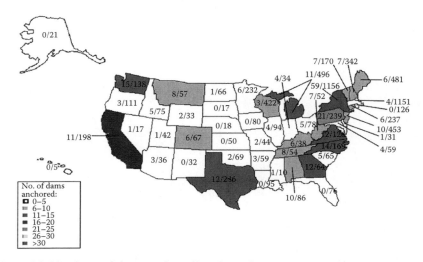

Figure 6.3 Numbers of dams anchored/number of masonry and concreted dams per U.S. state. Note: Statistics may be skewed based on how states classify their dams. (From Bruce, D. A., "Dam Remediation by Anchors and Cut-Offs: A Summary of Two National Research Programs," Ohio River Valley Soils Seminar, ORVSS XXXVIII, Louisville, KY, 2007. With permission.)

Figure 6.4 Numbers of dams anchored/number of large dams per Canadian province. (From Bruce, D. A., "Dam Remediation by Anchors and Cut-Offs: A Summary of Two National Research Programs," Ohio River Valley Soils Seminar, ORVSS XXXVIII, Louisville, KY, 2007. With permission.)

In the years since 2004, the actual number of dam anchor projects has dropped to approximately five projects per year. However, these have tended to be projects of very large scale on major private or municipal dams (e.g., Gilboa Dam owned and operated by the New York City Department of Environmental Protection) or federal dams (e.g., the USACE's Bluestone Dam, West Virginia) and have typically been highly challenging in regards to logistical and technical requirements.

6.2.3 The development and scope of the national recommendations

Recognizing the need for some type of national guidance and uniformity, the Post-Tensioning Division of the Prestressed Concrete Institute (PCI) formed an ad hoc committee that published in 1974 a 32-page document, *Tentative Recommendations for Prestressed Rock and Soil Anchors*. It is interesting to note (Table 6.1) that half of the document comprised an appendix of annotated project photographs intended to illustrate and

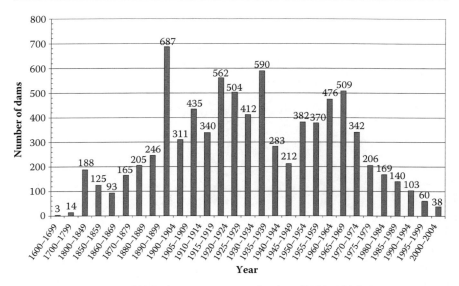

Figure 6.5 Histogram of U.S. dam construction by dam (1600–2004) for dams listed as concrete, gravity, buttress, arch, multiple arch, masonry, or dams listed as having a controlled spillway. Note the National Inventory of Dams lists 79,777 dams in total. This excludes approximately 9,500 dams with unreported/invalid data. The total number of dams shown in the histogram above is 8,178. (From Bruce, D. A., and J. Wolfhope, "Rock Anchors for North American Dams: The National Research Program Bibliography and Database," Paper presented at Institution of Civil Engineers, Ground Anchorages and Anchored Structures in Service, London, England, 2007. With permission.)

presumably promote anchor applications, including dam anchors at Libby Dam, Montana, and Ocoee Dam, Tennessee.

After publication of its document, the Post-Tensioning Division of PCI left to form the Post-Tensioning Institute (PTI) in 1976. Successive editions of "Recommendations" were issued in 1980, 1986, 1996, and 2004. As general perspective to the development of concepts, Table 6.1 provides an analysis of the relative and absolute sizes of the major sections in each successive edition. It is immediately obvious that the original documents stressed "applications"—in an attempt to promote usage—while the most recent edition provides very detailed guidance (and commentary) on the "big five" in particular (i.e., materials, design, corrosion protection, construction, and stressing/testing).

The structure of each successive edition of the PTI Recommendations has changed in the same way that the content has, although there are comparatively few structural differences between the 1996 and 2004 versions. The following detailed comparison by technical topic is based on the structure of the 2004 version of PTI.

Table 6.1 Number of pages in major sections of successive U.S. "Recommendations" documents

Aspect	1974	1980	1986	1996	2004
Materials	1	2	2	8	10
Site investigation	0	1	1	1	2
Design	2	6½	6½	12+ Appendix on grout/ strand bond	14
Corrosion protection	1	4	5	10	14
Construction	7	9	9	10	15
Stressing and testing	1	6	8	17	18
Bibliography/ references	0	1	1	1½	4
Applications	16	18	0	0	0
Recordkeeping	0	1	1	1½	1½
Specifications	0	1	1½	2	2
Epoxy-coated strand	0	0	Very minor reference	Frequent reference but no separate section	10 Separate sections
Total pages	32	57	41	70	98

Source: Bruce, D. A., and J. Wolfhope, "Rock Anchors for North American Dams: The Development of the National Recommendations (1974–2004)," Paper presented at Institution of Civil Engineers, Ground Anchorages and Anchored Structures in Service, London, England, 2007. With permission.

6.2.3.1 Scope and definitions (PTI Chapters 1 and 2)

The scope has remained relatively constant, and is limited to the anchors themselves (as components) as opposed to the analysis and design of the overall anchored system. A total of 72 technical terms are now defined, which represents a major expansion even over the 1996, edition; the first edition had 23 definitions, most of which, incidentally, remain valid and little changed. Table 6.2 presents the technical terms defined in the latest 2004 version of the PTI Recommendations. The bolded items identify those technical terms that were defined in the first edition published in 1974.

6.2.3.2 Specifications, responsibilities, and submittals

Whereas the 1974 edition provided no insight into specifications and the respective contractual responsibilities, certain records were required to be maintained on the grouting operations. By 1980, however, specifications had been addressed, reflecting the need to tailor procurement processes to "experienced" contractors "thoroughly experienced" and to match the innovation of the technique with alternative procurements methods. It is

Table 6.2 Terms defined in first and latest editions of PTI recommendations

Additive	Bondbreaker	Detensionable anchor head	Grout sock	**Primary grout**	**Temporary anchor**
Admixture	Cast	**Downward sloped anchor**	Holiday	**Proof test**	**Tendon**
Alignment load	Centralizer	Elastic movement	**Horizontal anchor**	Pulling head	Tendon bond length
Anchor	Coarse-grained soils	Encapsulation	Initial set	Relaxation	Test load
Anchor head	**Cohesive soils**	Epoxy-coating	**Lift-off**	Residual movement	Thixotropy
Anchor nut	**Consolidation grout**	F_{PU}	**Lock-off load**	Resin cartridge	Transition tube
Anchorage	Contractor	Final set	Memory	Restressable anchor head	Trumpet
Anchor cover	Corrosion-inhibiting compound	Fine-grained soils	**Noncohesive soils**	Rheology	**Unbonded anchor**
Apparent free tendon length	**Coupler**	Free stressing length	Patching material	Safety factor	Upward sloped anchor
Bearing plate	Creep movement	Fully bonded anchor	**Permanent anchor**	**Secondary grout**	Wedge
Bleed	Creep test	Gel time	Performance test	**Sheath**	Wedge plate
Bond length		Grit	Pressure filtration	Spacer	Wedge seating loss
Design load					

Note: Terms in bold text appeared in the first edition of PTI Recommendations (1980). Seven terms defined in the first edition no longer appear in the latest edition.

Sources: Prestressed Concrete Institute (PCI), *Tentative Recommendations for Prestressed Rock and Soil Anchors.* Phoenix, Arizona, 1974; Post-Tensioning Institute (PTI), *Recommendations for Prestressed Rock and Soil Anchors,* 4th ed., Phoenix, Arizona, 2004. With permission.

notable that the three types of specification outlined in 1980 (namely open, performance, and closed) have endured, although "closed" is now referred to as "prescriptive." Building on a 1996 innovation, the responsibilities to be discharged during a project—regardless of type of specification—were summarized in 2004 as shown in Table 6.3. Clear guidance is also provided on the content of preconstruction submittals and as-built records. The former also include the requirement for the contractor to prepare a construction quality plan. Emphasis remains on the need for "specialized equipment, knowledge, techniques and expert workmanship" and for "thoroughly experienced" contractors. The obvious, but often ignored, benefit of "clear communication and close co-operation," especially in the startup phase of a project, is underlined (PTI 2004, p. 7).

6.2.3.3 Materials (PTI Chapter 4)

The 1974 document very briefly refers to wire, strand, and bar, and to protective sheathing. In stark contrast, the 2004 version has built to ten pages providing definitive detail on the materials used in each of the ten major anchor components, with particular emphasis placed on steel, corrosion-inhibiting compounds, sheathings, and grouts (cementitious and polyester).

Table 6.3 Tasks and responsibilities to be allocated for anchor works

1. Site investigation, geotechnical investigation and interpretation, site survey, and potential work restrictions.
2. Decision to use an anchor system, requirements for a pre-contract testing program, type of specification and procurement method, and contractor prequalification.
3. Obtaining easements, permits, permissions.
4. Overall scope of the work, design of the anchored structure, and definition of safety factors.
5. Definition of service life (temporary or permanent) and required degree of corrosion protection.
6. Anchor spacing and orientation, minimum total anchor length, free anchor length, and anchor load.
7. Anchor components and details.
8. Determination of bond length.
9. Details of water pressure testing, consolidation grouting, and re-drilling of drill holes
10. Details of corrosion protection.
11. Type and number of tests.
12. Evaluation of test results.
13. Construction methods.
14. Requirements for QA/QC Program.
15. Supervision of the work.
16. Maintenance and long-term monitoring.

Source: Post-Tensioning Institute (PTI), Recommendations for Prestressed Rock and Soil Anchors, 4th ed., Phoenix, Arizona, 2004. With permission.

Strong cross-reference to relevant ASTM standards is provided as a direct guide to specification drafters. The use of multiple wire tendons ceased by the early 1980s in North American practice. The choice between strand tendons and bar tendons is now a practical matter of capacity and length. Bar tendons are used in applications requiring prestressing loads up to 200 kips per tendon whereas higher-design loads (up to 3,500 kips) are achievable only using multiple-strand tendons. Typically, bar tendons are only economically practical for relatively short tendon lengths as the diameter of the coupler requires oversizing the drilled hole. Experience has shown that 5-strand tendon anchors (design loads of about 200 kips) usually are more cost effective for tendon lengths greater than 80 feet than bar tendons.

6.2.3.4 Site investigation (included in PTI Chapter 6: Design)

Not referred to in 1974, recommended first in 1980 and completely rewritten and expanded in 1996 and 2004, this section now provides clear guidance on the goals and details of a site investigation program. "Minimum requirements" are recommended. However, this remains an area where the anchor specialist often has less "leverage" to exercise influence since the costs associated with such programs typically exercise strong control over the scope actually permitted by the owner.

6.2.3.5 Corrosion and corrosion protection (PTI Chapter 5)

Given the significance and relevance of this topic to the long-term service life of anchor systems, this subject is discussed separately, in Section 6.2.4, below.

6.2.3.6 Design (PTI Chapter 6)

Judging from the relatively short and simplistic coverage of this aspect in 1974, it is fair to say that not much was really then *known* of the subject. Core drilling was considered absolutely necessary and preproduction pullout tests were "strongly recommended." However, two enduring issues were addressed:

- The safety factor (on grout-rock bond) "should range from 1.5 to 2.5" (p. 5), with grout/steel bond not normally governing.
- A table of "typical (ultimate) bond stresses" (p. 5) was issued as guidance to designers.

Today, even despite superior and often demonstrated knowledge of load-transfer mechanisms (i.e., the issue of bond stresses *not* being uniform), the same philosophy prevails:

- The safety factor (reflecting, of course, the criticality of the project, rock variability, and installation procedures) is normally 2 or more.
- A table of "average ultimate" bond stresses is presented, which is basically identical (except for typographic errors) to the 1974 table.

However, the current edition does provide very detailed guidance on critical design aspects, including allowable tendon stresses; minimum free and bond lengths; factors influencing rock/grout bond stress development; anchor spacing; grout cover/strand spacing; and grout-mix design.

6.2.3.7 Construction (PTI Chapter 7)

As noted above, there was a strong bias in the 1974 document toward construction, largely because practice by far led theory. Furthermore, much of what was described in 1974 remains valid, especially with respect to issues relating to grouts, grouting, and tendon placement. Certain features, such as a reliance on core drilling, the use of a "fixed anchorage" (i.e., the use of a plate) at the lower end of multistrand tendons, and specific water take criteria to determine the need for "consolidation grouting," are, however, no longer valid.

The 2004 version expanded upon the 1996 guidance, itself a radical improvement over its two immediate predecessors, and is strongly permeated by an emphasis on quality control and assurance. Practical recommendations are provided on the fabrication of tendons (including the pregrouting of encapsulations) and storage, handling, and insertion. The following are examples of certain best practices that are described in the 2004 version of the PTI Recommendations:

- Drilling methods are best "left to the discretion of the contractor, wherever possible" (p. 52), although specifications should clearly spell out what is not acceptable or permissible.
- In rock, rotary percussion is favored. The drilling tolerance for deviation of 2 degrees is "routinely achievable," while finer tolerances may be difficult to achieve or to measure. Recent advances in directional and controlled drilling have allowed tighter tolerances to be achieved.
- It is recommended that holes left open for longer than 8 to 12 hours should be recleaned prior to tendon insertion and grouting.
- The acceptance criterion for water pressure testing is adjusted to 10.3 liters (2.7 gallons) in 10 minutes at 0.035 MPa (5 psi) for the entire hole. Technical background is provided on the selection of this threshold (based on fissure flow theory).
- Holes with artesian or flowing water are to be grouted and redrilled prior to water-pressure testing. The pregrout (generally water:cement

ratio = 0.5 to 1.0 by weight) is to be redrilled when it is weaker than the surrounding rock to help avoid hole deviation.

- When corrugated sheathing is preplaced before and separate from the tendon, a water test should be conducted on it also, prior to any grouting of its annulus, to verify it has not been damaged during placement.

The treatment of grouting is considerably expanded and features a new decision tree to guide in the selection of appropriate levels of QC programs. Holes are to be grouted in a continuous operation not to exceed one hour, with grouts batched to within 5 percent component accuracy. The value of testing grout consistency by use of specific gravity measurements is illustrated. Special care is needed when grouting large corrugated sheaths; multiple stages may be required to avoid flotation or distortion. The cutting of "windows" in the plastic (to equalize grout pressures) is strictly prohibited.

6.2.3.8 Stressing, load testing, and acceptance (PTI Chapter 8)

Given the professional experience and background of the drafting committee, it is surprising, in retrospect, to note the very simplistic contents of the 1974 document:

- "Proof test" every anchor to at least 115 percent "transfer" load (to a maximum of 80 percent of the guaranteed ultimate tensile strength of the tendon).
- Hold the load for up to 15 minutes (although no creep criterion is given).
- Lock-off at 50–70 percent guaranteed ultimate tensile strength (GUTS).
- The alignment load should be 10 percent of test load, with tendon extension only apparently recorded at the test load (115–150 percent transfer load). "If measured and calculated elongations disagree by more than 10 percent, an investigation shall be made to determine the source of the discrepancy" (p. 8).
- A lift-off test may be instructed by the engineer "as soon as 24 hours after stressing" (p. 8).

Despite significant advances in the 1980 and 1986 documents, reflecting heavily on European practice and experience, significant technical flaws persisted until the completely rewritten 1996 version. The 2004 document was little changed in structure and content, the main highlights being as follows:

- Practical advice is provided on preparatory and setup operations and on equipment and instrumentation including calibration requirements.
- Alignment load can vary from 5 to 25 percent of design load and 10 percent is common. This initial, or datum, load is the only preloading permitted prior to testing. On long, multistrand tendons, a monojack is often used to set the alignment load, to ensure uniform initial loading of the strands.
- Maximum tendon stress is 80 percent of its GUTS.
- Preproduction ("disposable," test anchors, typically 1–3 in number), performance, and proof tests are defined, the latter two covering all production anchors.
- For performance testing, the first 2 or 3 anchors plus 2–5 percent of the remainder are selected. The test is a progressive cyclic loading sequence, typically to 1.33 times working load. A short (10- or 60-minute) creep test is run at test load.
- Proof tests are simpler, requiring no cycling, and are conducted to the same stress limits. The option is provided to return to alignment load prior to lock-off (in order to measure the permanent movement at test load), otherwise this movement can be estimated from measurements from representative performance tests.
- Supplementary extended creep tests are not normally performed on rock anchors, except when installed in very decomposed or argillaceous rocks. A load cell is required and the load steps and reading frequencies are specified.
- Lock-off load shall not exceed 70 percent GUTS, and the wedges will be seated at 50 percent GUTS or more.
- The initial lift-off reading shall be accurate to 2 percent.
- There are three acceptance criteria for every anchor:
  - Creep at test load: less than 1 mm in the period 1 to 10 minutes, or less than 2 mm in the period 6 to 60 minutes.
  - Movement at test load: There is no criterion on residual movement, but clear criteria are set on the minimum elastic movement (equivalent to at least 80 percent free length plus jack length) and the maximum elastic movement (equivalent to 100 percent free length, plus 50 percent bond length plus jack length).
  - Lift-off reading: within 5 percent of the designed lock-off load.
- A decision tree guides practitioners in the event of a failure under any one criterion.
- The monitoring of service behavior is also addressed. Typically 3–10 percent of the anchors are monitored (if desired), by load cells or lift-off tests. Initial monitoring is at 1–3-month intervals, stretching to 2 years later.

6.2.3.9 Epoxy-coated strand (Supplement)

Epoxy-coated strand and its use was first discussed systematically in 1996, although minor references had been made in 1986. The 2004 document contains a separate supplement dealing with specifications, materials, design, construction, and testing, being a condensed and modified version of a document produced by the ADSC Epoxy-Coated Strand Task Force in November 2003. The Scope (Section 1) notes that anchors made from such strand "require experience and techniques beyond those for bare (i.e., uncoated) strand anchors." It supplements the recommendations provided in the overall document with respect to specifications/responsibilities/submittals, materials, design, construction, and stressing and testing.

6.2.4 Developments in tendon corrosion protection

6.2.4.1 1974

Figure 6.6 illustrates the very simple approach to tendon protection, that is, cement grout or nothing. "Permanent" is defined as "generally more than a 3-year service life." Sheathing is only discussed as a debonding medium, not a corrosion-protection barrier. For permanent anchors "protective corrosion seals over their entire length" are to be provided (but are not defined). For two-stage grouted tendons, *sheathing can be omitted*, the implication being that cement grout alone would be acceptable.

6.2.4.2 1980

The same Figure 6.6 is reproduced (as it was also in 1986). The term "permanent" is now reduced to 18 months or more, and growing attention is drawn to the requirements of permanent anchors: sheathing is for debonding "and/or to provide corrosion protection," as is secondary cement grout. Corrugated protection and epoxy coating for bars are discussed. The type and details of corrosion protection are to be based on longevity, anchor environment, and consequences of future and in-hole conditions/length of time before grouting. For the bond length, cement grout is considered "the first level of corrosion protection," and plastic corrugated sheathing ("for multiple corrosion protection schemes") or epoxy is permitted. Such protection is to extend at least 2 feet into the free length. The free length is to have, as a minimum, a sheath with cement grout or grease infill. A full-length outer sheath is regarded as "good practice."

6.2.4.3 1986

The emphasis is placed on first investigating the chemical aggressiveness of the soil and ground water: "Permanent anchors placed in environments

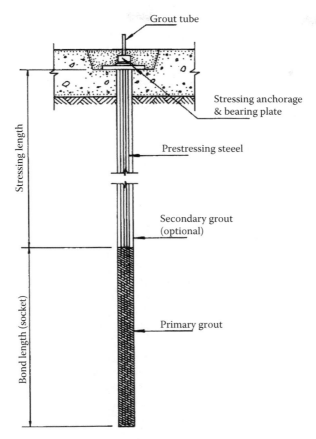

Figure 6.6 Rock anchor components. Note the lack of protection to the steel tendon other than cement grout. (From Prestressed Concrete Institute (PCI), *Tentative Recommendations for Prestressed Rock and Soil Anchors.* Phoenix, Arizona, 1974. With permission.)

where any one of these tests indicate critical values must be encapsulated over their full length." Thus, even up until the next set of Recommendations (1996), it was considered acceptable to allow anchors for dams to be installed without any protection for the bond length other than cement grout, depending on the results of laboratory tests on small samples. Encapsulation was not detailed.

6.2.4.4 1996

Permanence is now defined as a minimum of 24 months in a completely revised set of Recommendations. A wider spectrum of issues than simple chemistry now has to be considered when selecting corrosion protection principles. A major breakthrough was to identify two classes of protection (Class I and II) for permanent anchors to replace the poorly defined and loosely used "double" and "single" corrosion-protection systems offered by various tendon manufacturers. The details are summarized in Table 6.4 and a "decision tree" was provided for the guidance of designers (Nierlich and Bruce, 1997).

6.2.4.5 2004

The 1996 Recommendations were revalidated while it is stated that, for permanent anchors, "aggressive conditions shall be assumed if the aggressivity of the ground has not been quantified by testing" (p. 22). Table 6.4 was revised, as shown in Table 6.5, mainly to clarify the acceptable Class I status of epoxy protected steel in a "water proofed hole." The sophistication of contemporary tendons is shown in Figure 6.7. A long supplement is devoted to epoxy protected strand.

Overall, therefore, one is impressed that between 1974 and 2004 (a) extremely sophisticated corrosion protection systems were developed, and (b) the latitude offered to designers relative to choice of corrosion protection intensity and details was severely restricted; to install a permanent anchor in a dam without Class I protection in the United States is now not only impermissible, but unthinkable.

It must also be noted that the philosophy of pregrouting and redrilling the hole ("waterproofing") if it were to fail a permeability test was reaffirmed from 1974 onwards: indeed the early "pass-fail" acceptance criteria were, in fact, very rigorous and led to most anchors on most projects having to be pregrouted and redrilled several times. Although laudable, this was often, in fact, "extra work" since the criterion to achieve *grout* tightness is really much more lax than the criterion needed to provide the specified

Table 6.4 Corrosion protection requirements

Class	Protection requirements		
	Anchorage	Unbonded length	Tendon bond length
I	1. Trumpet	1. Grease-filled sheath, or	1. Grout-filled encapsulation, or
Encapsulated tendon	2. Cover if exposed	2. Grout-filled sheath, or 3. Epoxy for fully bonded anchors	2. Epoxy
II	1. Trumpet	1. Grease-filled sheath, or	Grout
Grout-protected tendon	2. Cover if exposed	2. Heat shrink sleeve	

Source: Post-Tensioning Institute (PTI), *Recommendations for Prestressed Rock and Soil Anchors*, 3rd ed., Phoenix, Arizona, 1996. With permission.

Table 6.5 Corrosion protection requirements

| Class | Corrosion protection requirements | | |
	Anchorage	Free stressing length	Tendon bond length
I Encapsulated tendon	Trumpet Cover if exposed	Corrosion inhibiting compound-filled sheath encased in grout, or Grout-filled sheath, or Grout-encased epoxy-coated strand in a successfully water-pressure tested drill hole	Grout-filled encapsulation, or Epoxy coated strand tendon in a successfully water-pressure tested drill hole
II Grout protected tendon	Trumpet Cover if exposed	Corrosion inhibiting compound-filled sheath encased in grout, or Heat shrink sleeve, or Grout-encased epoxy-coated bar tendon, or Polyester resin for fully bonded bar tendons in sound rock with non-aggressive ground water	Grout Polyester resin in sound rock with non-aggressive ground water

Source: Post-Tensioning Institute (PTI), *Recommendations for Prestressed Rock and Soil Anchors*, 4th ed., Phoenix, Arizona, 2004. With permission.

degree of *water* tightness. The saving grace of many of the early anchors was doubtless, therefore, the somewhat erroneous drill hole "waterproofing" criterion under which they were constructed.

An analysis of the case studies in the anchor database illustrates the evolution of systems and philosophies, as shown in Figure 6.8.

6.3 CASE HISTORIES

Three case studies on anchored dams are presented to highlight practices in U.S. dam anchoring. The first project is Tom Miller Dam located in central Texas, and the second project is Gilboa Dam located in upstate New York. Both dams were anchored for the primary purpose of increasing the stability of an existing concrete and masonry composite dam. A third project, John Day Dam, Washington, is presented as a salutary reminder of what can result from inappropriate practices.

6.3.1 Tom Miller Dam, Austin, Texas

Tom Miller Dam is located on the lower Colorado River, approximately six miles northwest of Austin, Texas (Wolfhope et al. 2005). The dam

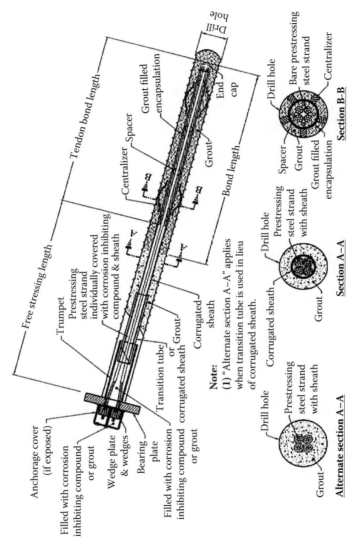

Figure 6.7 Class I Protection—Encapsulated Strand Anchor. (From Post-Tensioning Institute (PTI), *Recommendations for Prestressed Rock and Soil Anchors*, 4th ed., Phoenix, Arizona, 2004. With permission.)

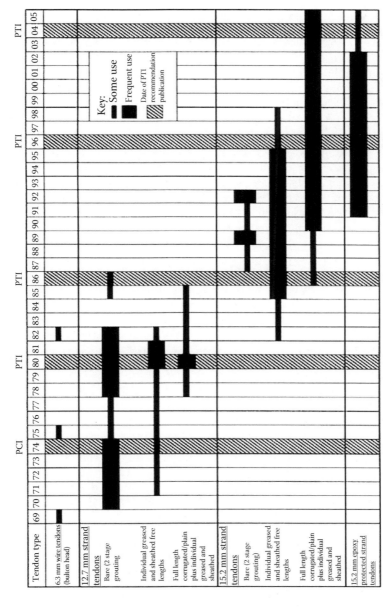

Figure 6.8 Illustrating how the types of corrosion protection evolved. (From Bruce, D. A., and J. Wolfhope, "Rock Anchors for North American Dams: The National Research Program Bibliography and Database," Paper presented at Institution of Civil Engineers, Ground Anchorages and Anchored Structures in Service, London, England, 2007. With permission.)

impounds Lake Austin, the last in a series of reservoirs referred to as the "Highland Lakes" of central Texas. The dam was originally constructed from 1890 to 1893 as an uncontrolled overflow gravity spillway. The original dam was a cyclopean structure built using a core of mortared irregular limestone rocks faced with granite masonry. The dam failed catastrophically in 1900 when a 500-foot-long section of the spillway slid downstream, as shown in Photo 6.1.

Following the first failure, a series of site investigations concluded there were undesirable conditions at the site to support reconstruction. A board of prominent engineers including the chief of the U.S. Reclamation Service examined the site and recommended the dam be rebuilt at a different location with improved geologic conditions. Despite the board's recommendations, the dam was rebuilt by the City of Austin in 1912 at its original location: the dam failed again in 1915. In a major engineering and reconstruction effort, the dam was rebuilt to its current configuration by the Lower Colorado River Authority from 1939 to 1941. The existing dam is therefore a composite structure that combines features of each of the previous dams, including a 500-foot-long uncontrolled overflow gravity spillway, a 600-foot-long flat slab-and-buttress dam, and a powerhouse with a gravity bulkhead intake structure (Photo 6.2). The present gravity spillway consists of a reinforced concrete cap over the original cyclopean masonry core.

As part of a ten-year program to modernize the Highland Lakes, the dam was upgraded in 2004 to meet current dam safety standards. Analysis of the dam revealed that the uncontrolled overflow and powerhouse intake gravity sections were unstable against floods exceeding the 500-year event and required stabilization to comply with state dam safety regulations. Design studies showed that post-tensioned anchors were clearly the most economical solution for increasing the dam's stability, notwithstanding the

Photo 6.1 The 1900 failure of Tom Miller Dam, Texas. (Courtesy of Lower Colorado River Authority archives.)

Photo 6.2 Ariel view of Tom Miller Dam, Texas. (Courtesy of Lower Colorado River Authority archives.)

significant challenges posed by the complicated geology and the composite construction of the dam.

The bedrock beneath Tom Miller Dam is the Edwards limestone formation, and specifically its basal 100–200 feet. It consists of moderately weathered limestone with highly weathered zones found in several horizontal planes below the structure. An extensive geotechnical investigation program identified zones of varying rock quality below the dam, ranging from a highly permeable fractured zone for the first 5–10 feet below the dam, a less-fractured zone exhibiting lower permeability, and a deep highly porous vuggy zone showing signs of significant water movement through the formation. Cavernous voids were found in many places, with individual openings exceeding 2 feet in diameter. Several caves and karst openings had been noted in previous geotechnical investigations and surveys near the dam, although no caves were encountered during the drilling program for this project.

The core of the dam comprises mortared irregular limestone blocks. The limestone was quarried from a site adjacent to the right abutment of the dam, and so the limestone is of a similar quality as the foundation rock beneath the dam. The site investigations identified this core to be highly permeable with poorly cemented areas and voids immediately beneath the concrete cap. The unconfined compressive strength of the foundation ranged from 480 to 22,400 psi, with an average value of 3,200 psi. The unit weight of the rock ranged from 98 to 168 pounds per cubic feet (pcf) and the rock quality designation ranged from 0 to 100 percent. The permeability of the foundation ranged from 0 to 957 Lugeons with an average value of 52 Lugeons. The unconfined compressive strength of the masonry dam core ranged from 1,090 to 9,460 psi, with an average value of 2,990 psi. The unit weight of the rock ranged from 124 to 153 pcf and the rock quality designation ranged from 0 to 90 percent. The permeability of the masonry core ranged from 0 to 979 Lugeons with an average value of 294 Lugeons.

Post-tensioned multistrand tendon anchors were used to provide an adequate factor of safety against sliding and overturning for the uncontrolled overflow spillway and powerhouse intake structure. The anchor design was carefully developed to handle difficult geologic and structural conditions that would often be considered detrimental to construction of post-tensioned stabilization systems including zones of poor quality, severely weathered, fractured, decomposed, cavitated, or highly permeable rock showing evidence of significant water movement through the dam core and foundation rock formation. The uncontrolled overflow spillway was stabilized using a single row of fifty-two vertical post-tensioned anchors spaced over the 500-foot spillway length (Figure 6.9).

The anchors were installed in 18 individual monoliths, and spacings were adjusted to provide clearance to the dam's internal drain system in the foundation and beneath the concrete overlay cap. The anchors were positioned

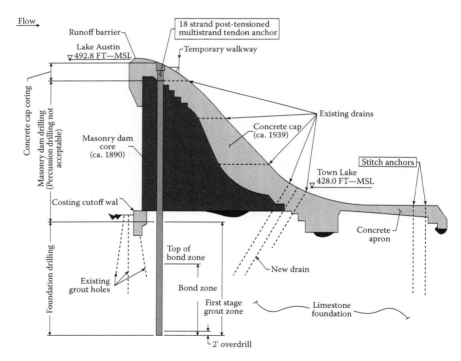

Figure 6.9 Schematic cross-section of the overflow section, Tom Miller Dam, Texas. (From Bruce, D. A., J. J. Jensen, and J. S. Wolfhope, "High Capacity Rock Anchors for Dams: Some Fundamental Observations on the Analysis of Stressing Data." Paper presented at USSD 2005 Conference, Salt Lake City, UT, June 6–10, 2005. With permission.)

immediately downstream of an existing concrete cutoff but upstream of existing foundation drains. The total drill-hole lengths alternated between 127 and 137 feet for adjacent anchors to vary the bond zone depths and avoid setting up a potential plane of failure in the foundation. The 18-strand tendons, with Class 1 corrosion protection, were installed in vertical, eight-inch diameter drilled holes. An additional three anchors were installed to stabilize the powerhouse intake structure, where hole lengths alternated between 143 and 153 feet for adjacent anchors (Figure 6.10).

The anchors were carefully positioned to allow installation through a narrow concrete wall section alongside and between the powerhouse intake penstocks. The design working load of each tendon was 60 percent of the guaranteed ultimate tensile strength (GUTS) and the test load stress was 80 percent GUTS.

A 30-foot-long bond zone was selected for the production anchors based on the results of a field-test anchor program conducted during final design. Two anchors were tested to failure and two were performance tested according to Post-Tensioning Institute (1996) procedures. The bond zone

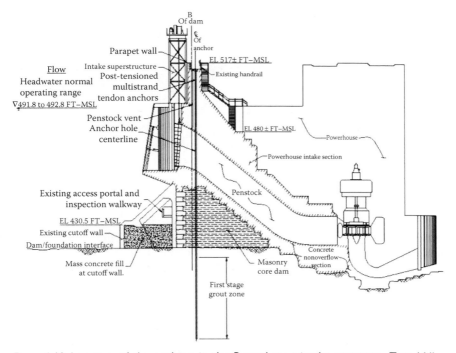

Figure 6.10 Location of the anchors in the Powerhouse intake structure, Tom Miller Dam, Texas. (From Bruce, D. A., J. J. Jensen, and J. S. Wolfhope, "High Capacity Rock Anchors for Dams: Some Fundamental Observations on the Analysis of Stressing Data." Paper presented at USSD 2005 Conference, Salt Lake City, UT, June 6–10, 2005. With permission.)

was carefully positioned to be in a region of relatively competent limestone, beneath the highly permeable zone identified immediately under the dam and above the deep vuggy zone. Despite this attention to positioning the bond zone, it was clear that the contractor would encounter random zones of poor rock and high permeability throughout the anchor holes.

In response to these difficult conditions for anchor drilling and grouting, each hole was mandated to be pretreated by gravity grouting using a sand-cement mixture. The holes were then redrilled, water-tightness tested, and neat-cement grouted (where necessary) to enable the tendons to be installed in accordance with the PTI watertightness criterion for the bond zone (2.5-gallons lost in 10 minutes at 5 psi excess head). A preplaced corrugated sheathing was then grouted (in several stages) in the hole prior to tendon insertion and grouting. The tendons were fabricated using a greased and sheathed free zone to ensure that post-tensioning forces were transferred into the foundation below the weaker highly fractured zone near the dam/foundation contact and away from the masonry core. A length of bare strand was provided at the top of the anchor beneath the anchorage to provide for fully grouted tendons bonded in the 13-foot-thick concrete cap. In order to improve the quality of the stressing data, the anchor contractor elected to apply performance test procedures to every anchor, and not only the 5 percent or so as requested in the specification.

From the risk management viewpoint, many lessons learned from previous anchoring projects on the Highland Lakes dams were implemented. For example, strict qualification requirements were developed to ensure the selection of a specialized anchor contractor experienced in similar post-tensioned stabilization projects. Two full-scale test anchors were first constructed to prove that the contractor could successfully install and test the production anchors. Vertical extensometers were installed adjacent to the test anchors to verify that the application of the post-tensioning forces did not have an adverse effect on the structure considering the irregular character of the masonry core.

By following a disciplined quality control program, the contractor successfully demonstrated that the holes could be drilled within the specified alignment tolerances, the foundation rock treated to meet the watertightness criterion, the tendons effectively installed and grouted, and the anchor tested and locked off in accordance with the acceptance criteria. Based on observations of the site-specific field conditions, several improvements were made to the fabrication and installation program to enhance the construction of the remaining 53 anchors. The two test anchors were accepted as production anchors and the remaining anchors were released for hole drilling and tendon fabrication. To gain a familiarity with the groutability of the masonry core and foundation, grouting procedures were refined based on a series of trial mix batches to enhance the effectiveness of the remaining grouting operations.

Sand-cement grout takes ranged up to 369 percent of the theoretical hole volume. In general, the volume of takes decreased as the project proceeded and the dam and its foundation progressively tightened. Subsequent neat-cement grouting was only required in 17 of the anchor holes, highlighting the effectiveness and benefits of the routine sand-cement pregrouting operation. Takes were generally moderate, with only four holes requiring more than one neat grout treatment. All other aspects of anchor construction (such as hole verticality, sheath testing and grouting, and tendon insertion and grouting) provided consistent, compliant, and acceptable results. The 28-day tendon grout strengths were typically well in excess of 8,000 psi.

6.3.2 Gilboa Dam, Gilboa, New York

Gilboa Dam is a major component of the New York City water supply system and is located in the Catskill Mountains approximately 120 miles north of New York City (Zicko, Bruce, and Kline 2007). Completed in 1927, the 180-foot-high dam consists of a 700-foot-long earth embankment and a 1,324-foot-long cyclopean concrete spillway. The Schoharie Reservoir, which is impounded by Gilboa Dam, can store up to 17.6 billion gallons and provides the city with a large percentage of its drinking water. The spillway was built on nearly horizontally interbedded layers of sandstone, mudstone, siltstone, and shale, with shale being the predominantly weak rock unit. The spillway is composed of cyclopean concrete and has a stepped downstream face (Photo 6.3). A cutoff wall of varying depth was constructed near the upstream face.

In early fall 2005, during the preliminary design phase for dam reconstruction, preliminary analyses showed that the sliding stability of the spillway structure did not meet current New York State dam safety criteria and was marginal for the 1996 record flood. Given the critical nature of the reservoir both in terms of public safety to over 8,000 residents living downstream and dependability for New York City's water supply, an interim stability improvement project was implemented for completion before the end of 2006, years prior to the foreseen major reconstruction. The design and construction phases of this interim project were completed in an unprecedented timeframe of 12 months.

To help ensure the successful completion of the job given the tight timeframe, bid packages for rock anchor installation were distributed to three specialty contractors judged to have the appropriate experience in high-capacity rock anchors for dams. In addition to conservative design assumptions, several construction measures were implemented to ensure the long-term performance of the anchors. These measures included the testing of preproduction anchors, the installation of "sentinel" anchors, corrosion protection of tendons, watertightness testing of the anchor hole and sheathing, performance testing of all production anchors, extended lift-off tests of select anchors, and anchor head encapsulation.

Photo 6.3 Gilboa Dam, New York, in summer 2003. (Modified from Zicko, K. E., D. A. Bruce, and R. A. Kline, Jr., "The Stabilization of Gilboa Dam, New York, Using High Capacity Rock Anchors: Addressing Service Performance Issues," Paper presented at Institution of Civil Engineers, Ground Anchorages and Anchored Structures in Service, London, England, 2007.)

The rock anchor system was designed in general accordance with criteria and guidance provided in the PTI Recommendations for Prestressed Rock and Soil Anchors (2004) for Class I tendons. Spillway stability analyses were conducted at seven cross-sections across the site to identify the required post-tensioned anchor loads. Based on these results, vertical anchors were required along the entire length of the spillway crest, from Monolith M1 to Monolith M17. Furthermore, inclined anchors were needed along the downstream face in the central portion of the spillway. These central monoliths are taller than the eastern monoliths, but they have shallower cutoff walls resulting in shorter failure surfaces compared to the higher western monoliths with deeper cutoff walls. For Monoliths M6 through M11, inclined anchors were angled 48° from horizontal and were located at the corner of step numbers 3–4; for Monoliths M12 through M14, the anchors were inclined 45° from horizontal and were located at the corner of step numbers 4–5. A plan view of the anchor layout is shown in Figure 6.11, and a schematic of the general anchor configuration is depicted in Figure 6.12. A total of 79 anchors were installed, of which 47 were vertical and 32 were inclined.

The anchor demands were grouped into four ranges of anchor size, groups A through D, based on their capacities and engineering judgment (Table 6.6). By grouping the anchors based on maximum required capacities and locking-off all the anchors to the same design load (DL) equal to

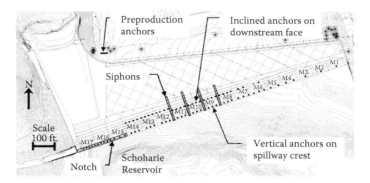

Figure 6.11 Plan view of Gilboa Dam, New York, showing the anchor layout. (From Zicko, K. E., D. A. Bruce, and R. A. Kline, Jr., "The Stabilization of Gilboa Dam, New York, Using High Capacity Rock Anchors: Addressing Service Performance Issues," Paper presented at Institution of Civil Engineers, Ground Anchorages and Anchored Structures in Service, London, England, 2007. With permission.)

60 percent GUTS of the steel tendons, some additional load was provided above that required to meet the minimum stability requirements.

Computations were performed to determine the minimum required corrugated sheathing diameter, hole diameter, and bond length. Based on the computed minimum diameters, commercially available products were selected to utilize readily available products minimizing lead times for material deliveries. Table 6.7 provides the design parameters as presented in the contract documents and also shows changes (in bold) implemented

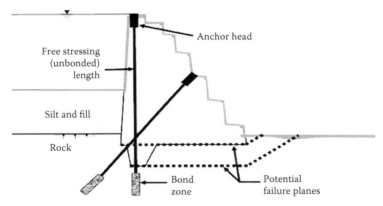

Figure 6.12 Typical section of the spillway with rock anchors, Gilboa Dam, New York. (From Zicko, K. E., D. A. Bruce, and R. A. Kline, Jr., "The Stabilization of Gilboa Dam, New York, Using High Capacity Rock Anchors: Addressing Service Performance Issues," Paper presented at Institution of Civil Engineers, Ground Anchorages and Anchored Structures in Service, London, England, 2007. With permission.)

Table 6.6 Anchor groups for Gilboa Dam, New York

Group ID	Range of number of strands	Design capacity range (kips)	Selected number of strands	Individual anchor design capacity (kips)
A	33–39	1160–1371	39	1371
B	40–45	1406–1582	45	1582
C	46–52	1617–1828	52	1828
D	53–58	1863–2039	58	2039

Source: Zicko, K. E., D. A. Bruce, and R. A. Kline, Jr., "The Stabilization of Gilboa Dam, New York, Using High Capacity Rock Anchors: Addressing Service Performance Issues," Paper presented at Institution of Civil Engineers, Ground Anchorages and Anchored Structures in Service, London, England, 2007. With permission.

by the contractor. These changes were made to minimize tool sizes and to ensure satisfactory performance during stressing.

The bond length was calculated based on the selected drill-hole diameter and the ultimate bond zone strength. To establish the ultimate bond stress, the interbedded site strata were presumed to be governed by sandstone and shale, which were the predominant strata in the bond zone. Typical ultimate bond strength values identified during preliminary design for shale ranged from 30 to 200 psi, and those for sandstone ranged from 100 to 250 psi. These values were significantly less than the ultimate bond stresses of 1,500 to 2,500 psi estimated from UCS testing. Therefore, an ultimate bond stress of 200 psi was selected for design of the rock anchors, which is the upper bound for shale, as given in PTI (2004). A working bond stress of 100 psi was selected providing a factor of safety of 2.0. These bond stresses were confirmed by UCS testing on rock samples downstream of the spillway, by conducting a site-specific preproduction test program early in construction, and by testing each installed anchor to a 33 percent overload to verify its load-carrying capacity.

The free stressing length was selected to locate the top of the bond zone at a depth at least 10 feet below the base of the existing cutoff wall. The distance of 10 feet was intended to account for uncertainty associated with

Table 6.7 Summary of minimum design parameters, Gilboa Dam, New York

Number of strands	Design load (kips)	Drill hole diameter (inch)		Sheathing diameter (inch)		Bond length (ft.)	
39	1,371	12	15	8	10	31	41
45	1,582	12	15	8	10	35	45
52	1,828	14	15	8	10	35	45
58	2,039	14	15	10	10	39	49

Source: Zicko, K. E., D. A. Bruce, and R. A. Kline, Jr., "The Stabilization of Gilboa Dam, New York, Using High Capacity Rock Anchors: Addressing Service Performance Issues," Paper presented at Institution of Civil Engineers, Ground Anchorages and Anchored Structures in Service, London, England, 2007. With permission.

the location of the actual concrete/rock interface. For design purposes, the location of this interface was based upon the original construction record drawings. The concrete/rock interface was encountered within 5 feet of its predicted location in preliminary borings drilled through the crest, thus documenting with reasonable accuracy the as-built drawings. PTI (2004) recommends that the free stressing length of anchors should extend a minimum distance of 5 feet beyond potential failure planes. Therefore, 10 feet was selected to provide additional assurance that the bond zone was at least 5 feet beyond the bottom of the cutoff wall. This additional anchor footage also allowed the contractor to order tendon materials prior to confirming the concrete/rock interface location during drilling.

Anchor group effects were evaluated to ensure that the interaction between anchors would not decrease the overall capacity of the anchored system. The pull-out resistance of an anchor was equated to the weight of an inverted cone of rock as presented in Littlejohn and Bruce (1977). This method assumes that a vertical plane develops where adjacent cones overlap and decreases the cone volume accordingly, and it also ignores the rock-shear strength along the edges of the cone. Although vertical and nearly vertical fractures were encountered in the borings drilled concurrently with the anchor design, these fractures were typically discontinuous and had surface roughness, or undulations, such that shear failure of the rock would be required to form a continuous vertical failure surface. Furthermore, grouting was completed in the dam foundation during the original construction, which would have further "locked" the rock together. Therefore, despite the presence of vertical fractures, the inverted cone method was considered appropriate since the rock shear strength was ignored.

Corrosion protection of the anchor tendons was addressed by specifying permanent (Class I) encapsulation. This protection included a grout-filled sheathing extending the full length of the strand, the trumpet welded to the bearing plate, and an overlap of the trumpet by the sheathing. The anchor holes and sheathing were subjected to extensive testing to ensure watertightness during the successive phases of anchor construction. After drilling, each anchor hole was required to pass a water test that limited water loss to 5.5 gallons per 10 minutes under a pressure head of 5 psi. If required due to failure of the water test, pregrouting and redrilling were performed until the anchor hole was sufficiently watertight. Furthermore, each anchor hole was videotaped to examine the sidewalls of the anchor hole. Consequently, the infiltration of water through rock joints into the anchor hole could be observed, whether as a trickle or a jet. Jets of water were more concerning since they would more likely wash anchor grout from the hole, reducing the degree of corrosion protection and the area of grout to rock contact within the bond zone.

Water testing was required for all sheathing, which consisted of full-length corrugated sheathing for vertical anchors, and of smooth sheathing in the unbonded zone and corrugated sheathing in the bond zone for inclined anchors. Regardless of the sheathing configuration, the criterion for water testing was that water loss would remain constant at less than 2.75 gallons per 10 minutes under a pressure head of 5 psi. For the vertical anchors, the corrugated sheathing was testing both prior to and after installation into the anchor hole. For the inclined anchors, the smooth and corrugated sheathing were heat-welded and then water-tested prior to installation into the anchor hole. The sheathing was further tested after its placement into the hole, both prior to and after tendon installation. This third additional water test was required because of the increased potential for damaging the sheathing during tendon installation on an angle since the annulus between the sheathing and wall of the anchor hole had not been pregrouted.

Future phases of dam reconstruction at Gilboa Dam include the removal of the deteriorated concrete and overlay stone masonry façade of the spillway to a depth of 6.5 feet. To minimize anchor constructability problems caused by setting the top of the encased anchor at this depth, a steel-reinforced concrete column was constructed prior to installation of the anchors. The column was built by removing the existing concrete to a depth of 10 feet and then refilling with steel-reinforced, high-strength concrete to the proposed bearing plate elevation approximately 2.5 feet below the existing face. This column was intended to protect the anchor during future construction and to transmit the anchor load to the structure below the limit of scaling. To facilitate removal of the deteriorated materials around the columns during reconstruction, a bond breaker was applied between existing and new concretes.

To provide a high level of corrosion protection of the anchor heads, the bearing plate, wedge plate, and tendon tails were coated with bitumastic material and then directly encased in concrete at all locations, except in the notch. Restressable anchors were considered in the design of the anchorage but were not utilized due to the increased potential for corrosion of restressable anchor heads as compared to traditional anchor heads. This was an especially important consideration at Gilboa Dam where the anchor heads are within the active spillway structure.

Although the dam was originally designed to be directly encased in concrete, questions regarding the final dam reconstruction instigated revision of the anchor heads in the notch during construction. The possibility of decommissioning anchors was realized, so access to the wedge plate was required in the 13 vertical anchors in the notch. The anchor heads in the 5.5-foot-deep notch were, therefore, modified to include a steel cap in-filled with corrosion inhibiting grease immediately around the wedge plate. This steel cap was encased in a larger steel cap bolted to the bearing plate that was filled with expandable closed-cell foam. Finally, the entire anchor head was encased in concrete. This configuration was used for the anchor heads to provide access, if necessary, without damaging the anchors during reconstruction. Finally, due to construction restraints and to ensure adequate concrete cover, the anchor tails were cut short, thereby not allowing for future restressing.

Although load cells are frequently incorporated on other projects into anchor heads to provide long-term monitoring of anchor loads, this was not feasible at Gilboa Dam due to the location of the anchors (in an active spillway) and future reconstruction activities (scaling of deteriorated concrete). In addition to the previously discussed design conservatism and anchor installation testing, extensive test methods were put in place to ensure satisfactory anchor performance including preproduction anchor tests, installation of "sentinel" anchors, performance testing all production anchors, and extended lift-off testing.

At the beginning of the anchor contract, an extensive preproduction anchor test program was performed to verify the bond stress and factor of safety used in design or establish the actual bond stress of the site strata, evaluate creep susceptibility of the site strata, and provide instrumented "sentinel" anchors at the site. Four preproductions anchors were installed downstream of the spillway, utilizing the same construction techniques and materials used in the production anchors but smaller in size. The preproduction anchors were incrementally loaded and unloaded until an ultimate rock-grout average bond stress of 200 psi was reached. The required 200-psi bond stress was achieved in each anchor, so the factors of safety were greater than the value of 2.0 used in design. It is important to note that no anchor could be failured.

To evaluate the creep susceptibility of the rock, a constant load equivalent to 80 percent GUTS was applied to each preproduction anchor. The anchors were monitored for up to 75 hours (4,500 minutes), which is significantly larger than in typical creep tests. Elongation of the tendons indicated creep less than 0.04 inches per log cycle of time, which is acceptable. Two preproduction anchors were equipped with permanent load cells after completion of the performance and creep testing and were locked off at 70 percent GUTS. Load cell readings for these "sentinel" anchors were recorded intermittently for one year, or approximately 6 months beyond completion of the anchor contract. The results are shown in Figure 6.13, a semi-log plot of the anchor loads over time. Assuming a 100-year service life for the dam, it is projected that 91–92 percent of the anchor lock-off load will be available at the end of that period, which corresponds closely to the design loads of the anchors within the spillway.

All 79 production anchors, vertical and inclined, installed in the spillway were subjected to performance testing. This exceeds the standard practice of proof testing the majority of the anchors and performance testing only the first 2 or 3 anchors and 2 percent thereafter. Due to the limited amount of subsurface information available during design, it was deemed prudent to apply this additional level of testing to ensure satisfactory anchor

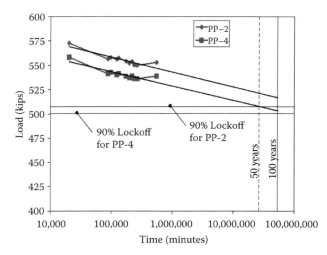

Figure 6.13 Results of load-cell monitoring of "sentinel" anchors, Gilboa Dam, New York. (From Zicko, K. E., D. A. Bruce, and R. A. Kline, Jr., "The Stabilization of Gilboa Dam, New York, Using High Capacity Rock Anchors: Addressing Service Performance Issues," Paper presented at Institution of Civil Engineers, Ground Anchorages and Anchored Structures in Service, London, England, 2007. With permission.)

performance. The performance testing was in accordance with PTI (2004), which included application of an alignment load to each strand (10 percent design load per strand), cyclic loading and unloading to a maximum load of 133 percent design load, and then creep testing at the maximum load.

With the exception of one anchor, all of the rock anchors were successfully stressed per the contract documents. Of these 78 anchors, 73 anchors performed adequately during performance testing and passed the 10-minute creep test. Due to excessive elongation during the 10-minute creep test, the 5 remaining anchors were subjected to the 60-minute creep test, which they subsequently passed with strand elongation not exceeding 0.08 inches. Typical performance test results are shown in Figure 6.14 in terms of total, elastic, and residual movement. One 58-strand inclined anchor had wires on 8 different strands break during performance testing. The anchor had been successfully loaded to 120 percent design load and experienced strands breaking as loading was cycled to 133 percent design load, at which time stressing was ceased. The cause of failure was attributed to either an uneven alignment load on individual strands or misalignment of the strands and jack. The remaining undamaged 50 strands were subsequently restressed to an appropriately reduced load.

Following lock-off of each anchor, an initial lift-off test was conducted to verify that load was successfully transferred to the anchor bond zone. Each of the anchors was within the contract limits of 5 percent of the specified

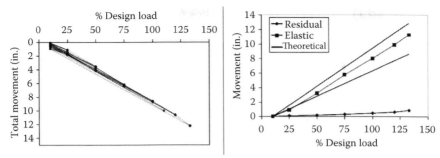

Figure 6.14 Movements recorded during performance testing of Anchor A40, a vertical 58-strand anchor, Gilboa Dam, New York. (From Zicko, K. E., D. A. Bruce, and R. A. Kline, Jr., "The Stabilization of Gilboa Dam, New York, Using High Capacity Rock Anchors: Addressing Service Performance Issues," Paper presented at Institution of Civil Engineers, Ground Anchorages and Anchored Structures in Service, London, England, 2007. With permission.)

lock-off load, which was 110 percent design load. Approximately 30 days after the initial lift-off test, 6 vertical and 4 inclined anchors were subjected to an extended lift-off test. The term "extended" simply refers to the time between the initial and second lift-off tests and does not imply that load was applied and subsequently held for an extended period of time. The lift-off loads were graphed similarly to the "sentinel" anchors (Figure 6.15). Extrapolation of the data indicates that the anchor load available at the end of the 100-year service life will be between 99 percent and 112 percent design load.

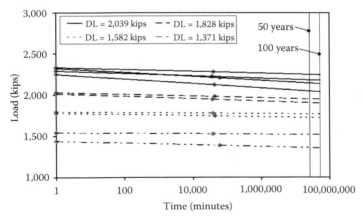

Figure 6.15 Results of initial and extended lift-off testing, Gilboa Dam, New York. (From Zicko, K. E., D. A. Bruce, and R. A. Kline, Jr., "The Stabilization of Gilboa Dam, New York, Using High Capacity Rock Anchors: Addressing Service Performance Issues," Paper presented at Institution of Civil Engineers, Ground Anchorages and Anchored Structures in Service, London, England, 2007. With permission.)

The fast-track interim improvement of Gilboa Dam provided several opportunities to "think outside the box" during rock anchor design and construction. Long-term performance of these anchors was crucial and was verified using design conservatism, extensive testing during construction, and instrumentation and monitoring. The success of the Gilboa Dam anchor project can be attributed to the following factors:

- Design conservatism was beneficial during construction.
- Preproduction anchors were installed in the same manner as the production anchors to provide insight into the construction procedure, load-carrying capacity of the rock, and tendon creep.
- Long-term monitoring of instrumented "sentinel" anchors was utilized to estimate the anchor load for the service life of the dam.
- Corrosion protection of the tendons and anchor heads was assured using thorough testing procedures and inspection at each construction step.
- Performance testing of all production anchors provided a very high quality of data on anchor behavior. This was especially important since the amount of subsurface information, including rock type at each anchor location and strengths, was limited.
- Extended lift-off tests performed on a small percentage of the production anchors were used to measure the anchor load several weeks after construction and to estimate the available load for the service life of the dam.

6.3.3 John Day Dam, Washington

John Day Dam is located on the Columbia River approximately 110 miles upstream of Portland, Oregon (Heslin et al. 2009). The Navigation Lock at John Day is located on the north side of the river between the spillway and a section of embankment dam, and measures 675 feet by 86 feet (chamber dimension). The maximum lift between forebay and tailwater is 113 feet. The north and south walls of the structure are symmetric and are shaped as shown in Figure 6.16. There is a filling/emptying culvert adjacent to the lock chamber in the base of the north and south walls. The walls were originally designed as full-gravity sections with constant foundation elevation. In an effort to reduce construction cost, the heel of each wall was founded on a layer of dense basalt at a higher elevation than the wall toe, where the filling/emptying culvert is located. The upper basalt layer is underlain by less competent flow breccia. Between the "lock full" and "lock empty" condition, horizontal movements up to one inch have been measured at the top of the south wall. This repetitive rocking resulted in cracks near the top of the filling/emptying culvert as shown in Figure 6.16.

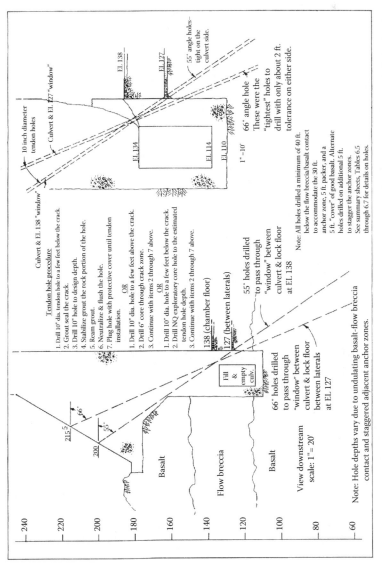

Figure 6.16 John Day Dam South Navigation Lock Wall, Washington. (From Heslin, G., D. A. Bruce, G. S. Littlejohn, and T. Westover, "Performance of Aging Post-Tensioned Rock Anchors in Dams," ASDSO Northeast Regional Conference, State College, PA, 2009. With permission.)

In 1981 the U.S. Army Corps of Engineers completed foundation grouting in the flow breccia and installed 73 ground anchors (design working load 1,518 kips) in an effort to stiffen the flow breccia and close the cracks in the filling/emptying culvert. The ground anchors were installed at the orientation shown in Figure 6.16 and were bonded into dense basalt underlying the floor of the lock chamber. After initial drilling, anchor holes were pregrouted and redrilled in an effort to develop watertight holes. Once the holes were redrilled, 37-strand tendons were inserted and grouted in the holes using two-stage grouting procedures. Anchor tendon details are shown in Figure 6.17.

An inflatable packer was used to separate the bond length and free length of the tendon: this was not an atypical detail of the time. Steel strands were bare below the packer and had individual sheaths surrounding each strand above the packer. There was post-tensioning grease inside the individual sheaths above the packer. Tendons were grouted into the structure in two stages: stage 1 involved grouting the bond zone and stage 2 involved grouting above the packer. Stage 2 grouting was completed after the anchors were stressed and locked off.

Fourteen of the 73 anchors were fitted with permanent load cells to monitor changes in load over time. These anchors would now be considered to have Class II corrosion protection by modern definition (PTI, 2004), and would not be considered adequate for permanent installations. Shortly after installation, USACE inspectors observed water seeping out of the strands of several anchors when the lock was full. This seepage (Photo 6.4) has continued and has resulted in corrosion at and below the anchor heads.

In an effort to track the rate of corrosion and the implications on monolith stability, the Corps of Engineers commissioned detailed inspections and liftoff tests in 2003 and 2008. The inspections showed that the number of anchors with visibly damaged strands increased by 11 percent between 2003 and 2008. Typical damage consisted of a missing center wire in the 7-wire strand. The center wires appeared to have corroded and ruptured some distance below the gripping wedges. This loss in steel area reduces the load locked into the anchors. Lift-off tests in 2008 had lift-off loads roughly 5 percent lower than the same anchors tested in 2003.

Typically, an anchor lift-off test is conducted by starting the hydraulic pump at a load somewhere between the alignment load and the lock-off load. The pump is allowed to run continuously and the rate of pressure increase is monitored by an inspector. When lift-off occurs, the pump begins to labor and the rate of pressure change decreases. Lift-off is confirmed by observing separation between the wedge plate and the bearing plate. For typical jacking systems, a lift-off load accuracy of 2 percent can be expected using this procedure.

Performing lift-off tests on anchors with damaged tendons can break corroded strands because the actual load in the anchor is unknown. During

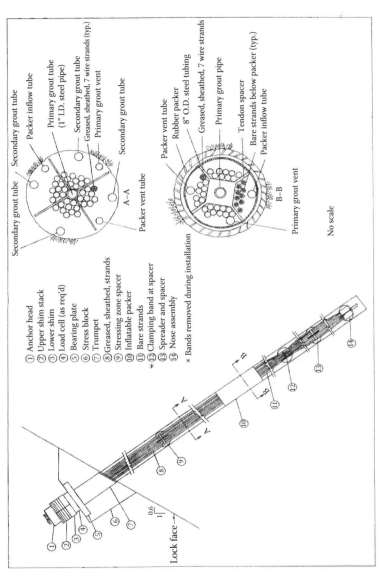

① Anchor head
② Upper shim stack
③ Lower shim
④ Load cell (as req'd)
⑤ Bearing plate
⑥ Stress block
⑦ Trumpet
⑧ Greased, sheathed, strands
⑨ Stressing zone spacer
⑩ Inflatable packer
⑪ Bare strands
⑫ Clamping band at spacer
⑬ Spreader and spacer
⑭ Nose assembly
✴ Bands removed during installation

Secondary grout tube
Secondary grout tube
Secondary grout tube
Packer inflow tube
Primary grout tube (1" I.D. steel pipe)
Secondary grout tube
Greased, sheathed, 7 wire strands (typ.)
Primary grout vent
Secondary grout tube
A–A
Packer vent tube

Packer vent tube
Rubber packer
8" O.D. steel tubing
Greased, sheathed, 7 wire strands
Primary grout pipe
Tendon spacer
Bare strands below packer (typ.)
Packer inflow tube
B–B
Primary grout vent

No scale

Lock face

0.6
1

Figure 6.17 Anchor tendon details, John Day Dam, Washington. (From Heslin, G., D. A. Bruce, G. S. Littlejohn, and T. Westover, "Performance of Aging Post-Tensioned Rock Anchors in Dams," ASDSO Northeast Regional Conference, State College, PA, 2009. With permission.)

Photo 6.4 Typical corroded anchor head in 2008 at John Day Dam, Washington. (Note water spraying out of strands.) (From Heslin, G., D. A. Bruce, G. S. Littlejohn, and T. Westover, "Performance of Aging Post-Tensioned Rock Anchors in Dams," ASDSO Northeast Regional Conference, State College, PA, 2009. With permission.)

the test, the applied load can quickly overshoot the structural capacity of the corroded steel tendon and rupture strands. To minimize the chance of breaking strands, a three-stage procedure was developed for conducting lift-off tests at John Day. In stage 1, the anchor was loaded in increments of 5–10 percent of the anticipated lift-off load. Anchor head deflection was measured and plotted versus load (or pump pressure) in the field. When the slope of the deflection plot increased drastically, lift-off had likely occurred. This was confirmed by observing separation between the stressing head and bearing plate. The load was then reduced below lift-off and stage 2 loading was performed.

The objective of stage 2 loading was to more accurately define the point where the slope of the load-deflection plot changed. Stage 2 involved loading the anchor in increments of 1 percent to 2 percent of the lift-off load determined in stage 1. Anchor head deflection was measured and plotted as in stage 1. Once lift-off was confirmed by observing separation between the stressing head and bearing plate, the load was again reduced below lift-off.

Stage 3 involved performing the standard lift-off test where the pump was allowed to run continuously until a change in the rate of pressure increase was observed. The risk of damaging the tendon with the stage 3, or standard, lift-off procedure was lessened because the lift-off load was known based on the stage 1 and 2 tests. With a known target, the risk of overshooting the lift-off load and damaging the anchor tendon was greatly reduced.

Typical lift-off test data from stages 1 and 2 are shown in Figure 6.18. By fitting the prelift and postlift portions of the curve with straight lines, the lift-off load can typically be determined to less than 1 percent of the lock-off load. While this is more accurate that the 2 percent figure typically associated with the conventional lift-off test procedure (stage 3 loading), the primary advantage was that the risk of damaging anchor tendons was reduced. A second advantage of the three-stage lift-off procedure was that the post liftoff anchor stiffness could be quantified.

The anchors at John Day Dam can be grouped into damaged and undamaged tendons based on visual inspections. For anchors with apparently undamaged tendons, the 2008 lift-off loads are typically 85 percent to 90 percent of the original lock-off load. The largest component of the load loss appears to be due to stress relaxation in the steel tendons. For anchors with damaged tendons, lift-off loads are roughly proportional to the number of visually intact strands at the anchor head. Anchors with damaged strands have measured lift-off loads ranging from 0 percent to 89 percent of the original lock-off load.

Fourteen of the 73 anchors at John Day Dam have permanent load cells. The USACE monitored the load cells for a period of nearly one year following installation. The data from these instruments have a large scatter

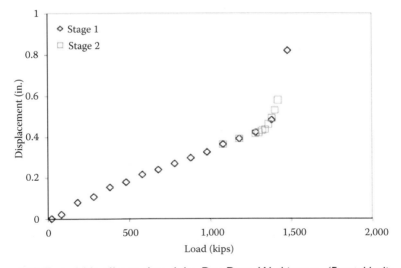

Figure 6.18 Typical lift-off test data, John Day Dam, Washington. (From Heslin, G., D. A. Bruce, G. S. Littlejohn, and T. Westover, "Performance of Aging Post-Tensioned Rock Anchors in Dams," ASDSO Northeast Regional Conference, State College, PA, 2009. With permission.)

and occasionally data trends are unexplainable. Plots of data collected from "good" and "bad" load cells are shown in Figure 6.19. These show that the "good" load cell registers a gradual loss of load over time and that the rate of load loss decreases over time (log-linear relationship). Data from a "bad" load cell, as shown in Figure 6.19, have a very large scatter and no discernible trend. The "bad" load cell appears to show an increase in load over time.

Post-tensioned rock anchors could lose load over time for several reasons, including (1) bond zone creep, (2) creep of navigation lock concrete, (3) strand slippage at the anchor head, (4) stress relaxation of the tendon steel, and (5) loss of steel section due to corrosion. For the anchors at John Day Dam, one would expect the largest change in load over time to be due to stress relaxation in the steel tendon because the anchors are bonded into dense basalt (negligible bond zone creep), the navigation lock concrete has not deformed appreciably (negligible concrete creep), and the amount of strand above the locking wedges has not changed since 1981 (no observed strand slippage).

The steel used for the tendons in the John Day anchors consisted of "stress-relieved" steel rather than the current standard "low-relaxation" steel. Stress relaxation involves a gradual reorientation of the steel fabric. The process is a function of temperature and stress level. Higher

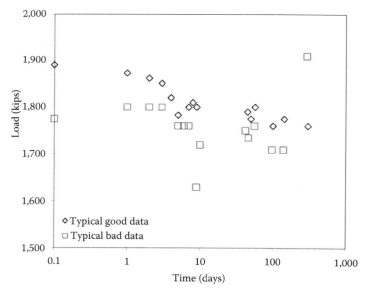

Figure 6.19 Typical load cell data, John Day Dam, Washington. (From Heslin, G., D. A. Bruce, G. S. Littlejohn, and T. Westover, "Performance of Aging Post-Tensioned Rock Anchors in Dams," ASDSO Northeast Regional Conference, State College, PA, 2009. With permission.)

temperatures and stress levels induce higher relaxation losses. Load loss due to stress relaxation follows a log-linear relationship. By ignoring load cell data that do not follow a log-linear trend, assuming all losses were due to stress relaxation, and projecting the relationship to present, the load cells would predict current anchor load between 83 percent and 93 percent of original lock-off load. Lift-off tests on undamaged anchors with load cells showed that actual lift-off load was between −2 percent and +5 percent of the load predicted from load cell data. This appears to support the theory that stress relaxation is the largest component of the load loss, other than corrosion.

Visual inspections can determine the number of visually intact strands at the anchor head. However, the actual number of intact strands, or aggregate area of remaining steel, contributing to the anchor load is always less than this number. This is presumably due to corrosion below the anchor head. Lift-off test results can be evaluated to determine the number of effective strands remaining by making some assumptions about load losses. For anchors with visually undamaged tendons at the anchor head, measured lift-off loads were 84 percent to 90 percent of original lock-off load. If it is assumed that stress relaxation is the sole source of load loss, the present load can be predicted using the log-linear relationship for stress relaxation. The ratio of the measured lift-off load to the predicted load can be set equal to the ratio between the number of "effective" strands and the original number of strands. Using this procedure, the number of effective strands is one-half to 5 strands less than the number of intact strands observed at the anchor

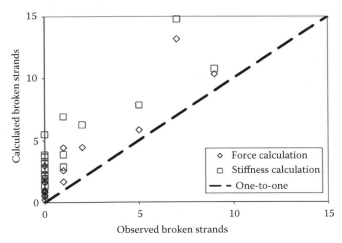

Figure 6.20 Visual inspection versus calculated number of damaged strands, John Day Dam, Washington. (From Heslin, G., D. A. Bruce, G. S. Littlejohn, and T. Westover, "Performance of Aging Post-Tensioned Rock Anchors in Dams," ASDSO Northeast Regional Conference, State College, PA, 2009. With permission.)

head. A plot of this "force-based" relationship is shown in Figure 6.20. These data show that actual conditions are worse than what is indicated by visual inspection.

The number of effective strands can also be calculated based on the post lift-off anchor response. This approach involved evaluating the stiffness, or slope of the post lift-off plots (stage 1 and stage 2 tests). By assuming the modulus of the steel is unchanged by stress relaxation, the effective area of steel can be determined from Hooke's law. The number of effective strands is then calculated as the effective area of steel divided by the area of a single, intact strand. Using this "stiffness" procedure, the number of effective strands is 2 to 7 strands less than the number of visually intact strands observed at the anchor head.

The anchors at John Day were fabricated and installed prior to the advent of modern Class I corrosion protection systems, or encapsulated tendons, defined by the Post-Tensioning Institute. The anchor deterioration is the result of corrosion of the steel tendons. Sealing anchor holes by pregrouting did not protect many of the anchors from exposure to groundwater, even in the short term, presumably due to difficulties posed to the operation by the ground conditions. The bond zone steel was not protected.

Visual inspections are an important tool for assessing long-term anchor performance. However, actual conditions can be substantially worse than those indicated by visual inspection of the anchor head. Generally, anchor failures occur near the head, but this is highly dependent on the tendon fabrication details and anchor environment. Visual inspections should be used to monitor changes in anchor condition, but lift-off testing must be performed to truly assess the performance of the anchors over time.

Monitoring for signs of corrosion in the anchor system is particularly important for anchor tendons fabricated before the advent of modern Class I–encapsulated tendons. Visual inspections can be used with anchor lift-off tests to quantify the rate of anchor deterioration and load loss. Load cells on the John Day anchors were not helpful for monitoring loads over time, since instrument performance was erratic and did not follow normal trends in a majority of cases.

REFERENCES

ADSC Epoxy-Coated Strand Task Force. 2003. "Supplement for Epoxy-Coated Strand." *ADSC: The International Association of Foundation Drilling,* November.

Bruce, D. A. 1988. "Practical Aspects of Rock Anchorages for Dams." Paper presented at Association of State Dam Safety Officials 5th Annual Conference, Manchester, NH, September 25–28, 26 pp.

Bruce, D. A. 1997. "The Stabilization of Concrete Dams by Post-Tensioned Rock Anchors: The State of American Practice." Pp. 508–21 in *Ground Anchorages and Anchored Structures*, edited by G. S. Littlejohn. London: Thomas Telford.

Bruce, D. A. 2002. "A Historical Review of the Use of Epoxy Protected Strand for Prestressed Rock Anchors." Paper presented at Dam Safety 2002, Annual Conference of the Association of State Dam Safety Officials, Tampa, FL, September 8–11, available on CD, 21 pp.

Bruce, D. A. 2007. "Dam Remediation by Anchors and Cut-Offs: A Summary of Two National Research Programs." Paper presented at Ohio River Valley Soils Seminar, ORVSS XXXVIII, Louisville, KY, November 14, 35 pp.

Bruce, D. A., J. J. Jensen, and J. S. Wolfhope. 2005. "High Capacity Rock Anchors for Dams: Some Fundamental Observations on the Analysis of Stressing Data." Paper presented at USSD 2005 Conference, Salt Lake City, UT, June 6–10, 2 pp.

Bruce, D. A., and J. S. Wolfhope. 2005. "Rock Anchors for Dams: The Preliminary Results of the National Research Project." Paper presented at ASDSO Dam Safety Conference, Orlando, FL, September 23–26, 6 pp.

Bruce, D. A., and J. S. Wolfhope. 2006a. "Rock Anchors for Dams: The National Research Project: The Evolution of Post-Tensioned Anchors on Hydropower Dams." Paper presented at Hydrovision 2006, Portland, OR, July 31–August 4, 11 pp.

Bruce, D. A., and J. Wolfhope. 2006b. "Rock Anchors for Dams: The National Research Project: The (Semi) Final Results of the Phase 1 Study." Paper presented at ASDSO Dam Safety, Boston, MA, September 10–14.

Bruce, D. A., and J. Wolfhope. 2007a. "Rock Anchors for North American Dams: The Development of the National Recommendations (1974–2004)." Paper presented at Institution of Civil Engineers, Ground Anchorages and Anchored Structures in Service, November 26–27, London, England, 11 pp.

Bruce, D. A., and J. Wolfhope. 2007b. "Rock Anchors for Dams: Evolution and Analysis of Corrosion Protection Systems and Construction Costs." Paper presented at Dam Safety '07, Annual National Conference, September 9–13, Austin, TX, 20 pp.

Bruce, D. A., and J. Wolfhope. 2007c. "Rock Anchors for North American Dams: The National Research Program Bibliography and Database." Paper presented at Institution of Civil Engineers, Ground Anchorages and Anchored Structures in Service, November 26–27, London, England, 11 pp.

FIP Commission on Practical Construction. 1986. *Corrosion and Corrosion Protection of Prestressed Ground Anchorages*. London: Thomas Telford.

Heslin, G., D. A. Bruce, G. S. Littlejohn, and T. Westover. 2009. "Performance of Aging Post-Tensioned Rock Anchors in Dams." Paper presented at ASDSO Northeast Regional Conference, June 14–16, State College, PA, 12 pp.

Littlejohn, G. S., and D. A. Bruce. 1977. *Rock Anchors—State of the Art*. Essex, England: Foundation Publications. (Previously published in *Ground Engineering* in 5 parts, 1975–76.)

National Rock Anchor Research Program. 2005. "High Capacity Rock Anchors for Dams: Thirty Years of Recommendations for Practice." Co-Principal Investigators D. A. Bruce and J. W. Wolfhope. Pp. 88–105 in *Geo3 GEO Construction Quality Assurance/Quality Control Conference Proceedings*, edited by D. A. Bruce and A. W. Cadden. Dallas, TX: ADSC.

Nierlich, H., and D. A. Bruce. 1997. "A Review of the Post-Tensioning Institute's Revised Recommendations for Prestressed Rock and Soil Anchors." Pp. 522–30 in *Ground Anchorages and Anchored Structures*, edited by G. S. Littlejohn. London: Thomas Telford.

Post-Tensioning Institute (PTI). 1986. *Recommendations for Prestressed Rock and Soil Anchors*. 2nd ed. Phoenix, AZ: PTI.

Post-Tensioning Institute (PTI). 1996. *Recommendations for Prestressed Rock and Soil Anchors*. 3rd ed. Phoenix, AZ: PTI.

Post-Tensioning Institute (PTI). 2004. *Recommendations for Prestressed Rock and Soil Anchors*. 4th ed. Phoenix, AZ: PTI.

Prestressed Concrete Institute (PCI). 1974. *Tentative Recommendations for Prestressed Rock and Soil Anchors*. Phoenix, AZ: PCI (now PTI).

Wolfhope, J. S., D. A. Bruce, G. Forbes, and M. L. Boyd. 2005. "Post-Tensioned Rehabilitation of Hydropower Dams: Continuously Improving on an Anchor Program." Paper presented at Waterpower XIV Conference, Austin, TX, July 18–22, 12 pp.

Zicko, K. E., D. A. Bruce, and R. A. Kline Jr. 2007. "The Stabilization of Gilboa Dam, New York, Using High Capacity Rock Anchors: Addressing Service Performance Issues." Institution of Civil Engineers, Ground Anchorages and Anchored Structures in Service, November 26–27, London, England. 10 pp.

Chapter 7

Instrumented performance monitoring

Marcelo Chuaqui

7.1 INTRODUCTION

This chapter focuses upon the instrumented monitoring that should be conducted during the remediation of a dam or levee. Geotechnical and structural instrumentation is highly specialized and therefore the material presented is not intended to be a comprehensive guide for implementation of the monitoring program, but rather an overview of this mandatory operation.

In this section, objectives for instrumented monitoring with respect to dam remediation are defined and discussed. In Section 7.2 some of the key aspects of planning an instrumented monitoring program are discussed. In Section 7.3 the instruments themselves are examined, by looking at the different instrument types and then reviewing the instruments most commonly used to measure specific parameters. Later sections focus on data management, field staffing, and process automation. The material presented in this chapter has been compiled from a variety of referenced sources and for the reader who is either involved in the selection or implementation of a monitoring program the referenced publications are strongly recommended.

Remediation can be considered to be one phase in the life of a dam or levee. The geotechnical monitoring program for one of these structures ideally starts prior to construction and continues throughout the life of the structure. Monitoring of performance can include visual examination/ inspection, video/photographic/audio surveillance, dam/reservoir operations data and reports, water quality sampling programs, geotechnical and/ or materials testing of samples, self-potential testing, thermal monitoring, resistivity surveys, seismic reflection/refraction studies, ground-penetrating radar and geologic exploration programs. Piezometers, crack meters, flow meters, and inclinometers represent fundamental instrumentation.

Additionally, specific devices can be installed to measure a particular parameter of interest.

Instruments provide quantitative data on such parameters as groundwater pressure, deformation, total stress, temperature, seismic events, leakage, and water levels. A wide variety of instruments may be utilized in a comprehensive monitoring program to ensure that all critical parameters for a given project are covered sufficiently. The most commonly used geotechnical instrumentation is described in Section 7.3. These devices provide quantitative data and can be installed to measure at locations where visual observations are not possible. Quantitative data permit databases to be created that permit subtle trends in the performance of the structure to be detected.

The importance of implementing monitoring programs for dam and levee safety cannot be challenged. There are many historical cases of dam failures where early warning signs of failure might have been detected if a good dam safety monitoring program had been in place (Myers and Stateler 2008), and indeed Peck was quoted (Myers 2008) at the International Commission on Large Dams (ICOLD) Conference in September 2000 as stating, "Monitoring of every dam is mandatory because dams change with age and may develop defects. There is no substitute for systematic and intelligent surveillance."

Instrumentation and monitoring are risk-management tools. Monitoring programs provide the data required to determine if a dam or levee is performing as expected, if that performance is changing, and if the performance remains acceptable. The ability to detect a change in this performance is critical to managing the risk involved with potential failure of the structure. A well-planned and carefully implemented dam safety monitoring program should be a key part of every dam owner's risk management program (Myers and Stateler 2008), and within the context of remediation, monitoring should be used to manage any potential risks associated with the implementation, and to confirm the effectiveness of the remediation. The golden rule of instrumentation notes that every instrument should be installed to answer a specific question (Dunnicliff 1993). If there is no question, then there should be no instrument.

Through the use of computers, dataloggers, and automated instruments, it is now possible to automate measurements (USCOLD 1993). Data can be collected in practically real-time at high frequencies and automated monitoring allows for the use of alarms for sudden or unexpected changes in data values. These systems permit the monitoring systems to be used more efficiently for prevention and warning.

Because the monitoring needs for a particular dam change over time, it is important to reevaluate the monitoring program on a regular basis. This reevaluation is needed to determine if the correct information is being provided to effectively and accurately monitor the ongoing performance of the structure (Myers and Stateler 2008). The monitoring requirements for a remediation program are themselves the result of a specific performance issue. Additionally, during remediation, it is likely that the normal operational parameters of the dam or levee may be changed either temporarily or permanently. The remediation itself may involve utilizing a process that introduces changes to the normal conditions the structure experiences. Furthermore, the remediation may result in modified operational parameters for the structure. The revised monitoring program must specifically address each of these changes.

Monitoring efforts that focus on parameters that are no longer important or valid can provide a false sense of security by distracting from more critical parameters relating to the remediation and future safe performance of the dam/levee. In this regard, the process of conducting a potential failure modes analysis (PFMA) is becoming increasingly popular and valuable throughout the dam engineering community. As a basic principle, the nature of the instrumentation program (including the number and type of instruments, their frequency of reading, and their permissible response levels) are dictated by the outcome of the PFMA process. The process applies not only to failure modes that might occur under normal operating conditions, but also to failure modes that might occur under the extreme loading conditions of remediation, floods, and earthquakes as examples. Therefore, in performing the PFMA, it is vital to measure the effects that the construction means and methods will have on the structure and any effects the completed remediation will have on the structure. In effect, the PFMA process drives the scope of the instrumentation to be installed and monitored. The PFMA process involves the following successive steps:

Step 1: Identifying the potential failure modes for the dam or levee that warrant attention at the present time (based on all currently available information and data). Some common failure modes include:

- Overtopping occurs when the water level rises above the crest of the dam, sometimes by displacement of water by a large amount of material from a landslide upstream, but more typically as a result of intense rainfall events.
- Instability refers to an unbalance of the forces acting upon the dam so that resistance to shearing along a surface, for example, is overcome by the hydrostatic pressures of the reservoir. This may occur through incorrect design, or increases in pore water pressure as a result of geological changes.
- Sliding can occur within the embankment or around the abutment or foundation depending on the strength of the materials.
- Piping through, under, or around an embankment occurs when seepage flows are great enough to carry material downstream. This internal erosion can weaken the foundation and limit the effectiveness of the structure.
- Traverse cracking can occur with nonuniform settlement of the foundation, which again will weaken the structure and potentially create a flow pathway.

Step 2: Defining the monitoring necessary to manage the risks associated with each potential failure mode. As with any risk-management program, it is necessary to consider monitoring costs in relation to the likelihood of each failure mode and associated consequences and impacts.

Step 3: Defining expected and unexpected performance associated with each monitoring effort, and to provide a framework for personnel charged with carrying out the routine dam safety monitoring to identify developing conditions of concern.

Instrumented monitoring that is included to detect potential failure modes that may develop during remediation effectively addresses the issues of providing an early detection of unusual or unexpected performance. It may not directly address the other issues or needs.

While performing a remediation, instrumentation should also be used for:

- Verification of design parameters, assumptions, and construction techniques. It is likely that monitoring of the structure's performance assisted in defining the need for remediation and some of the parameters required for determining the type and schedule for the remediation. In order to select and design the remediation process, field investigation work is typically the key information-gathering method; characterization of the geology and materials at and around the site is essential. Instrumentation can verify design parameters with observations of actual performance, thereby enabling engineers to determine the suitability of the design. Instrumentation also aids the modification of designs by incorporating the effects of actual field conditions. The remediation may require responsive design modifications, as it is being performed, rendering critical the ability to continuously monitor the parameters of interest for detecting and correcting problems that may arise.

- Analysis of adverse events, i.e., the outcome of the PFMA process.

- Verification of apparent satisfactory performance. It is just as necessary to confirm satisfactory performance of the remediation as it is to identify areas of concern.

- Prediction of future performance. Instrumentation data can be used to make informed, valid predictions of future behavior of the remediated structure.

- Legal evaluation. Quantitative documentation of some parameters may be needed for legal reasons, for example, claims related to construction activities or changes in groundwater levels.

- Research and development. Analysis of the performance of the remedial process can be used to advance the state-of-the-art of design and construction of these methods.

7.2 PLANNING

An effective geotechnical instrumentation program is a complete program of systematically installing instruments, collecting data, evaluating the results and taking timely action. It is not just putting a few instruments in the ground and having a technician take some readings on an irregular basis (Marr 2001).

From this concept of an instrumentation program, two golden rules (Dunnicliff 1993) follow:

- every instrument must have a purpose (every instrument should provide data to answer a question)
- every program must be planned and executed in a systematic way

Development of the instrumentation system should begin with the definition of an objective (as related to the results of the PFMA process) and then proceed through a comprehensive series of logical steps that include all aspects of the system. Without a carefully executed plan, the geotechnical instrumentation program will more than likely fail to meet those objectives.

A systematic planning approach is recommended and described by Dunnicliff in "Geotechnical Instrumentation for Monitoring Field Performance" (1993). Each of these sequential steps is listed in Table 7.1, which constitutes a most useful guideline.

The persons designing the instrumentation program must be familiar with the existing dam or levee and they must understand the existing instrumentation program. They must also have an awareness of the construction methods that will be used for the remediation. They should also be experienced with the design and implementation of geotechnical instrumentation. On most projects, planning the instrumentation will require a team approach between the different specialists or groups, each skilled in different aspects of the project.

During the planning process, it is critical to identify the most important geotechnical questions that will arise during the design, construction, and operation phases of the remediation. Special attention should be given to the effects on the dam or levee by the construction means and methods, as well as changes brought about by the completed remediation. Instrumentation can then be selected to help answer each of these specific questions. Many different parameters can be monitored, but it is essential that engineers are selective as to which parameters are relevant with respect to the remediation, and the underlying geotechnical challenges.

Locations for instruments should be determined based on the predicted behavior of the site. The locations should be compatible with the geotechnical concerns and the method of analysis that will be used when interpreting the data. Care must, of course, be taken in selecting

Table 7.1 Systematic approach to planning monitoring programs using geotechnical instrumentation

Step	Action
1	Define the project conditions
2	Predict mechanisms that control behavior
3	Define the geotechnical questions that need to be answered
4	Define the purpose of the instrumentation
5	Select the parameters to be monitored
6	Predict magnitudes of change
7	Devise remedial action
8	Select instruments
9	Select instrument locations
10	Plan recording of factors that may influence measured data
11	Establish procedures for ensuring reading correctness
12	Prepare budget
13	Plan installation
14	Plan regular calibration and maintenance
15	Plan data collection, processing, presentation, interpretation, reporting, and implementation

Source: From Dunnicliff, J., *Geotechnical Instrumentation for Monitoring Field Performance*, John Wiley and Sons, Inc., New York, 1993.

instrument locations and the installation means and methods so as to not introduce a potential failure mechanism. For example, casing should not be installed vertically through the core material, and horizontal tubes and cables should not fully run through the core and should exit downstream. When selecting instrument locations, it is important to consider the benefits of the installation versus potential detrimental effects. Just as instruments can introduce failure mechanisms, instruments can also affect the parameter that is being measured. For example, a grouted borehole extensometer must be surrounded by grout with similar characteristics to the surrounding soil; otherwise, it will not behave in the same manner had the instrument not been present (U.S. Army Corps of Engineers 1995).

Dams and levees present relatively harsh and demanding environments for sophisticated instruments. It is therefore a good idea to use instruments that have a history of proven performance rather than testing new technology. In planning the instrumentation, it is important to have back-up or redundancy for measurement of critical parameters, and different instrument types have specific strengths and weaknesses. Whenever possible, it is advisable to use instruments with different characteristics or principles to provide better measurement of a critical parameter. For example, tiltmeters can provide very accurate data, resolving very small movements, but provide a relative movement (tilt), while optical surveys offer a geodetic measurement that, in most instances, is less accurate. Utilizing both to measure the deformation at a critical location will help provide very accurate movement data that can be correlated to an overall geodetic location (Chuaqui, Ford, and Janes 2007). Expense and reliability must be factored to provide appropriate coverage and redundancy.

Another key component of the planning is performing some detailed numerical analysis in order to define data ranges for instruments, that is, maxima, minima, and how and if change represents a concern. During the planning phase, warning levels for the parameters of interest should be defined. Every measurement includes a degree of uncertainty and several factors should be examined when selecting instruments.

Accuracy and precision are concepts that are sometimes thought to be interchangeable, but are distinct factors. Accuracy refers to the closeness to a true value, whereas precision is the reproducibility or repeatability of the measurement (Bevington 1969). Accuracy is expressed as a ± number, such as ±1 mm, indicating the measured value is within 1 mm of the true value. Precision is also stated as a ± number with the number of significant digits indicating the degree of precision such that ±1.00 is more precise than ±1.0. The readings should take this into account because readings taken to three significant digits are unwarranted for an instrument with a precision of ±1.0 (Dunnicliff 1993). Instruments should be both accurate and precise. The "noise" of an instrument is an external random factor that causes a decrease in accuracy and precision. Resolution refers to the smallest division of the readout scale and will determine how accurate a reading can be. Sensitivity is the amount of output in response to an amount of input, so a small change in pressure, for example, will produce a larger voltage change in an instrument with a higher sensitivity. Linearity is often a factor to be considered as the indicated values are directly proportional to the actual values being measured and, assuming linear calibration and interpolation, is more accurate. Hysteresis is a similar concept in cyclic changes where an increasing value is not the same as a decreasing value over the same range (U.S. Army Corps of Engineers 1995).

Assessing monitoring data requires the ability to correlate changes in readings to specific events in the field. All factors that might cause changes in the measured parameters should be recorded, including construction details and progress, geology and subsurface conditions, reservoir and tailwater levels, rainfall amounts, ambient and water temperatures, barometric pressure, and seismic events. The value of visual observation should not be underestimated. For example, correlating the drilling of a hole at a specific location to the presence of sediment in the tailwater is a critical piece of information when trying to identify potential seepage paths. It is important that the need for detailed documentation of the construction process be identified as a requirement of the monitoring program.

It is important to include a baseline monitoring period to determine instrument performance and normal instrument values prior to the start of remediation. It is best to make the baseline monitoring period long enough to capture a full cycle of each parameter that affects the readings, although sometimes this is not practical given the construction schedule (Chuaqui, Ford, and Janes 2007).

Part of the planning includes writing procedures for collecting, processing, interpreting, and reporting the data. Resources for these tasks must be allocated; failure to process, interpret, and report data is a common shortcoming of monitoring programs. It is all too common for there to be weeks, months, and even years of data that have been captured but not interpreted, collecting in archives (U.S. Army Corps of Engineers, 1995).

A plan should be formulated for the procedures of collecting and processing the data, the frequency of readings, extent of analysis, and reporting requirements. A schedule for each instrument should be constructed based on the importance of the parameter to be measured or specific activities on site. It is beneficial to have consistency in personnel and equipment with multiple readings, if necessary, to confirm readings. Communication between all personnel involved in the data collection, analysis, and reporting needs to be very precisely planned to ensure that all information flows, especially if an unexpected event occurs. The plan should reflect how the data will be used; a format easily imported into a database or for graphical presentation should be considered.

Automation of data collection, processing, reporting, and notification of readings exceeding predefined thresholds has paradoxically made instrumented monitoring both simpler and more complicated. It is simpler in the fact that data are collected regularly with less manpower, but complicated by the increased amount of data and the complex systems requiring expertise. Computerized systems cannot replace sound engineering judgment, and engineers must make a special effort to ensure that measured effects are correlated with probable causes.

Other miscellaneous but important items to be considered during the planning phase include calibration, maintenance, replacement of instruments, and lightning protection/grounding of the instruments.

7.3 INSTRUMENT OPERATING PRINCIPLES

This section describes the most commonly used borehole instruments as well as surveying methods and the Global Positioning System (GPS). To illustrate a typical instrumentation system, cross-sections showing where different instruments are located in typical earth and concrete dams are provided.

Most instrumentation measurement methods consist of three components: a transducer, a data acquisition system, and a linkage between these elements. A transducer is a component that converts a physical change into a corresponding electrical output signal. Transducers are typically limited to directly measuring temperature, change in dimensions, force (mass), and color (or frequency). However, transducers based on these primary measurements can provide data on many geotechnical parameters, such as total stress, pore water pressure, deformation, strain, tilt, acceleration, velocity, temperature, sound intensity, and light intensity. Data acquisition systems range from simple portable readout units to complex automatic systems.

The technologies of instrumentation fall into three general categories: pneumatic, vibrating wire, and electrical/electronic (Dunnicliff 1993). Surveying with the use of total stations and GPS are included in the electrical/electronic category.

Pneumatic devices measure gas pressure applied to a diaphragm by equalizing this unknown pressure with a gas supply that flows through an inlet tube past the diaphragm and passes through an outlet tube to a pressure gauge. Pneumatic devices include piezometers, pressure cells, and settlement gauges. Figure 7.1 is a schematic of a pneumatic piezometer.

Vibrating wire devices have a tensioned wire that vibrates at its natural frequency. When there are small movements between the clamped ends of the wire, the frequency at which it vibrates changes and this can be used to construct many different instruments. For example, the wire can be used as a pressure sensor as shown in Figure 7.2. The wire is plucked magnetically by an electrical coil attached near the wire at its midpoint, and either this same coil or a second coil is used to measure the period or frequency of vibration. Frequency (f) is dependent on the bending of the diaphragm, hence on the pressure (P). With vibrating wire transducers, undesirable effects involving signal-cable resistance, contact resistance, electrical signal seepage to ground, or length of signal cable are negligible. Very long cable lengths are practical. Vibrating wire devices include piezometers, pressure cells, load cells, settlement, and deformation gauges.

Electrical and electronic instruments involve a wide variety of technologies. Table 7.2 summarizes some of the more common electrical instruments and how they operate, and Figure 7.3 shows a schematic of an electrical resistance strain gauge.

The following comments pertain to other families of instruments:

Temperature sensors: Temperature monitoring is used to detect seepage within a dam and for correction factors for data from instruments sensitive to temperature changes. Thermistors are composed of semiconductive materials that change in resistance with temperature. Thermocouples are composed of wires of different materials and the voltage generated between the two wires is proportional to the temperature read. Resistance temperature devices measure a voltage that is dependent on the resistance of a wire that is proportional to temperature.

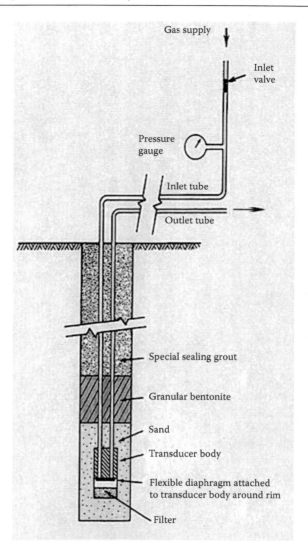

Figure 7.1 Schematic of normally closed pneumatic piezometer. (From Dunnicliff, J., *Geotechnical Instrumentation for Monitoring Field Performance*, John Wiley and Sons, Inc., New York, 1993. With permission.)

pH sensors: pH is measured using a potential difference between a reference electrode and a sensing electrode. The voltage is proportional to the hydrogen ion activity in the water sample and is converted to a pH value. Both the temperature and pH sensors can be manually read using handheld devices or via integrated multiparameter monitoring and logging devices that are installed into boreholes ("trolls").

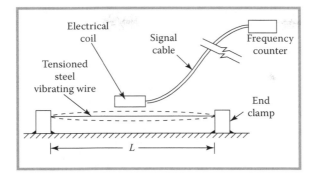

Figure 7.2 Schematic of vibrating wire transducer. (From Dunnicliff, J., *Geotechnical Instrumentation for Monitoring Field Performance*, John Wiley and Sons, Inc., New York, 1993. With permission.)

Flow and velocity meters: Seepages in a dam or levee occur through, under, and/or around the structure. Changes in seepage rate are an indication of problems, and seepage monitoring illustrates the effectiveness of the structure and drainage. Flow rates can be measured in weirs that have regular-shaped overflow openings. Reference tables can determine the rate

Figure 7.3 Schematic of electrical resistance strain gauge. (From Dunnicliff, J., *Geotechnical Instrumentation for Monitoring Field Performance*, John Wiley and Sons, Inc., New York, 1993. With permission.)

Table 7.2 Description of selected electrical/electronic monitoring instruments

Instrument	Operating principle
Electrical resistance strain gauges	A change in resistance is caused by a change in length. Figure 7.3 shows a schematic of a micro-circuit bonded to a medium that is attached to a structure. With stress applied to the structure, the associated strain changes the length of the circuit and the resistance so a voltage can be read to measure these changes.
Linear variable differential transformers (LVDTs)	Linear variable differential transformers measure linear displacement and consist of a movable magnetic core passing through wire coils. When an AC voltage is applied to the primary coil a voltage is induced in two secondary coils with a magnitude dependent on the distance from the core to each secondary coil. Typical instruments include crackmeters and jointmeters.
Potentiometers	A movable slider makes contact with a resistance strip. As this slider moves, the voltage across the strip changes and can be related to the movement.
Force balance accelerometers	Currents are induced by movements of a mass subjected to gravity suspended in a magnetic field. This technology is typically used in tilt sensors, in tiltmeters, and in inclinometers.
Induction coil transducers	A powered electrical coil induces a magnetic field inside a steel wire ring. When a voltage is applied to the ring, a current can be measured that is proportional to the distance between the coil and the ring. These transducers are used in probe extensometers.
MicroElectroMechanical Systems (MEMS)	MicroElectroMechanical Systems are the integration of mechanical elements, sensors, actuators and electronics on a common silicon substrate through microfabrication technology. There are numerous types of MEMS. This is a large and rapidly growing high-technology area. The first to appear in the geotechnical field are accelerometers (similar in principle to the sensors used in most inclinometers), which are being used as tilt sensors.
Total station	An optical surveying instrument combined with an electronic distance measurement (EDM).
Global Positioning System (GPS)	The system consists of satellites in orbit around the globe where receivers can receive signals from a minimum of four satellites in order to calculate the receivers' position.

by a measurement of the water level. Flow can also be measured through collection in drains, wells, channels, and ditches using timed or calibrated buckets. Flow and velocity meters are available using electromagnetic devices. Measurement of seepage is often bedeviled by access restraints, or the simple practicality of measuring its rate in a tailwater area already submerged.

Total station: The reference benchmarks are installed at locations outside the area to be monitored. A total station is used to optically survey at least three of the reference points and the location of the total station is then determined by a least-square method. From this location, the monitoring target locations are surveyed. Readings are compared to previous values and movements are calculated. Figure 7.4 shows a schematic of total station-monitoring prisms. With conventional surveying, reference targets are necessary in a stable area outside the zone of influence and can be a significant distance away. A line of sight is required and atmospheric conditions can affect readings (Chuaqui and Hope 2007).

GPS: The Global Positioning System (GPS) provide three-dimensional monitoring of surface displacements. The system consists of satellites in orbit around the globe where receivers can receive signals from a minimum of four satellites in order to calculate the receivers' position. Basic GPS receivers are accurate to 10 m horizontally and 15 m vertically. However, greater accuracy—to within several millimeters—can be achieved with more complex but expensive equipment and applying a technique termed "relative positioning" (Hudnut and Behr 1998). With this technique, a receiver is located at a point of reference and another is positioned at the location of interest. Both GPS units receive information from the same satellites and compares them so as to eliminate common errors to both the receivers and satellites. The result is a three-dimensional vector between the two receivers, and with the known reference point the coordinates of the second receiver can be calculated.

Piezometers: Readings are taken using a water-level indicator to sound the water level or by installing a pressure transducer in the standpipe below the lowest piezometric level. There are many types of piezometers including observation wells, open standpipe piezometers, twin-tube hydraulic piezometers, pneumatic piezometers, vibrating wire piezometers, and electrical resistance piezometers (U.S. Army Corps of Engineers 1995). Piezometers can be used to monitor the pattern of water flow, that is, determining piezometric pressure conditions prior to construction, monitoring seepage, and effectiveness of drains, relief wells, and cutoffs curtains/walls. Piezometers can also be used to provide an index of soil strength by monitoring the pore water pressure, an estimate of effective stress can be made, and allow an assessment of strength. Multipoint piezometers can be installed within a single borehole so that several points can be read. The piezometers are usually separated by packers or grout to isolate the instruments at a given depth and movable probes are an option to provide a practically unlimited number of measurement points.

Inclinometers: An inclinometer casing is grouted into a borehole or attached to a structure. The probe has wheels that track in grooves that

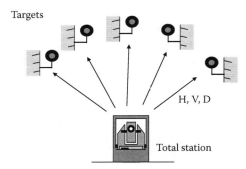

Targets

H, V, D

Total station

Figure 7.4 Principle of total station movement monitoring: Measurement of H (horizontal angle), V (vertical angle), and D (distance) allows for the position of a target to be measured in three dimensions.

are machined into the casing so that the probe does not rotate within the casing. The grooves are oriented to be parallel to the expected direction of maximum deformation. The probe measures tilt over its gauge length and readings are taken at intervals equal to the gauge length. The readings are summed to calculate the movement along its length. The casing must be installed to sufficient depth to ensure the bottom is a fixed point or the top of the casing must be surveyed to determine how much translation movement is occurring. Figure 7.5 shows the casing, grooves, and how the readings are summed to calculate a deformation. These inclinometer measurements create a profile and subsequent readings show changes in this profile over time, as shown in Figure 7.6.

In-place inclinometers (IPI): These are long-term or permanent installations consisting of multiple accelerometers at defined intervals joined by articulating rods so that a full set of readings are more readily available rather than taking readings at each interval. IPI permit remote, real-time readings. It is not possible to remove the probe and take a second set of readings with the probe rotated by 180 degrees as is normal practice with a manual inclinometer probe. This makes it much harder to separate probe drift from real movement in the data.

Extensometers: Extensometers are used to measure movements of soil and rock along a single axis. There are many different types of extensometers that work by different operating principles ranging from rods or wires anchored to different elevations and read with micrometers, vibrating wire displacement transducers, LVDTs, or potentiometers (Dunnicliff 1993). When an extensometer is installed in a borehole, it is referred to as a borehole extensometer. When multiple elevations are read within one hole, these are referred to as multiple-position borehole extensometers (MPBX). The rod extensometer consists of anchors set at specified depths, rods inside protective tubing, and a reference head. Measurements are taken at the reference head by micrometer or by an

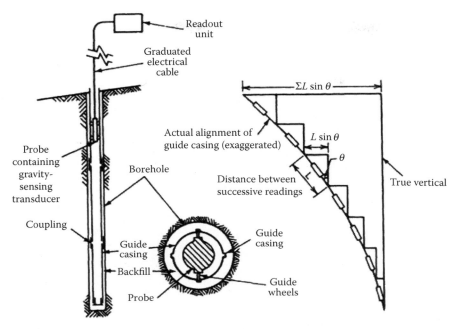

Figure 7.5 Inclinometer schematic. (From Dunnicliff, J., *Geotechnical Instrumentation for Monitoring Field Performance*, John Wiley and Sons, Inc., New York, 1993. With permission.)

electric sensor. The magnet extensometer consists of a series of magnets that are installed with an access pipe. The magnets are anchored at specified depths. Measurements are taken by lowering a probe through the access pipe to detect the depth of the magnets. The Sondex system consists of a series of rings attached to a flexible corrugated pipe. Measurements are taken by lowering a probe through an inner access pipe to detect the position of the rings. The Borros anchor settlement point is used to monitor settlement of soil under an embankment. It consists of an anchor and two concentric riser pipes that are extended up through the embankment. Measurements are made with a graduated tape and optical survey.

Settlement points: These points are used to measure settlement in the ground beneath embankments or heave during grouting. They consist of a steel pipe that is anchored to the soil via a concrete plug or pronged anchor. The pipe is within a steel or PVC sleeve that allows the pipe to move freely. The top of the anchored pipe is optically surveyed to determine changes in elevation from baseline readings.

Vibration monitoring: A recent consideration is seismic instrumentation in seismic zones and traditionally nonseismic areas as earthquakes are unpredictable. Triaxial accelerometers contain a magnet that moves

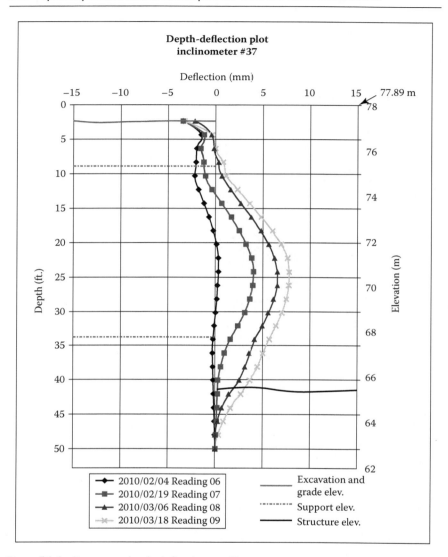

Figure 7.6 Inclinometer depth-deflection profile.

relative to a surrounding coil. Any motion in any of the vertical, longitudinal, or transverse axes induces a voltage that is proportional to standard parameters of velocity or acceleration. Strong motion monitoring using accelerographs or seismographs measures the response of the dam to ground shaking and can assist in guiding the inspection and repair efforts. Ideal locations are the base and crest of the dam, downstream slope, and nearby rock outcrop if possible.

Table 7.3 Common parameters, instruments, and technologies

Parameter	Instrument	Technologies
Groundwater pressure	Piezometer	Open standpipe, pneumatic, vibrating wire, hydraulic
Seepage	Weirs, calibrated containers, flow meters	
Deformation	Pendulum, survey, extensometers, settlement gauges, inclinometers	Optical, induction coil, LVDT, strain gauge, potentiometer, vibrating wire, accelerometer
Stress	Pressure cells	Pneumatic, strain gauge, vibrating wire
Seismic	Seismographs, accelerographs	Accelerometer
Temperature	Thermistors, thermocouples, resistance temperature devices	

Once the PFMA process is completed and the potential failure modes are identified, the monitoring requirements for each mode are defined. The common factors that are evaluated for a dam or levee and the instruments to measure them are summarized in Table 7.3 along with examples of different technologies that are available for these instruments (U.S. Army Corps of Engineers 1995).

The array of monitoring instruments that are installed in an embankment is illustrated in Figure 7.7 and those in a concrete dam are shown in Figure 7.8. The monitoring program is dynamic—parameters are changing and new questions require answers. Instruments are added, but also must be decommissioned when they serve little purpose so that the data are focused on relevant factors.

7.4 DATA MANAGEMENT

Data management and analysis are crucial to achieving the goals identified in Section 7.1. Data management includes the collection, reduction and processing, presentation, and storage of data. Data analysis includes review of the reported data, making the appropriate conclusions and taking the necessary actions.

There is no value in collecting data if they are not going to be analyzed and actions not taken in response to the data collected. Analysis is required for assessing the effectiveness and progress of the remediation, for detecting unsafe developments, and for determining the performance of the instrument systems. A project-specific schedule for data collection is necessary. The frequency of readings should be adjusted to instrument characteristics, site conditions, construction activity, or the occurrence of unusual events. The schedule must be responsive to changes and instrument readings.

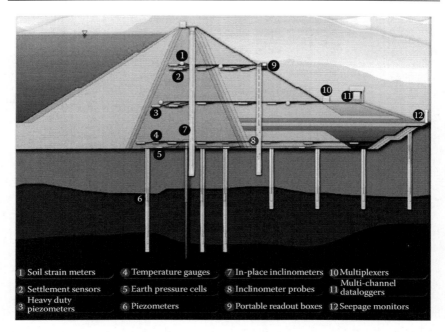

1 Soil strain meters	4 Temperature gauges	7 In-place inclinometers	10 Multiplexers
2 Settlement sensors	5 Earth pressure cells	8 Inclinometer probes	11 Multi-channel dataloggers
3 Heavy duty piezometers	6 Piezometers	9 Portable readout boxes	12 Seepage monitors

Figure 7.7 Section of earth dam showing some potential instrument types and locations. (Courtesy of Geokon, Inc., www.geokon.com, 2010.)

For efficient data collection, consistency can be enhanced from any specific instrument by employing a designated person and equipment, if feasible, using standardized procedures. Data can also be verified with multiple readings and redundancy with data from other monitoring instruments or means and immediate comparison with previous readings. Any unusual conditions, activities, observations, or results should be communicated between the team as part of the analysis and a warning issued if threshold levels are exceeded. Alarms can be raised and a contingency plan should be implemented.

After collection, the data must be expeditiously converted into meaningful values through processing. The raw data are often voltages that require mathematical manipulation using calibration factors of the instruments. The results should be reported in a format easily understandable and summarized, such as graphs, for analysis and interpretation. Throughout the process, careful attention should be paid in real time to errors and anomalous readings so that the readings can be repeated.

Many of these tasks can be simplified through the use of computers and simple programs to reduce, convert, and display the readings. It is important to note that vigorous testing of the programs and periodic manual checking should be implemented beforehand to verify results (U.S. Army Corps of Engineers 1995).

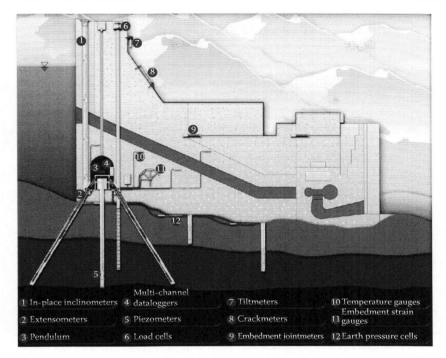

1 In-place inclinometers	4 Multi-channel dataloggers	7 Tiltmeters	10 Temperature gauges
2 Extensometers	5 Piezometers	8 Crackmeters	11 Embedment strain gauges
3 Pendulum	6 Load cells	9 Embedment jointmeters	12 Earth pressure cells

Figure 7.8 Section of concrete dam showing some potential instrument types and locations. (Courtesy of Geokon, Inc., www.geokon.com, 2010.)

For ease of analysis, graphical representation of the data is usually the most suitable in the form of plots that can quickly show comparisons and trending. Typical formats include time-history plots with one or two parameters on the axes versus time, shown in Figure 7.9, and positional plots showing changes in a parameter relative to a position indicating movement as in Figure 7.10. Guidelines for effective presentation include an appropriate scale, standardized formats, inclusion of location sketches, related conditions on multiple graphs, and predicted or limit values.

The analysis involves evaluation and interpretation of the data. As part of the process, the data are assessed for validity and possible errors in calculations. For dam safety monitoring, the analysis focuses on the performance of the instrument system and the performance of the dam structure or elements that are being observed. Data analysis pertaining to a monitored feature of a higher degree of importance should be treated as such and have priority with detection of potential problems within hours and meaningful analyses within a day or two at most.

The analysis process will review the current data along with initial data and the last readings to detect changes, irregularities, and possible malfunction of instruments. Consideration of predicted performance should

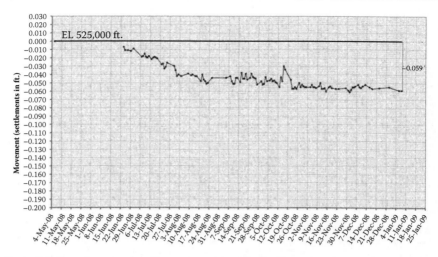

Figure 7.9 Time-history plot of movement monitoring data.

be included and an important aspect of the analysis is performance over a significant time period to reveal consistencies and trends. The predicted design trends can be compared to the behavioral trends and act as a basis for future prediction. With a large amount of data, statistical analysis is meaningful in determining average values, variances, and deviations (Bevington 1969), and so can help with setting acceptable threshold and action levels.

The results of the analysis can reveal the effectiveness of instruments and their requirement for calibration or replacement, need for changes in

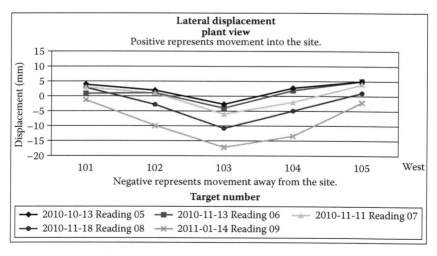

Figure 7.10 Positional plot of movement monitoring data.

the monitoring frequency, re-evaluation of priority areas, need for further information or study, and verification or contradiction of predictions.

There are potential problems (U.S. Army Corps of Engineers, 1995) to be avoided related to data analysis:

- insufficient data for field comparison;
- delays in data entry, processing, analysis, and distribution;
- assuming correct calculations and valid data;
- assuming changes in data are invalid or a concern;
- assuming no changes are satisfactory;
- not incorporating significant factors that affect the readings; and
- assuming computer-generated plots are accurate without review.

Presentation of the data not only includes the creation of plots for ease of analysis, but also summarizing, combining data from different sources, and revealing trending to compare predicted and actual behaviors. Conclusions can be made to the effectiveness of the monitoring program, and the need for improvements or adjustments.

Reporting should focus on an aspect of the dam monitoring, such as seepage, and only include the data and instrumentation relevant to that feature with values in meaningful and consistent units. The information should be clear and pertinent with discussions of changes, rate of change, and trends. Wherever possible, some degree of graphical representation or synthesis should be incorporated such as piezometric cross-section, or changes with time as related to construction activities, or reservoir, or river levels.

7.5 PROJECT STAFFING

In order to accomplish the tasks required for a monitoring program, sufficient dedicated staff must be made available. The range of tasks involved is large with the need for individuals with diverse skill sets who are delegated specific responsibilities or can perform multiple tasks. The overall project and program should be supervised by an experienced senior-level technician or engineer. This person or his or her staff would study the structure routinely and would play a role in planning the monitoring program. Other possible staff that could fill the role would be project engineers, superintendents, or foremen. Inspections include looking for seepage downstream of the structure or areas of movement in the embankment or other types of physical distress. The manager would aid in the planning, design, and procurement of the system as well as in the training and guidance of all staff, analysis of the data, and overall supervision. It is important to have close cooperation between the owner and the contractor for the flow of information and a clear understanding of the responsibilities for each group to avoid misunderstanding and disputes.

Once the array of instruments has been carefully planned and acquired, the instrumentation installation personnel must test them before installation. Coordination with other personnel or site activities is vital. Installation of each instrument requires specific knowledge and a procedure. Protection of the instruments must also be kept in mind. These personnel also play a ˙ role in maintenance and replacement as required.

The data-collection staff is an important part of the team with proper training and clear understanding of the data that are being acquired. Often, this task falls on temporary or contract, inexperienced employees who bear the physical labor with limited knowledge of the concerns. A more suitable option is permanent employees who can also retrieve automated data if available and cover a project or area on a regular basis to allow for consistency.

Once the data are collected from the field, data entry staff can input the information. This position can be handled by nontechnical personnel, but they should still be familiar with the data in order to be aware of potential errors and invalid readings. The management of data, reporting, and plotting staff requires strong computer skills and a more technical background, awareness of the data-collection procedures, and familiarity with the instrumentation. It may be necessary to use various software programs and retrieve data from remote sites.

After processing of the data, experienced personnel must review and analyze the reports. Multiple people should be involved to offer differing perspectives and comprehensiveness.

Of course, the number of staff required depends on the number of instruments and the frequency required for readings. Many of these tasks can also be accomplished by one single, dedicated and well-trained individual.

7.6 AUTOMATION

With the advancement of technology and computer systems, automated data acquisition systems (ADAS) are available as an alternative or as a supplement to manual readings (USCOLD 1993). Automated tasks include data acquisition, processing, presentation, and reporting. Initial costs are relatively high, but the overall long-term costs of the program are competitive and should be considered on major dam/levee remediation programs. With reduced or limited resources, the automated alternative can aid in accomplishing the tasks required. The components generally include an electronic sensor or transducer connected to a datalogger or computer with a communication interface to transmit data locally or remotely. Existing systems can be evaluated for feasibility of retrofitting with automation. Table 7.4 lists the advantages and disadvantages of an automated system.

A combination of conventional power with battery, solar, or emergency generator backup power is ideal for the system in case of power outages. Communication from the instruments to the datalogger and datalogger to the computer and possibly to another office can be via cable, radio, or cell phone. A central computer with special software can provide the interface to the internet for remote monitoring. These systems have the potential to provide alarms via cell phones or e-mails if set threshold limits are exceeded (U.S. Army Corps of Engineers 1995).

Trending of historical data is a key to the data analysis as it provides indication of performance and early warning of potential failure. Once alarms are triggered, a predetermined contingency plan should be in place and executed. There should be several stages for the alarms on vital, reliable instruments in which the response will escalate as the threat is greater.

When selecting instruments for automation, the factors for consideration are acceptable performance and longevity, as well as the ability to install them in the chosen location. Also, the compatibility for automation and ease of reverting to a manual backup are considerations. Table 7.5 summarizes some frequently used dam instrumentation and their relative ease of automation. With relative ease of automation, Category 1 is fairly straightforward, whereas Category 2 has more specialized requirements, and Category 3 has further limitations and is not available in certain environments.

Table 7.4 Advantages and disadvantages of automation

Advantages	Disadvantages
Overall cost of monitoring	Initial cost of installation expensive
Regular, immediate, or real-time monitoring	Higher maintenance costs
Accessibility to limited areas	Produces large volumes of data; overtaxes storage
Increased frequency—more data, less system error	Removes personal attention from the field
Increased accuracy—reduced human error	Lightning; variable voltage potential is destructive
Increased data reliability and consistency	Excessive downtime with complex integrations
Data and system validity checks enhance quality	Computer and electronics expertise required
Reduces manpower	Requires constant electrical power source
Alarms for exceeding thresholds and system health	Requires use of electronic transducers with least long-term reliability of instrument type
Remote diagnostics, calibrations, and programming	

Table 7.5 Instruments and ease of automation

Instrument	Category of relative ease	Automation method	Manual backup
Piezometers			
Open standpipe	2	Pressure transducer	Straightforward by measuring to water surface
Pneumatic	2	Pressure actuating	Straightforward w/manually operated readout
Vibrating wire	3	Frequency counter	Straightforward w/manually operated readout
Deformation gauges			
Tiltmeter	1, 2	Depends on transducer	Straightforward w/manually operated readout
Embankment extensometer	1, 2	Depends on transducer	Straightforward w/manually operated readout
Inclinometer, in-place	2	Depends on transducer	Straightforward w/manually operated readout
Earth pressure cells			
Pneumatic	3	Pressure actuating	Straightforward w/manually operated readout
Strain gauge	2	Strain gauge circuitry	Straightforward w/manually operated readout
Temperature			
Thermistor	1, 2	Resistance readout	Straightforward w/manually operated readout
Thermocouple	2	Voltage readout	Straightforward w/manually operated readout
RTD	1, 2	Resistance readout	Straightforward w/manually operated readout
Survey			
Robotic total station	2	Digital output	Straightforward w/manually operated instrument
GPS	2	Digital output	Straightforward w/manually operated instrument

Source: U.S. Army Corps of Engineers, "Engineering and Design—Instrumentation of Embankment Dams and Levees," *Engineer Manual*, Publication EM 1110-2-1908, Washington, DC, 1995.

With the potential for large amounts of data, reduced human interaction, and potential threshold alarms in place, it is vital to have a manager and/or support team dedicated to the instrumentation and data to oversee the monitoring program and implement the contingency plan throughout the project.

7.7 CASE HISTORIES

7.7.1 Saluda Dam

Saluda Dam is 200 feet high, 7,800 feet long, and is located in South Carolina (Sossenkina et al. 2004). This earth dam was constructed in 1930 and consists primarily of silty and clayey sand but is somewhat heterogeneous, and its construction featured little quality control, buried railroad tracks, and networks of drains and tunnels. For its most recent remediation for a seismic upgrade, a large berm was constructed immediately downstream of the dam as shown in Figure 7.11. The challenge was that, for a dam of such complex nature, excavation at the toe could cause instability and failure. In order to minimize the risk, the excavation was performed in short stages or cells so that each cell would be backfilled before excavation progressed to the next cell. Based on a slope-stability analysis, the phreatic surface had to be lowered by a major dewatering system. To provide early warning of potential problems and to ensure safety, an intensive monitoring program was employed.

The potential failure mechanisms were identified as sliding, piping, internal erosion, and bottom heave. The parameters to monitor these mechanisms were changes in pore pressure and deformations, both vertical and lateral, within the embankment and foundation. After baseline readings were taken, appropriate threshold levels and a response plan were established. Detailed records of activities such as dewatering, drawdown, excavation, and rainfall were kept in order to correlate these with the data.

To monitor slope failure and bottom heave, inclinometers and piezometers were installed with additional data provided by surface surveys. The vibrating wire piezometers provided information on pore water pressure where increases could indicate shear in the soil resulting in a sliding slope failure due to seepage. They were selected due to their quick response time, ability to measure negative pore pressure, automating capability, and ease of

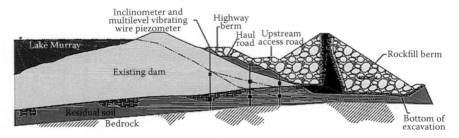

Figure 7.11 Existing dam and back-up berm cross-section. (From Sossenkina, E., M. Glunt, J. Mann, S. Newhouse, and Paul C. Rizzo Associates. "Listening to the Dam— Instrumentation and Monitoring Program Saluda Dam Remediation." Paper presented at ASDSO Regional Conference, April 2004. With permission.)

installation. Two piezometers were installed per borehole and over 100 were installed in total within the dam and foundation. High hydraulic gradient seepage could also cause internal piping and bottom heave. Open standpipe piezometers were also installed as redundancy and verification. Stable conditions were observed as well as during the dewatering program and drawdown of lake level in order to evaluate deviations in the data and performance.

Three rows of inclinometers (eighty in total) were installed above each cell extending down to weathered rock or competent foundation soil. Analysis revealed distinctive shapes of the inclinometer profiles relating to dewatering-induced settlement, surcharge response, soil relaxation, and shear sliding planes as shown in Figure 7.12. In order to evaluate the quickly changing conditions, the monitoring system was automated using a datalogger and radio-link transmission or wireless transmission from field computers.

Additional data were provided by vibrating wire tiltmeters to monitor important structures, shear strips on the slopes, and pipe lasers to monitor a line of targets along the cut slope. Survey monuments monitored using GPS technology correlated well with the inclinometer data and the expected movements of the ground surface.

The detailed record of activities, inspections, and visual observations were essential in interpreting the data and making sound decisions. Figure 7.13

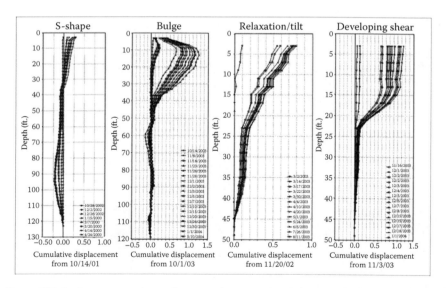

Figure 7.12 Inclinometer plots showing dewatering-induced settlement, surcharge response, soil relaxation, and shear sliding planes. (From Sossenkina, E., M. Glunt, J. Mann, S. Newhouse, and Paul C. Rizzo Associates. "Listening to the Dam—Instrumentation and Monitoring Program Saluda Dam Remediation." Paper presented at ASDSO Regional Conference, April 2004. With permission.)

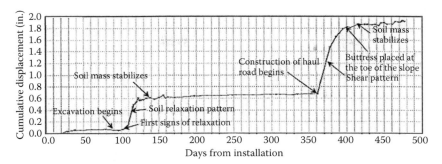

Figure 7.13 Inclinometer plot correlated with construction activities. (From Sossenkina, E., M. Glunt, J. Mann, S. Newhouse, and Paul C. Rizzo Associates. "Listening to the Dam—Instrumentation and Monitoring Program Saluda Dam Remediation." Paper presented at ASDSO Regional Conference, April 2004. With permission.)

demonstrates the correlation between inclinometer results and construction activities. A small local slide went undetected by instruments upslope, but was discovered by routine visual observations of developing tension cracks that indicated a developing larger failure. A buttress at the toe of the slope was constructed to halt the movement. In another instance, murky seepage was undetected by piezometers but discovered by inspections. The dewatering system was found to be operational and not the cause. Review of old drawings and records revealed a large drain tunnel and statements from workers recalling similar seeps in the area of old drainage features that would cease in short periods of time. The observed seepage did abate and a filter blanket was installed to prevent the loss of fines. This record and experience prevented potential costly and unnecessary remedial actions.

7.7.2 New Orleans levee test sections

Test sections to improve levee construction using geosynthetic reinforcement were created with the deep mixing method (DRE system) to research potential new design methodologies (Varuso and Grieshaber 2004). Previous tests indicated that foundation soils, consisting predominantly of soft fat clays with high water content, gained significant shear strength due to consolidation of the soft materials during and immediately following embankment construction.

To monitor the performance of the test sections, the instrumentation program consisted of settlement plates to measure vertical movement of the embankment and foundation, inclinometers measuring lateral movement, piezometers monitoring pore pressure, and strain gauges and extensometers to monitor the location and magnitude of stresses in the geosynthetics.

The piezometer and settlement plate data determined the rate at which consolidation occurred. The inclinometer results indicated potential failure surfaces and determined global strain. The strain gauges and extensometers monitored local strains in the reinforcement caused by the embankment loading. The data aided in developing a levee-design methodology incorporating geosynthetics to minimize the levee cross-section to reduce costs of construction, material, and real estate. Figure 7.14 shows some of the instrumentation locations.

Inclinometer data indicated stability of the embankment with lateral spreading of the fill material in both directions as opposed to a movement in one direction indicative of a failure mode. The twelve piezometers provided data that showed increases in pore pressure during embankment construction. When the pore water pressure dissipated, the soft clay layers underwent primary consolidation. The settlement data revealed a similar trend with increased settlement during lift placement of the embankment. The settlement was less than predicted due to the use of the geosynthetics. Also, the rate of settlement decreased after completion of primary consolidation. This period of increase in shear strength would occur rapidly during levee construction and was confirmed with post-construction borings and testing. Strain in the geosynthetics was measured with 108 total strain gauges and indicated a gradual increase in strain over time, but less than expected due to the increase in shear strength of the foundation soils. Data were verified with the installation of 18 extensometers.

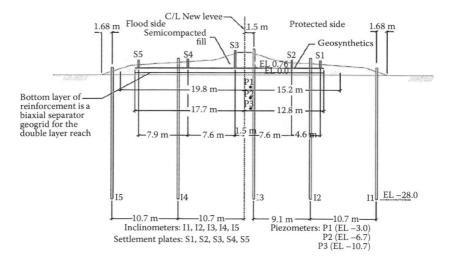

Figure 7.14 Instrumentation locations. (From Varuso, R., and J. Grieshaber, *Geosynthetic Reinforced Levee Test Section on Soft Normally Consolidated Clays*, U.S. Army Corps of Engineers, New Orleans District, Elsevier Ltd., 2004. With permission.)

The findings allowed a new design methodology to be recommended for the construction of levees in New Orleans that reduced costs by decreasing the cross-section required.

7.7.3 Unnamed dam, United States

A current dam remediation project established an instrumentation and monitoring plan with specific thresholds and responsibilities that were distributed to both the owner and contractor. The plan relied on judgment in data evaluation, application of the established threshold criteria, and responsiveness. In this plan, each group had a data manager responsible for coordinating data collection and distribution between the parties. Collaboration and communication were key to their roles. Data were shared through an FTP site (to store the data), as well as a web-based monitoring site to view automated piezometer data.

The threshold levels were established by advisory panels for piezometers, inclinometers, extensometers, settlement monuments, and crack pins. Alert levels were also created with increasing levels associated with increasing risk. The highest alert level was triggered by visual observations of distress and failure-in-progress indicators such as a whirlpool in the lake, sinkholes, depressions, settlement, cracking, muddy or grout discharge downstream, increased seepage, or boils with piping and slope instability.

Threshold exceedance response begins with evaluation of the cause and risk involved. The monitoring system is checked for proper functionality, manual readings are taken, and possible contributing construction or climatological activities are assessed. Visual inspections are made and monitoring frequencies may be adjusted. Exceedances trigger verbal notification up the command chain for each group to initiate response. During the evaluation, the data managers for both the owner and contractor work together and share their findings along with project officials and advisory panels. The outcome assesses the threat level and determines the action required whether construction can continue, must be modified, or should be temporarily stopped for further review. Afterward, an incident review of the process must be completed to determine if the monitoring plan requires revision.

7.8 FINAL REMARKS

There are major and inherent risks associated with performing the remediation of a dam or levee as the structure senses changes related to the construction process and/or modified operational parameters. These changes may unexpectedly and adversely affect its performance. Risks can be reduced by implementing an instrumented monitoring program, as driven by the outcome of the PFMA process.

An effective monitoring program must be executed systematically and consists of an integrated process encompassing the installation of the instruments, collection of the data, evaluation of the results, and reaction to the evaluations. In order to implement such a program, the program requires early and continuous collaboration and communication between the owner and contractor.

REFERENCES

Bevington, P. 1969. *Data Reduction and Error Analysis for the Physical Sciences.* New York: McGraw-Hill.

Chuaqui, M., and C. J. Hope. 2007. "Precision Monitoring of Shoring and Structures." *Field Measurements in Geomechanics (FMGM) Symposium Proceedings.* Boston, MA, September. Reston, VA: ASCE.

Chuaqui, M., S. Ford, and M. Janes. 2007. "Field Instrumentation for an Innovative Design-Build Excavation Adjacent to Heritage Structures." *Field Measurements in Geomechanics (FMGM) Symposium Proceedings.* Boston, MA, September. Reston, VA: ASCE.

Dunnicliff, J. 1993. *Geotechnical Instrumentation for Monitoring Field Performance.* New York: John Wiley and Sons.

Hudnut, K., and J. Behr. 1998. "Continuous GPS Monitoring of Structural Deformation at Pacoima Dam." *Seismological Research Letters* (July/August): 299–308.

Marr, A. 2001. "Why Monitor Geotechnical Performance?" Paper presented at the 49th Annual Geotechnical Engineering Conference, University of Minnesota, February.

Myers, B., and J. Stateler. 2008. "Why Include Instrumentation in Dam Monitoring Programs?" U.S. Society on Dams, A White Paper prepared by the USSD Committee on Monitoring of Dams and Their Foundations, November, 2008.

Sossenkina, E., M. Glunt, J. Mann, S. Newhouse, and Paul C. Rizzo Associates. 2004. "Listening to the Dam—Instrumentation and Monitoring Program Saluda Dam Remediation." Paper presented at ASDSO Regional Conference, April.

U.S. Army Corps of Engineers. 1995. "Engineering and Design—Instrumentation of Embankment Dams and Levees, Engineer Manual, Publication EM 1110-2-1908, Washington, DC, June.

U.S. Committee on Large Dams (USCOLD). 1993. "General Guidance and Current U.S. Practice in Automated Performance Monitoring of Dams." Committee on the Monitoring of Dams and Their Foundations of the United States Committee on Large Dams, May.

Varuso, R., and J. Grieshaber. 2004. *Geosynthetic Reinforced Levee Test Section on Soft Normally Consolidated Clays.* U.S. Army Corps of Engineers, New Orleans District, Elsevier Ltd., November.

Chapter 8

A distant mirror and a word of warning

Donald A. Bruce

8.1 HALES BAR DAM, TENNESSEE

The concrete structures comprising Hales Bar Dam, Tennessee, were built by the Chattanooga and Tennessee River Power Company between 1905 and 1913 on the Tennessee River about 33 miles downstream of Chattanooga (Rogers 2011). This was the first occasion a private power company had constructed a major dam across a navigable channel in the United States, and this undertaking required congressional approval. When completed it contained the world's highest single-lock lift (41 feet) although at 265 feet in length, it would soon become the shortest on the Tennessee River. On the other—left—abutment, the powerhouse required a 98-foot-by-240-foot excavation extending 75 feet below the original river bed.

The dam was located across the narrowest reach of the river for many miles; the river channel at this site was narrow, crooked, and shallow, incised into Mississippian Age Bangor limestone. At the time of design in 1904, it was assumed that the topography of the site (the narrow channel) meant that the bedrock there was especially resistant. However, the limestone that dips moderately toward the left abutment was later found to contain numerous steeply inclined fault zones striking under the dam (Figure 8.1), which had in fact controlled the river course. Karstification associated with these systems was extremely well developed, the cavities being both clay-filled and open.

Between 1905 and 1910, four different contractors failed to complete the project due to problems with excavation and dewatering and three were bankrupted. However, in 1910, another contractor was found who agreed to work only on a time-and-materials basis and, during the next three years, was able to complete the project. This involved the drilling (by diamond coring) of a series of secant, reinforced piles: 40 each 45 feet deep on the upstream side, and 30 piles each 32 feet deep on the downstream side of the powerhouse excavation. Extensive grouting was also conducted, extending from 30 to 50 feet into the bedrock. Upon completion, the dam was 2,315 feet long, with the central overflow section comprising 1,300 feet. It stood from 59 to 113 feet above the riverbed, compared to the original design

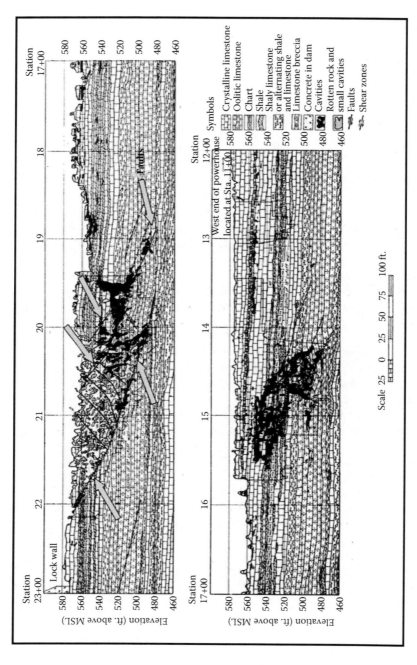

Figure 8.1 Developed geological section beneath Hales Bar Dam, Tennessee. (From Rogers, J. D., "Hales Bar Dam and the Potential Pitfalls of Constructing Dams on Karst Foundations," PowerPoint Lecture, Missouri University of Science and Technology, Rolla, MO, 2011.)

assumption of 45 to 65 feet. Wooden flashboards were tacked onto the crest to increase the operating pool by 3 feet soon after construction.

Excessive seepage was noted upon first filling, especially around the left abutment, and attempts in 1914 and 1915 to stem the leaks with dumped rock, concrete, and other materials were unsuccessful, merely redistributing the flow. The project pioneered the use of divers to locate leaks, both upstream and downstream, and to try to seal them. In 1919 an asphalt grouting project was undertaken from holes drilled from inside the dam's inspection gallery. Over 78,000 cubic feet of hot asphalt were injected into 68 holes averaging 92 feet deep. By 1922 victory was declared.

However, by 1929, a progressive increase in seepage had resulted in total flows reaching 1919 levels, and analyses indicated that the grouting had only been effective in filling voided karst in the uppermost 15 feet of the formation but, predictably, had had no benefit in the lower clay-filled features, which continued to be eroded. Between 1930 and 1931, an intense program of investigation was undertaken, using dyes and oils to identify specific flow conduits in the karst. Estimates of flow ranged from 100 to 1,200 cubic feet per second (cfs). Little appears to have been done for many years thereafter, partly due to the impact of the Great Depression, but mainly as a result of a 5-year lawsuit between the owner and the federal government, which eventually reached the U.S. Supreme Court. The Tennessee Valley Authority (TVA) finally took ownership of the project in August 1939, and seepage flow investigations by the TVA and the U.S. Geological Survey were quickly conducted. Consequently, it was found that the total leakage rate was between 1,650 and 1,720 cfs, equivalent to about 10 percent of the Tennessee River's normal flow. Figure 8.2 shows the mapped flow and indicates the 13 boils that had formed in the gravel bar on the right abutment.

TVA took over the dam in August 1939 when it acquired the assets of the Tennessee Electric Power Company. They began further investigations in November 1940 by drilling 3-inch core holes vertically near and along the upstream face of the dam. This led to the design in 1941 of a multirow, multimaterial cutoff centered on the dam's centerline, just downstream of the core holes. The three steps were as follows (Figure 8.3):

1. Drilling 750 each 18-inch-diameter calyx holes on 24-inch centers in a staggered, secant fashion to a maximum depth of 163 feet (i.e., 25 to 103 feet below top of rock). These holes each accommodated a 6-inch, asbestos-cement pipe and were backfilled with concrete.
2. Coring 3-inch-diameter holes at 10-inch centers just upstream of the secant wall, and grouting with hot asphalt.
3. Installation of 13-inch-diameter holes at 10-inch centers, filled with concrete, and 3-inch core holes midway between these 13-inch holes, injected with cement grout. This composite row was located just downstream of the secant wall.

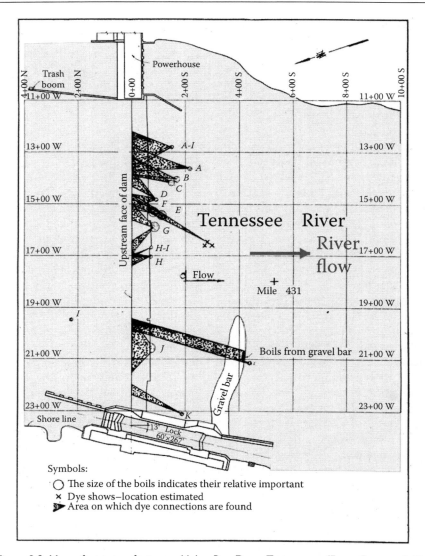

Figure 8.2 Map of seepage features, Hales Bar Dam, Tennessee. (From Rogers, J. D., "Hales Bar Dam and the Potential Pitfalls of Constructing Dams on Karst Foundations," PowerPoint Lecture, Missouri University of Science and Technology, Rolla, MO, 2011.)

A 25-foot-long, movable template was used to locate accurately the 18-inch holes with the downstream holes being primaries. The holes in each row were at 24-inch centers and were also installed in a split-spaced sequence, presumably to avoid disturbance to recently placed concrete in completed piles. After removal of the calyx cores, a water-current meter

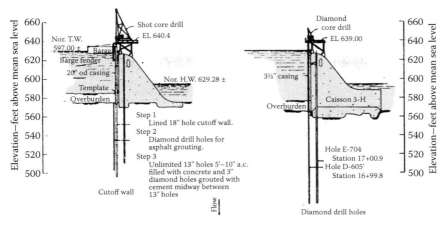

Figure 8.3 Section showing composite cutoff, Hales Bar Dam, Tennessee. (From Rogers, J. D., "Hales Bar Dam and the Potential Pitfalls of Constructing Dams on Karst Foundations," PowerPoint Lecture, Missouri University of Science and Technology, Rolla, MO, 2011.)

was inserted into each hole to record the seepage velocities. These proved to be as high as 4.5 feet per second and provided information on grout-mix design for use in the step 3 holes. All work was completed in 1945 after which the addition of 20 tainter gates permitted a further rise in pool elevation.

Evaluation of the extremely detailed and plentiful drilling and grouting records confirmed the extremely erratic nature of the rock, both parallel and transverse to the dam's axis. Many low-angle faults were discerned, parallel to the strike of the rock, i.e., orthogonal to the dam's axis. Extremely large and interconnected cavities were found to depths of almost 100 feet beneath rockhead: the secant wall was terminated at variable depths, several feet beneath these features, and into (presumably) competent and intact rock (Figure 8.4).

However, the seepage problems gradually worsened during the 1950s and, by the early 1960s, "suck holes" and vortices were being observed just upstream of the dam, with foaming boils noted just downstream of the dam.

The story ends in April 1963 when the TVA announced it was abandoning Hales Bar Dam, and replacing it with a new lock and dam 6 miles downstream: the new site was chosen primarily for geological, not topographical, reasons. Engineering studies had shown that improvements at Hales Bar would be more extensive than previously indicated and their success in completely sealing and stabilizing the dam could not be assured. In addition to the underseepage problem, which had resumed, the old lock was less than half the size of those used elsewhere along the upper Tennessee River (110 by 600 feet). Most of the structure was demolished between 1965 and 1968

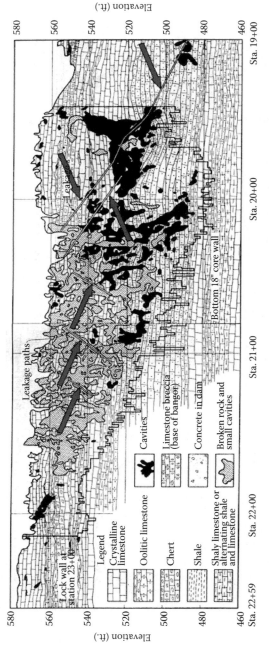

Figure 8.4 One of two major faulted areas as mapped during the remediations, Hales Bar Dam, Tennessee. (From Rogers, J. D., "Hales Bar Dam and the Potential Pitfalls of Constructing Dams on Karst Foundations," PowerPoint Lecture, Missouri University of Science and Technology, Rolla, MO, 2011.)

Photo 8.1 Demolition of the spillway overflow section of the old Hales Bar Dam, Tennessee, in 1967–68, as seen from the right bank. The powerhouse (far side) was gutted of all equipment, but left in place, as was the navigation lock (foreground). (Courtesy Rogers, J. D., "Hales Bar Dam and the Potential Pitfalls of Constructing Dams on Karst Foundations," PowerPoint Lecture, Missouri University of Science and Technology, Rolla, MO, 2011.)

(Photo 8.1), except for the powerhouse, and the old lock was converted into a coal-barge terminal. These remnants can still be seen in the new Nickajack Lake. Hales Bar Dam was the first dam owned by a governmental agency to be removed because of engineering problems.

So why choose this sad case history to end a book detailing successes and triumphs with a tale of a dispiriting inevitable repetitive failure? The answer is quite simple: as the Good Book says: "Pride goeth before destruction, and a haughty spirit before a fall" (Proverbs 16:18). As a profession, we should not let our technological progress obscure the magnitude of the challenges posed by nature. Nor should we relax the standards of quality and attention to detail that are essential during the execution of each and every process and phase of each and every project. We must not let good experiences foster complacency.

These admonitions apply to any engineering project. It is particularly relevant to dam and levee remediation where flaws in design or mistakes during construction can trigger the direst risks to the structure, to the population, to the environment, and to the banks.

REFERENCES

Rogers, J. D. 2011. "Hales Bar Dam and the Potential Pitfalls of Constructing Dams on Karst Foundations." PowerPoint Lecture, Missouri University of Science and Technology, Rolla, MO.

Index